CONTENTS

1. LENGTH

2. ANGLES

3. 2D SHAPES

4. 3D OBJECTS

5. AREA

CONTENTS

6. VOLUME AND CAPACITY

7. MASS

8. TIME

9. POSITION

ANSWERS

  ISBN: 9781925726190

ABOUT CATCH-UP MATHS

The Catch-Up Maths series enables students to start from scratch when they are struggling with their year level maths. Each book takes maths back to the foundation and ensures that all basic concepts are consolidated before moving forwards. Lots of revision and opportunities to practise and build confidence are provided before moving on to new topics.

Each new strand and sub-strand of the primary maths curriculum is introduced clearly with simple explanations, examples and trial questions (with answers), before children move to the Practice section. To ensure concepts are understood fully, videos of the author working through and explaining new concepts are included in every chapter.

A QR code on the topic page provides access to the subtitled video.

This book has 9 chapters divided between the Year 6 Measurement and Space strands of the Australian Maths Curriculum. The chapters are:

1 Length
2 Angles
3 2D Shapes
4 3D Objects
5 Area
6 Volume & Capacity
7 Mass
8 Time
9 Position

★ A Review section that can be used as assessment and to check students' progress is included at the end of each chapter.

★ Answers are at the back of the book.

How to use this book

Children can work through the pages from front to back, or choose individual topics to reinforce areas where they are struggling.

The topics are introduced with:

- clear instructions, using simple language
- completed examples and incomplete examples for students to tackle before moving on to the **Your Turn** section
- a video linked by QR code that shows the incomplete examples and has the author giving extra instruction about the page.

HOW TO USE THE QR CODES IN CATCH-UP MATHS

A unique aspect of the **Catch-Up Maths** series is the **instructional video** created by the author for every new topic.

The videos give a step-by-step explanation of the examples and work through them to provide a helpful lesson for the student. The videos are simply accessed via the QR code on the same page and can be watched on a phone, tablet, computer or interactive whiteboard.

Access the video using a QR code reader app

SCAN to watch video

Each video shows the page from the book. The author talks through the concepts and examples, and demonstrates what students need to do. Her words are shown as easy-to-read captions so that the audio can be turned down in noisy classrooms, and for students with hearing difficulties. The solutions to the examples are presented before students are expected to tackle the 'Your Turn' section. This careful instruction ensures that students can confidently move on to the following Practice questions. Those assisting students should encourage them to check their 'Your Turn' answers before moving on.

Over 60 instructional videos!

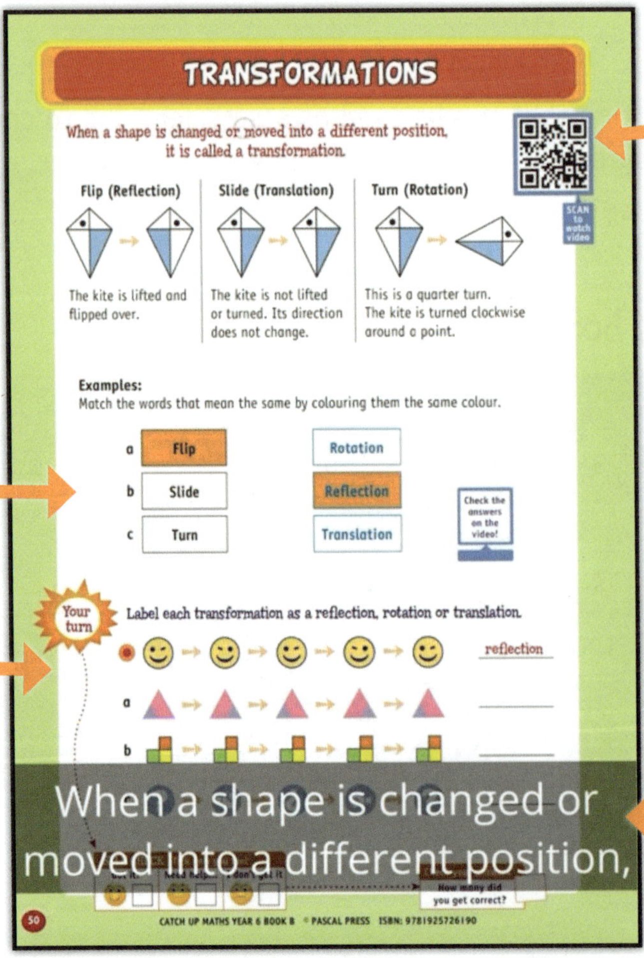

Scan this to access the video.

The answers to these examples can only be found on the video.

After watching the video, students can confidently complete the Your turn section.

Clear and easy-to-read captions

TRANSFORMATIONS

When a shape is changed or moved into a different position, it is called a transformation.

Flip (Reflection) — The kite is lifted and flipped over.

Slide (Translation) — The kite is not lifted or turned. Its direction does not change.

Turn (Rotation) — This is a quarter turn. The kite is turned clockwise around a point.

Examples:
Match the words that mean the same by colouring them the same colour.

a	Flip	Rotation
b	Slide	Reflection
c	Turn	Translation

Check the answers on the video!

Your turn — Label each transformation as a reflection, rotation or translation.

reflection

a

b

When a shape is changed or moved into a different position,

CATCH UP MATHS YEAR 6 BOOK B © PASCAL PRESS ISBN: 9781925726190

AUSTRALIAN CURRICULUM CORRELATIONS

Australian Curriculum: Mathematics F–6 Version 9.0

ACARA CODE	Content Description	Pages
1. Length		**PAGES 1–19**
AC9M4M01 Year 4	interpret unmarked and partial units when measuring and comparing attributes of length, mass, capacity, duration and temperature, using scaled and digital instruments and appropriate units	**1, 2, 3, 4, 5, 6, 12, 13**
AC9M4M02 Year 4	recognise ways of measuring and approximating the perimeter and area of shapes and enclosed spaces, using appropriate formal and informal units	**9, 10, 11**
AC9M5M01 Year 5	choose appropriate metric units when measuring the length, mass and capacity of objects; use smaller units or a combination of units to obtain a more accurate measure	**7, 8**
AC9M6M01 Year 6	convert between common metric units of length, mass and capacity; choose and use decimal representations of metric measurements relevant to the context of a problem	**1, 2, 3, 4, 5, 6, 7, 8**
2. Angles		**PAGES 20–39**
AC9M4M04 Year 4	estimate and compare angles using angle names including acute, obtuse, straight angle, reflex and revolution, and recognise their relationship to a right angle	**20, 21, 22, 23**
AC9M5M04 Year 5	estimate, construct and measure angles in degrees, using appropriate tools including a protractor, and relate these measures to angle names	**24, 25, 28, 29, 32, 33**
AC9M6M04 Year 6	identify the relationships between angles on a straight line, angles at a point and vertically opposite angles; use these to determine unknown angles, communicating reasoning	**20, 21, 26, 27, 30, 31**
3. 2D Shapes		**PAGES 40–61**
AC9M1SP01 Year 1	make, compare and classify familiar shapes; recognise familiar shapes and objects in the environment, identifying the similarities and differences between them	**46, 47, 48, 49**
AC9M2SP01 Year 2	recognise, compare and classify shapes, referencing the number of sides and using spatial terms such as "opposite", "parallel", "curved" and "straight"	**40, 41, 42, 43, 44, 45, 46, 47, 48, 49**
AC9M4SP03 Year 4	recognise line and rotational symmetry of shapes and create symmetrical patterns and pictures, using dynamic geometric software where appropriate	**52, 53, 54, 55**
AC9M5SP03 Year 5	describe and perform translations, reflections and rotations of shapes, using dynamic geometric software where appropriate; recognise what changes and what remains the same, and identify any symmetries	**50, 51, 52, 53**
AC9M6SP03 Year 6	recognise and use combinations of transformations to create tessellations and other geometric patterns, using dynamic geometric software where appropriate	**52, 53**
4. 3D Objects		**PAGES 62–81**
AC9M3SP01 Year 3	make, compare and classify objects, identifying key features and explaining why these features make them suited to their uses	**62, 63, 64, 65, 66, 67, 68, 69, 74, 75**
AC9M5SP01 Year 5	connect objects to their nets and build objects from their nets using spatial and geometric reasoning	**72, 73**
AC9M6SP01 Year 6	compare the parallel cross-sections of objects and recognise their relationships to right prisms	**70, 71**
5. Area		**PAGES 82–103**
AC9M4M02 Year 4	recognise ways of measuring and approximating the perimeter and area of shapes and enclosed spaces, using appropriate formal and informal units	**82, 83, 84, 85, 86, 87, 88, 89, 95, 96, 97**
AC9M5M01 Year 5	choose appropriate metric units when measuring the length, mass and capacity of objects; use smaller units or a combination of units to obtain a more accurate measure	**82, 83, 84, 85**
AC9M5M02 Year 5	solve practical problems involving the perimeter and area of regular and irregular shapes using appropriate metric units	**90, 91, 92, 93, 94**
AC9M6M02 Year 6	establish the formula for the area of a rectangle and use it to solve practical problems	**90, 91, 92, 93, 94**

Australian Curriculum Correlations continued

ACARA CODE	Content Description	Pages
6. Volume and Capacity		**PAGES 104–123**
AC9M3M01 Year 3	identify which metric units are used to measure everyday items; use measurements of familiar items and known units to make estimates	**112,**
AC9M3M02 Year 3	measure and compare objects using familiar metric units of length, mass and capacity, and instruments with labelled markings	**104, 105, 106, 107, 108, 109**
AC9M4M01 Year 4	interpret unmarked and partial units when measuring and comparing attributes of length, mass, capacity, duration and temperature, using scaled and digital instruments and appropriate units	**104, 105, 106, 107**
AC9M5M01 Year 5	choose appropriate metric units when measuring the length, mass and capacity of objects; use smaller units or a combination of units to obtain a more accurate measure	**110, 111, 112, 113, 116, 117**
AC9M6M01 Year 6	convert between common metric units of length, mass and capacity; choose and use decimal representations of metric measurements relevant to the context of a problem	**106, 114, 115**
7. Mass		**PAGES 124–143**
AC9M3M01 Year 3	identify which metric units are used to measure everyday items; use measurements of familiar items and known units to make estimates	**124, 125, 126, 127, 128, 129**
AC9M4M01 Year 4	interpret unmarked and partial units when measuring and comparing attributes of length, mass, capacity, duration and temperature, using scaled and digital instruments and appropriate units	**130, 131, 134, 135**
AC9M6M01 Year 6	convert between common metric units of length, mass and capacity; choose and use decimal representations of metric measurements relevant to the context of a problem	**126, 127, 132, 133, 136, 137,**

ACARA CODE	Content Description	Pages
8. Time		**PAGES 144–173**
AC9M1M03 Year 1	describe the duration and sequence of events using years, months, weeks, days and hours	**164, 165**
AC9M2M03 Year 2	identify the date and determine the number of days between events using calendars	**154, 155, 156, 157**
AC9M2M04 Year 2	recognise and read the time represented on an analogue clock to the hour, half-hour and quarter-hour	**144, 145, 146, 147, 148, 149**
AC9M3M03 Year 3	recognise and use the relationship between formal units of time including days, hours, minutes and seconds to estimate and compare the duration of events	**144, 145, 146, 147**
AC9M3M04 Year 3	describe the relationship between the hours and minutes on analog and digital clocks, and read the time to the nearest minute	**144, 145, 146, 147, 148, 149**
AC9M4M03 Year 4	solve problems involving the duration of time including situations involving “am” and “pm” and conversions between units of time	**148, 149, 150, 151, 158, 159, 160, 161, 162, 163**
AC9M5M03 Year 5	compare 12- and 24-hour time systems and solve practical problems involving the conversion between them	**152, 153**
AC9M6M02 Year 6	establish the formula for the area of a rectangle and use it to solve practical problems	**158, 159, 160, 161, 162, 163**
AC9M6M03 Year 6	interpret and use timetables and itineraries to plan activities and determine the duration of events and journeys	**158, 159, 160, 161, 162, 163**
9. Position		**PAGES 174–193**
AC9M1SP02 Year 1	give and follow directions to move people and objects to different locations within a space	**174, 175, 176**
AC9M3SP02 Year 3	interpret and create two-dimensional representations of familiar environments, locating key landmarks and objects relative to each other	**178, 179, 187, 188**
AC9M4SP02 Year 4	create and interpret grid reference systems using grid references and directions to locate and describe positions and pathways	**180, 181**
AC9M5SP02 Year 5	construct a grid coordinate system that uses coordinates to locate positions within a space; use coordinates and directional language to describe position and movement	**182, 183**
AC9M6SP02 Year 6	locate points in the 4 quadrants of a Cartesian plane; describe changes to the coordinates when a point is moved to a different position in the plane	**184, 185, 186**

METRES

A metre is a unit of measurement used to measure long things.

One metre is one-thousandth ($\frac{1}{1000}$) of a kilometre.

There are 100 cm in one metre.

1 m = 1000 mm

Examples: Use blue to colour the boxes of the items that would measure approximately one metre or more, use red for $\frac{1}{2}$ metre and green for $\frac{1}{4}$ metre.

a ☐ book
b ☐ truck
c ☐ ruler
d ☐ boogie board
e ☐ small dog
f ☐ surfboard
g ☐ newborn baby
h ☐ car
i ☐ shoe

Your turn

List five more things that match the length.

Less than 1 m	About 1 m	More than 1 m
a shoe	coffee table	dining table

SELF CHECK Tick how you feel

Got it!	Need help...	I don't get it
☐	☐	☐

Check your answers
How many did you get correct? ☐

  ISBN: 9781925726190

PRACTICE

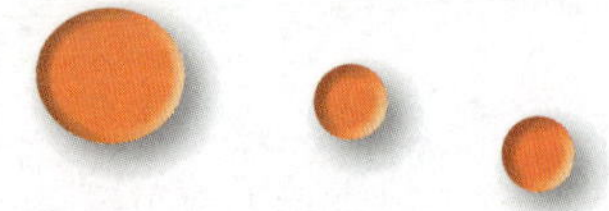

1 Write these lengths in centimetres.

- 2 m 36 cm 236 cm
- a 3 m 27 cm ______
- b 8 m 64 cm ______
- c 6 m 42 cm ______
- d 4 m 54 cm ______
- e 9 m 92 cm ______
- f 5 m 38 cm ______
- g 7 m 9 cm ______
- h 12 m 4 cm ______
- i 15 m 28 cm ______

2 Write in metres and centimetres.

- 526 cm 5 m 26 cm
- a 143 cm ___ m ___ cm
- b 156 cm ___ m ___ cm
- c 481 cm ___ m ___ cm
- d 204 cm ___ m ___ cm
- e 267 cm ___ m ___ cm
- f 902 cm ___ m ___ cm
- g 6751 cm ___ m ___ cm
- h 2895 cm ___ m ___ cm
- i 3086 cm ___ m ___ cm

3 Write the lengths in Question 1 in order from shortest to longest.

236 cm ______________________________

4 Write the lengths in Question 2 in order from longest to shortest.

6751 cm ______________________________

5 Complete the tables.

	cm	m and cm	Decimal m
	369 cm	3 m 69 cm	3.69 m
a		4 m 38 cm	
b	906 cm		
c			8.10 m
d		6 m 49 cm	
e			4.70 m
f		8 m 51 cm	

	cm	m and cm	Decimal m
g		5 m 60 cm	
h			1.03 m
i	215 cm		
j		7 m 72 cm	
k	1042 cm		
l			24.49 m
m	1503 cm		

  ISBN: 9781925726190

CENTIMETRES

Smaller lengths are measured in centimetres (cm).

One centimetre is one-hundredth ($\frac{1}{100}$) of a metre.

This ruler is 15 centimetres (15 cm) long.

0 CM 1 2 3 4 5 6 7 8 9 10 11 12 13 14 15

This is 1 cm.

Always measure from the 0.

The purple string measures 10 cm long.

1 m = 100 cm

Examples: List 10 items that would be measured in centimetres.

your hand ______ ______

______ ______

______ ______

______ ______

______ ______

Check the answers on the video!

Your turn

Use a ruler to measure each piece of string in centimetres.

Make sure you measure with your ruler starting at 0.

- ● 12 cm
- **a** ___ cm
- **b** ___ cm
- **c** ___ cm
- **d** ___ cm

SELF CHECK Tick how you feel

Got it!	Need help...	I don't get it
☐	☐	☐

Check your answers

How many did you get correct? ☐

  ISBN: 9781925726190

PRACTICE

1 Estimate the lengths of the lines and then measure them.

	Line	Estimate (cm)	Actual (cm)
●	blue	2	$2\frac{1}{2}$
a	green		
b	purple		
c	orange		
d	red		
e	yellow		
f	pink		
g	dark green		
h	black		
i	brown		

2 Draw lines of these lengths.

● 12 cm

a $3\frac{1}{2}$ cm

b 5 cm

c 1 cm

d $6\frac{1}{2}$ cm

e $10\frac{1}{2}$ cm

3 Convert the lengths.

	cm	m
●	62 cm	0.62 m
a	37 cm	
b		0.04 m
c	15 cm	
d	30 cm	

	cm	m
e		0.1 m
f		0.08 m
g	143 cm	
h	208 cm	
i		3.5 m

CATCH UP MATHS YEAR 6 BOOK B © PASCAL PRESS ISBN: 9781925726190

MILLIMETRES

Very small things are measured in millimetres (mm).
A millimetre is one-tenth ($\frac{1}{10}$) of a centimetre.

This ruler is 10 cm long. 10 cm = 100 mm

This is 10 mm.
There are 10 markings because there are 10 mm in 1 cm.

The small marks on the ruler are millimetres.
Each small mark is 1 mm.

Examples: Write the measurements in millimetres.

a 5 cm = 50 mm
b 2 cm = ____ mm
c 8 cm = ____ mm
d 10 cm = ____ mm
e 14 cm = ____ mm
f 19 cm = ____ mm
g 23 cm = ____ mm
h 38 cm = ____ mm
i 47 cm = ____ mm

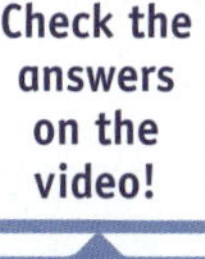

Your turn

1 Write the lengths marked on the ruler in millimetres.

● A = 2 mm **a** B = ___ mm **b** C = ___ mm **c** D = ___ mm

2 Mark and label these lengths on the ruler below.

● E = 28 mm **a** F = 41 mm **b** G = 103 mm **c** H = 144 mm

Check your answers
How many did you get correct?

PRACTICE

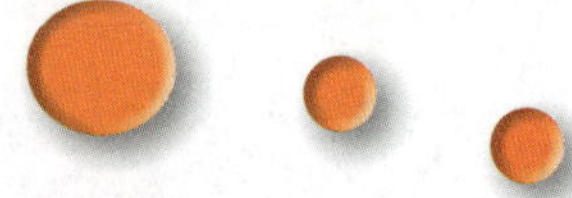

1 Write mm or cm next to each length.

- ● length of a pencil: 10 cm
- a width of a thumbtack: 1 ___
- b width of a baby's finger: 6 ___
- c width of the head of a pin: 1 ___
- d length of a paperclip: 3 ___
- e length of a crayon: 11 ___

2 Draw lines of these lengths.

- ● 8 mm ——
- a 6 mm
- b 18 mm
- c 43 mm
- d 109 mm
- e 156 mm

3 Measure the lines in mm and write the length.

- ● 24 mm ————
- a ______ ————————————
- b ______ ———————————————
- c ______ ——————————————————
- d ______ ———————————————————
- e ______ ——————————————————————

4 Convert the measurements.

	mm	cm	m
●	100 mm	10 cm	0.10 m
a		5 cm	
b			0.03 m
c	200 mm		
d		60 cm	
e			0.70 m
f	800 mm		

	mm	cm	m
g	3000 mm		
h		250 cm	
i			4.35 m
j	7250 mm		
k	9500 mm		
l		890 cm	
m			6.82 m

CATCH UP MATHS YEAR 6 BOOK B © PASCAL PRESS ISBN: 9781925726190

KILOMETRES

Kilometres (km) are used to measure long distances.

A highway

The distances between towns

SCAN to watch video

Examples: Use green to colour the boxes of the items you would measure in millimetres, use red for centimetres, blue for metres and yellow for kilometres.

- **a** ☐ A plane journey
- **b** ☐ The width of a baby's toenail
- **c** ☐ The distance to Qld from NSW
- **d** ☐ A handspan
- **e** ☐ The width of a pen tip
- **f** ☐ The distance around your room
- **g** ☐ A cubit
- **h** ☐ The length of a finger
- **i** ☐ The width of a staple
- **j** ☐ The height of a mobile phone
- **k** ☐ The length of an eyelash
- **l** ☐ The distance from Australia to New Zealand
- **m** ☐ The length of a shoe
- **n** ☐ A train journey

Check the answers on the video!

1 How many metres?

- ● 4 km = 4000 m
- **a** 7 km = ________
- **b** 15 km = ________
- **c** 28 km = ________
- **d** 44 km = ________
- **e** 274 km = ________

2 How many kilometres?

- ● 3000 m = 3 km
- **a** 6000 m = ________
- **b** 13 000 m = ________
- **c** 37 000 m = ________
- **d** 24 000 m = ________
- **e** 62 000 m = ________

SELF CHECK Tick how you feel

Got it!	Need help...	I don't get it
☐	☐	☐

Check your answers

How many did you get correct? ☐

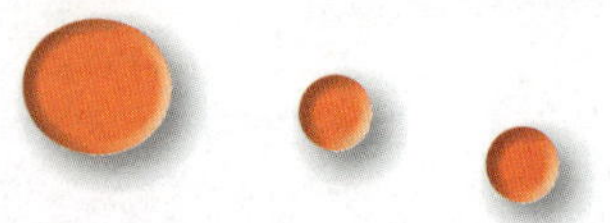

1 Write these lengths in metres.

- ● 3 km 245 m = 3245 metres
- a 5 km 420 m = ______ metres
- b 6 km 62 m = ______ metres
- c 8 km 8 m = ______ metres
- d 15 km 210 m = ______ metres
- e 72 km 43 m = ______ metres
- f 437 km 3 m = ______ metres
- g 624 km 25 m = ______ metres
- h 580 km 10 m = ______ metres
- i 804 km 1 m = ______ metres

2 Write these lengths in kilometres and metres.

- ● 5492 m = 5 km 492 m
- a 2515 m = ____ km ____ m
- b 5381 m = ____ km ____ m
- c 8620 m = ____ km ____ m
- d 6494 m = ____ km ____ m
- e 7730 m = ____ km ____ m
- f 96 003 m = ____ km ____ m
- g 10 780 m = ____ km ____ m
- h 125 680 m = ____ km ____ m
- i 654 964 m = ____ km ____ m

3 Convert to kilometres.

- ● 223 m = 0.223 km
- a 940 m = ______ km
- b 32 m = ______ km
- c 105 m = ______ km
- d 7314 m = ______ km
- e 5606 m = ______ km
- f 1140 m = ______ km
- g 8175 m = ______ km
- h 65 075 m = ______ km
- i 416 410 m = ______ km

4 Complete the tables.

	cm	m	km
●	300	3	0.003
a		8.25	
b			4
c		7	
d	1500		
e			6.25
f	613		

	cm	m	km
g		5000	
h	200		
i	900 000		
j		1125	
k			0.005 15
l		54 370	
m		1	

  ISBN: 9781925726190

PERIMETER

The distance around a 2D shape is called the perimeter (P).

SCAN to watch video

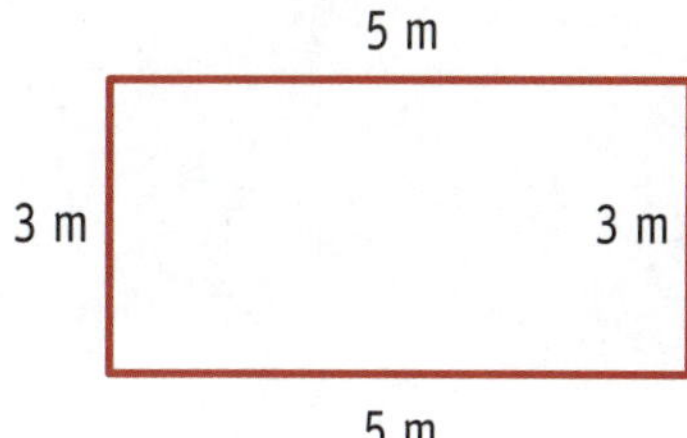

Perimeter = 3 m + 5 m + 3 m + 5 m

P = 16 m

The perimeter (P) of the rectangle is 16 m.

3 cm

Perimeter = 3 cm + 3 cm + 3 cm + 3 cm

P = 12 cm

The perimeter (P) of the square is 12 cm.

Examples: Find the perimeter (P) of each shape.

Not all the sides are labelled. Use what you know about shapes to work out the other lengths.

a

P = 3 cm + 2 cm + 3 cm + 2 cm

= 10 cm

b 4 m

P = 4 m + 4 m + 4 m + 4 m

= ___ m

c 6 cm, 4 cm

P = 6 cm + 4 cm + 6 cm + 4 cm

= ___ cm

Check the answers on the video!

Your turn Colour the correct answer.

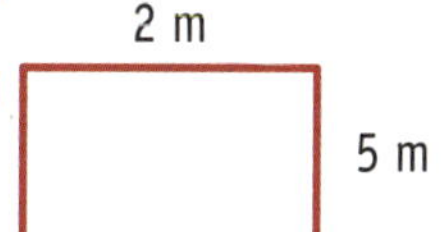

P = 2 m + 5 m + 2 m + 5 m

P = 13 m	P = 14 m	P = 14 cm

a

P = ____ + ____ + ____ + ____

P = 10 cm	P = 15 cm	P = 20 cm

SELF CHECK Tick how you feel

Got it!	Need help...	I don't get it
☐	☐	☐

Check your answers

How many did you get correct? ☐

PRACTICE

Calculate the perimeter of these 2D shapes.

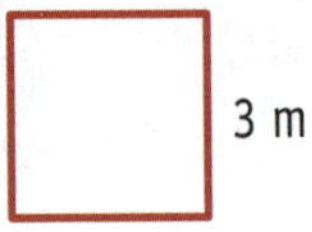

● (square, 3 m) P = 3 m + 3 m + 3 m + 3 m
= 12 m

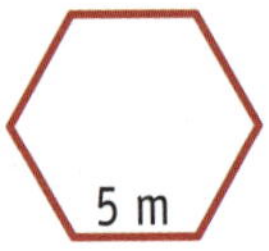

a (hexagon, 5 m) P = ________________
= ________

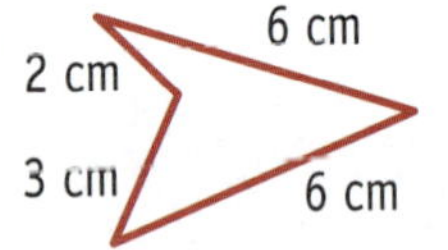

b P = ________________
= ________

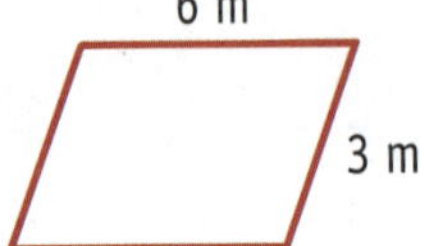

c P = ________________
= ________

Draw three different shapes with a perimeter of 18 cm.

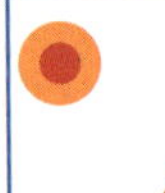

Complete the tables for rectangles with these measurements.

	Height	Width	Perimeter
●	24 m	11 m	70 m
a	5 m		20 m
b		7 m	32 m
c	16 cm	13 cm	
d	12 m		60 m
e	32 mm	44 mm	
f	7 m		32 m
g		15 m	64 m
h	4.5 cm		30 m

	Height	Width	Perimeter
i		6.5 mm	50 mm
j	15 m	3 m	
k		12 cm	40 cm
l		10 mm	28 mm
m	28 m		58 m
n	16 cm	19 cm	
o		35 mm	100 mm
p	24 cm		64 cm
q	27 m	15 m	

  ISBN: 9781925726190

Calculate the perimeter.

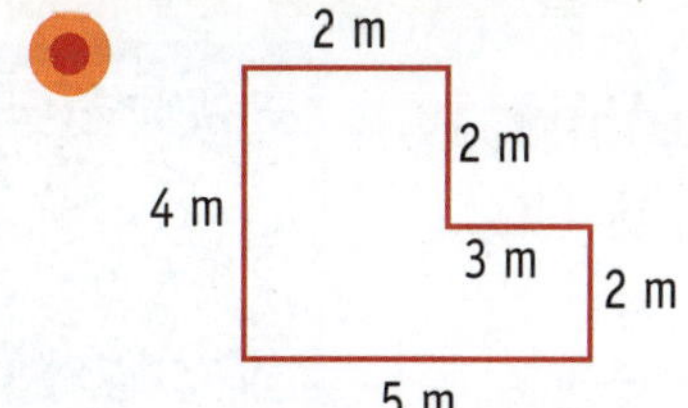

P = 2 m + 2 m + 3 m + 2 m + 5 m + 4 m
= 12 m

a

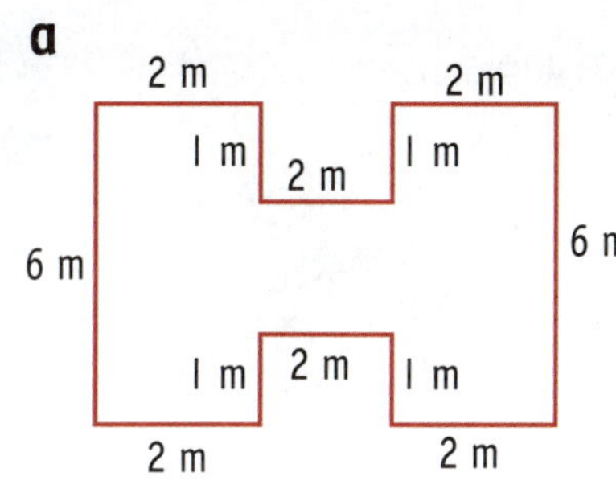

P = ____________________
= __________

b

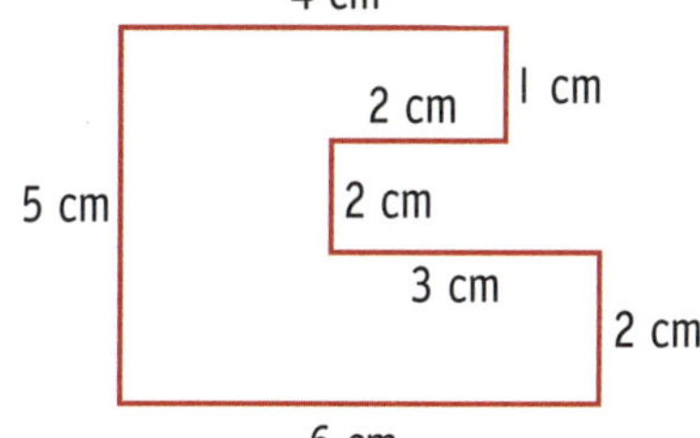

P = ____________________
= __________

c

4 m
5 m
3 m
10 m

P = ____________________
= __________

d

1 m
1 m
1 m
1 m
6.5 m
3.5 m

P = ____________________
= __________

e

4 m
5 m
5 m
15 m
4 m
10.5 m

P = ____________________
= __________

TEMPERATURE

Temperature is the measurement of how hot or cold something is. The units for measuring temperature are degrees Celsius (°C).

SCAN to watch video

The temperature on this thermometer is 35 °C.

35 °C would be the temperature on a hot summer day.

Example:
Draw arrows to show where the temperatures would appear on the thermometer.

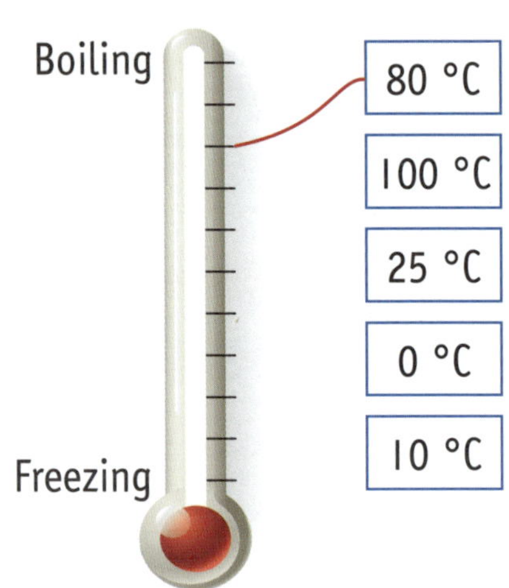

Check the answers on the video!

Your turn

Colour the thermometer sections to match the labels.

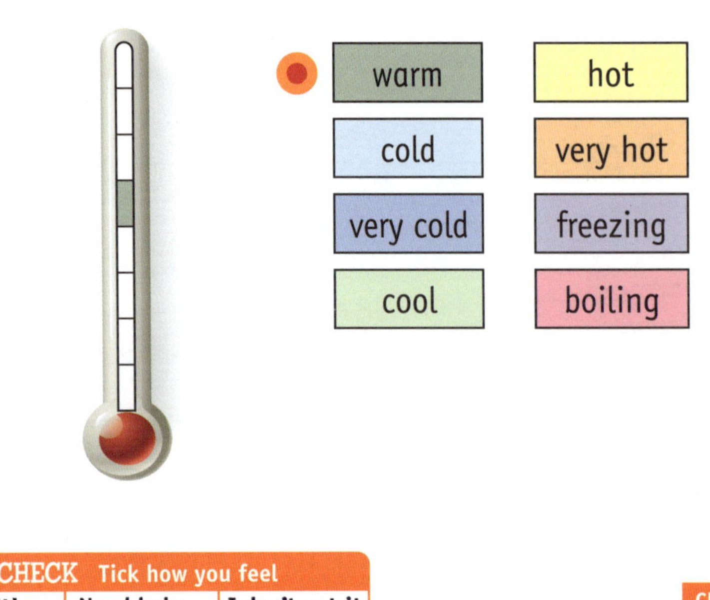

SELF CHECK Tick how you feel

Got it!	Need help...	I don't get it
☐	☐	☐

Check your answers
How many did you get correct? ☐

CATCH UP MATHS YEAR 6 BOOK B © PASCAL PRESS ISBN: 9781925726190

PRACTICE

1 Write the temperatures in words.

- 37 °C thirty-seven degrees Celsius

a 10 °C ______________________

b 46 °C ______________________

c 105 °C ______________________

d 490 °C ______________________

2 Write the temperatures in short form.

- twenty-four degrees Celsius 24°C

a sixty-three degrees Celsius _____

b seventy degrees Celsius _____

c nineteen degrees Celsius _____

d three hundred and two degrees Celsius _____

3 What is the temperature shown on each thermometer?

- 32 °C

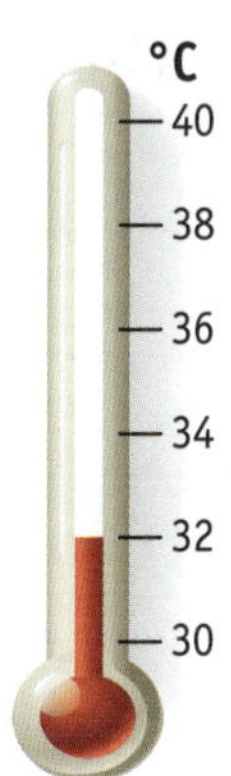

a _____

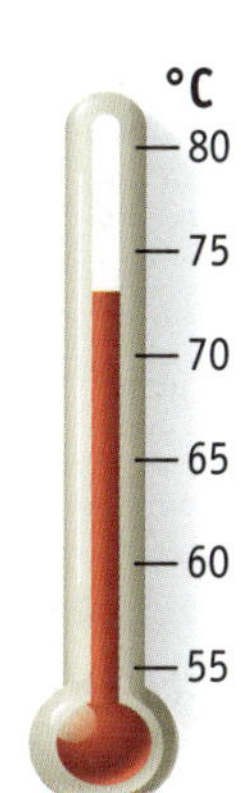

b _____

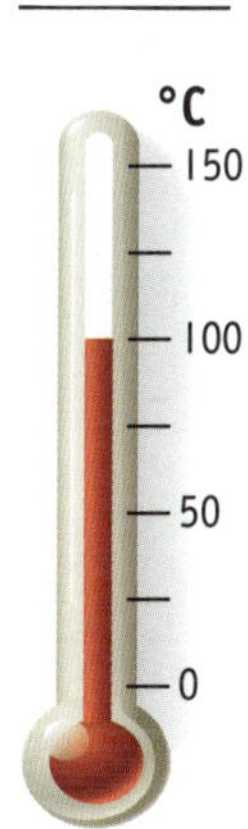

c _____

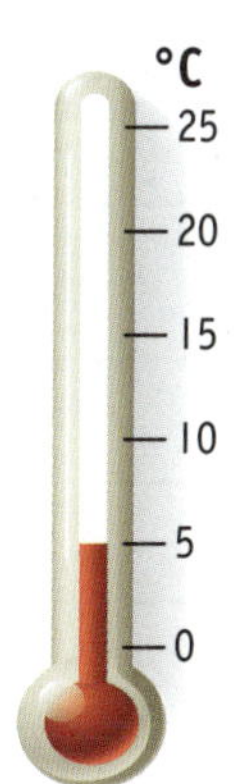

4 Colour each thermometer to show the temperature.

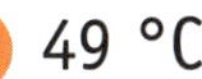

- 49 °C

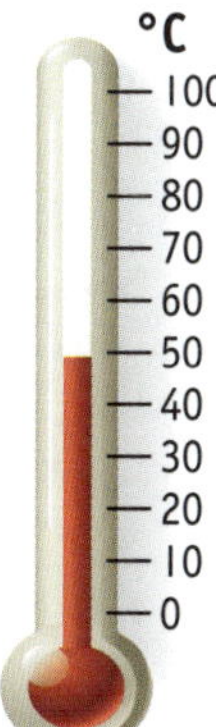

a 81 °C

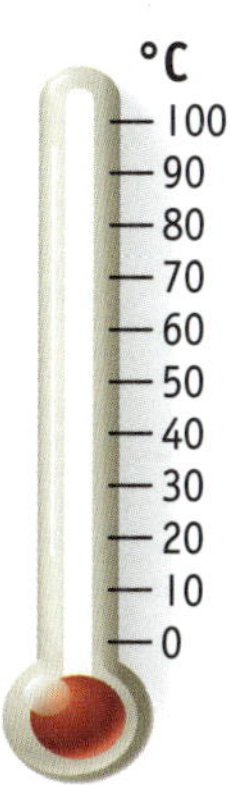

b 53 °C

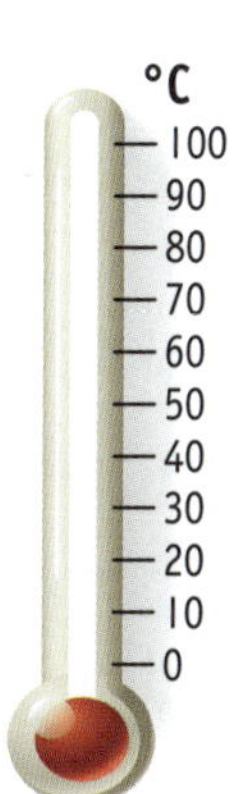

c 95 °C

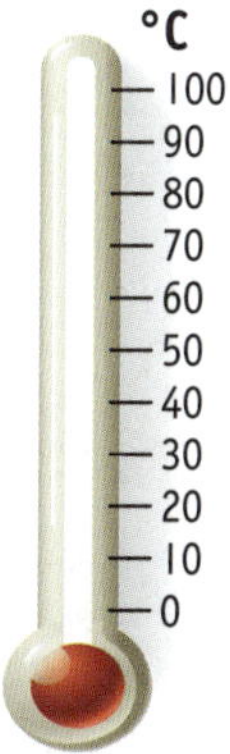

LENGTH REVIEW

1 How many centimetres?

a 3 m 26 cm ______

b 1 m 44 cm ______

c 2 m 2 cm ______

d 9 m 20 cm ______

e 10 m 20 cm ______

f 6 m 67 cm ______

g 4 m 5 cm ______

h 5 m 45 cm ______

i 84 m 14 cm ______

2 Write in m and cm.

a 249 cm ___ m ____ cm

b 480 cm ___ m ____ cm

c 802 cm ___ m ____ cm

d 734 cm ___ m ____ cm

e 175 cm ___ m ____ cm

f 5602 cm ___ m ____ cm

g 3023 cm ___ m ____ cm

h 9414 cm ___ m ____ cm

i 65 cm ___ m ____ cm

j 4384 cm ___ m ____ cm

3 Complete the tables.

	cm	m and cm	Decimal m
a	529 cm		
b		6 m 48 cm	
c			7.35 m
d			6.2 m
e		4 m 15 cm	
f	108 cm		
g			8.05 m

	cm	m and cm	Decimal m
h	342 cm		
i		1 m 6 cm	
j			5.85 m
k		2 m 66 cm	
l	689 cm		
m			11.27 m
n		18 m 38 cm	

4 Measure the lines and record the lengths in cm in the table.

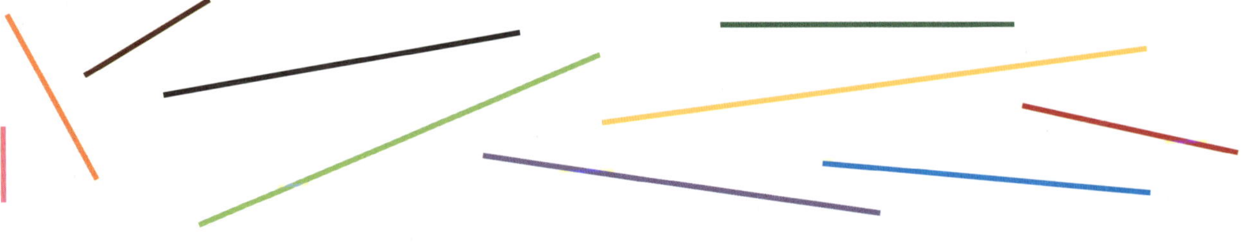

Line colour	a	b	c	d	e	f	g	h	i	j
Measurement (cm)										

CATCH UP MATHS YEAR 6 BOOK B © PASCAL PRESS ISBN: 9781925726190

5 Draw lines of these lengths.

a 5 cm

b $3\frac{1}{2}$ cm

c 10 cm

d $12\frac{1}{2}$ cm

6 Convert the lengths.

cm	32	78	5				422		950
m				10	3	8.9		8.03	

7 Complete the sentence.

One millimetre is _____-tenth of one centimetre.

8 Write these measurements in millimetres.

a 4 cm ________

b 12 cm ________

c 28 cm ________

d 49 cm ________

e 98 cm ________

f 8 cm ________

g 16 cm ________

h 36 cm ________

i 58 cm ________

j 64 cm ________

9 Write in millimetres the measurements marked by the arrows.

a A ______ **b** B ______ **c** C ______ **d** D ______ **e** E ______ **f** F ______

10 Mark and label these measurements on the ruler above.

a G: 17 mm

b H: 58 mm

c I: 82 mm

d J: 90 mm

e K: 121 mm

f L: 166 mm

REVIEW

11 Measure these lines in mm.

a ______

b ______

c ______

d ______

e ______

12 Fill in the tables.

	mm	cm	m
a	120		
b		15	
c		5	
d			0.70
e	360		

	mm	cm	m
f	4000		
g		125	
h			5.98
i	6350		
j		990	

13 Write the missing number.

1 kilometre = ______ metres

14 How many metres?

a 291 km ______

b 601 km ______

c 137 km ______

d 512 km ______

e 8 km ______

f 415 km ______

g 359 km ______

h 774 km ______

15 How many kilometres?

a 4000 m ______

b 32 000 m ______

c 49 000 m ______

d 805 000 m ______

e 1500 m ______

f 83 000 m ______

g 70 000 m ______

h 99 000 m ______

CATCH UP MATHS YEAR 6 BOOK B © PASCAL PRESS ISBN: 9781925726190

16 Write these lengths in metres.

a 10 km 258 m = ________

b 4 km 59 m = ________

c 7 km 34 m = ________

d 3 km 10 m = ________

e 19 km 345 m = ________

f 25 km 4 m = ________

g 8 km 246 m = ________

h 3 km 753 m = ________

17 Write these lengths in km and m.

a 3565 km = ___ km _____ m

b 4858 km = ___ km _____ m

c 3002 km = ___ km _____ m

d 7130 km = ___ km _____ m

e 9267 km = ___ km _____ m

f 3749 km = ___ km _____ m

g 5931 km = ___ km _____ m

h 2478 km = ___ km _____ m

18 Convert to kilometres.

a 39 m _______ km

b 215 m _______ km

c 437 m _______ km

d 608 m _______ km

e 590 m _______ km

f 8269 m _______ km

g 1401 m _______ km

h 7268 m _______ km

19 Complete the table.

	cm	m	km
a	400		
b			3
c		6.25	
d		2	
e			8.75
f	512		
g	600 000		
h		4000	
i		23	
j	950 000		

20 Calculate the perimeter.

a

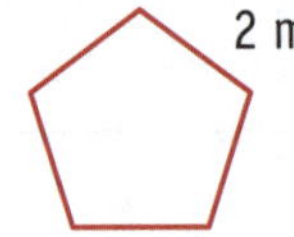

P = ____________________

= __________

b

2 cm
3 cm
1 cm
1 cm
8 cm
1 cm
1 cm
8 cm
1 cm
1 cm
2 cm
3 cm

P = ____________________

= __________

c

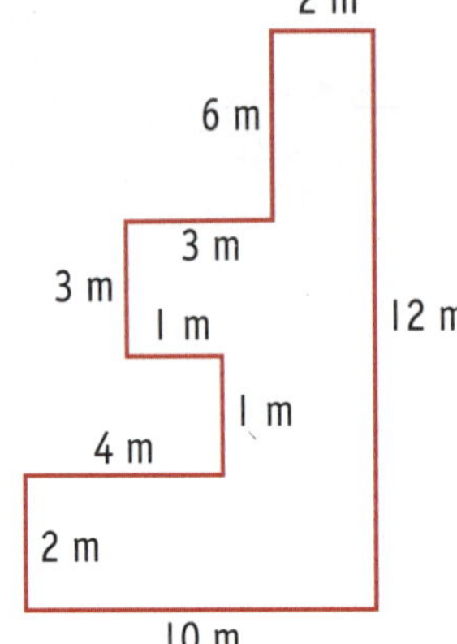

P = ____________________

= __________

d

4.3 m
5.8 m
3.2 m
6.3 m

P = ____________________

= __________

21 Complete the tables for rectangles with these measurements.

	Height	Width	Perimeter
a		25 cm	60 cm
b	30 mm		120 mm
c		21 m	80 m
d	55 cm	90 cm	
e		21 mm	88 mm
f	27 m	45 m	

	Height	Width	Perimeter
g	39 cm	63 cm	
h		35 mm	85 mm
i	15.5 m		91 m
j		6.8 cm	22 cm
k	12.7 mm	17.3 mm	
l	24.2 m	13.3 m	

CATCH UP MATHS YEAR 6 BOOK B © PASCAL PRESS ISBN: 9781925726190

22 Draw three different shapes with a perimeter of 20 cm.

23 What is the temperature shown on each thermometer?

a ______ b ______ c ______ d ______ e ______

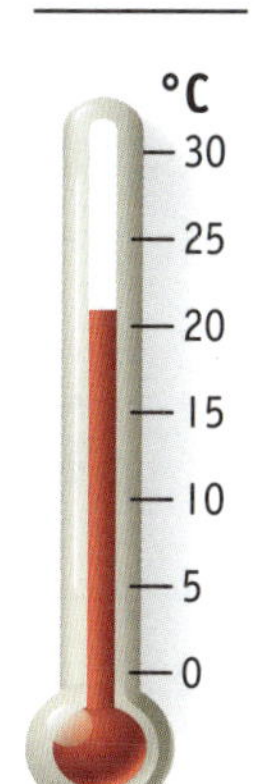

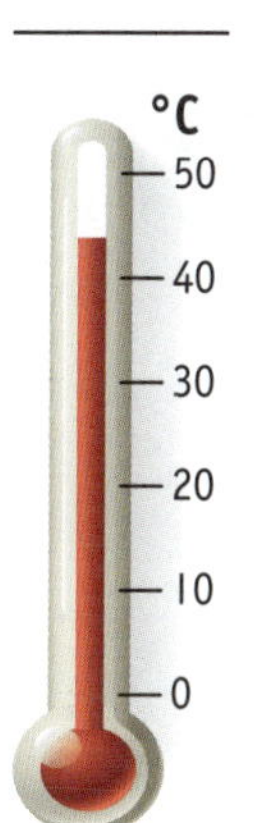

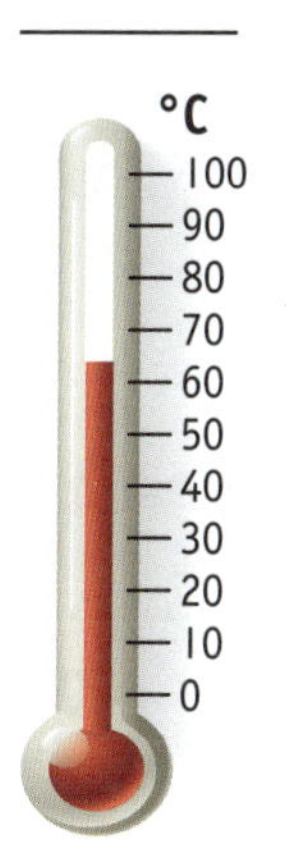

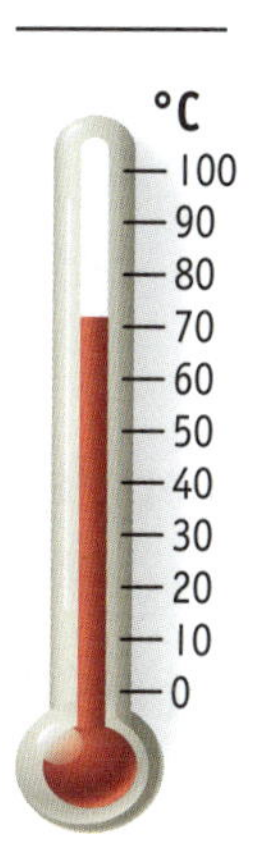

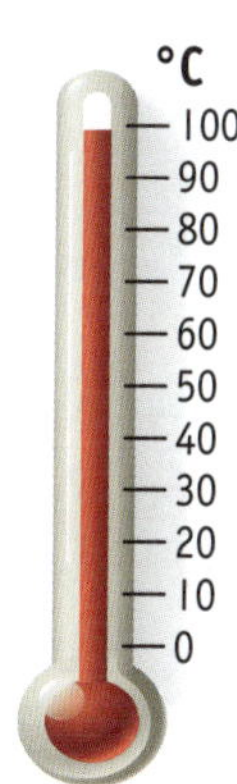

24 Colour each thermometer to show the temperature.

a 15 °C b 38 °C c 59 °C d 84 °C e 92 °C

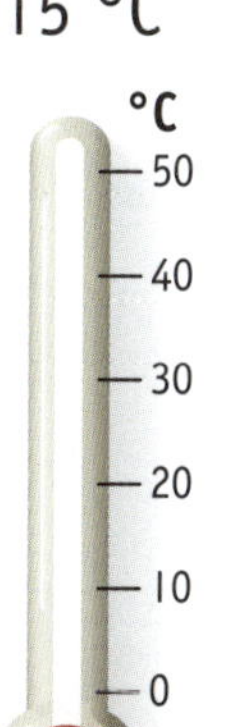

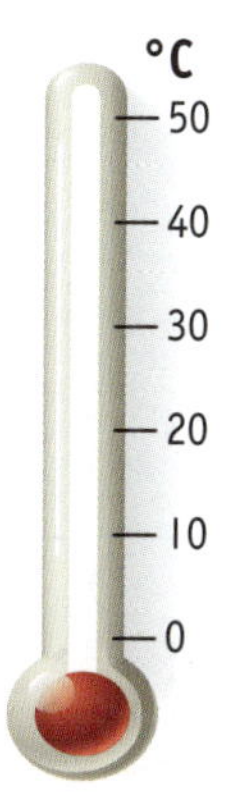

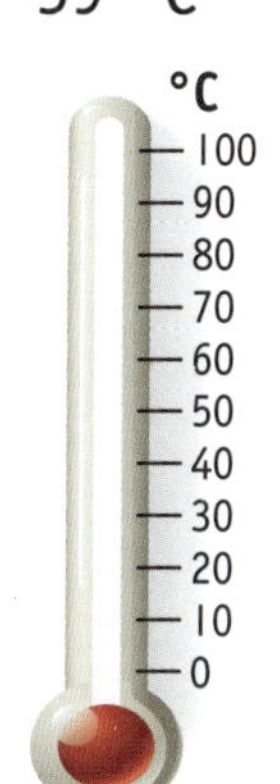

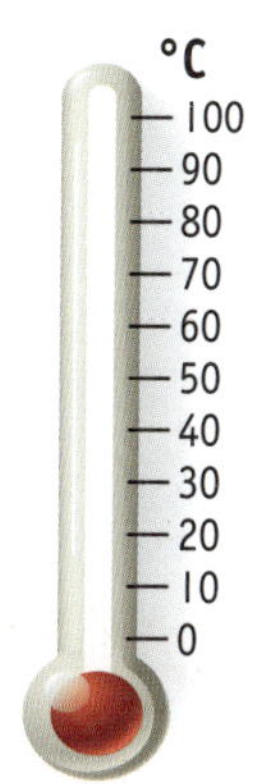

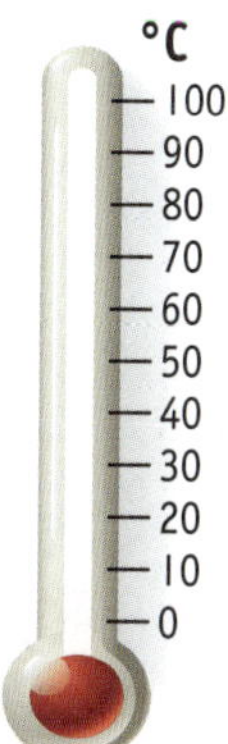

RIGHT ANGLES AND COMPLEMENTARY ANGLES

A right angle measures 90°.

This corner symbol shows the angle is 90°.

Two angles that add up to 90° are called complementary angles.

25°

65°

Complementary angles:
25° + 65° = 90°

SCAN to watch video

Complementary angles make 90° – the same size as a right angle.

Examples: Turn each ray into a right angle.
Mark each right angle with the corner symbol.

a b c d e

Check the answers on the video!

Your turn Draw five angles that match each description.

Smaller than 90°	90°	Larger than 90°

SELF CHECK Tick how you feel

Got it!	Need help...	I don't get it
☐	☐	☐

Check your answers
How many did you get correct? ☐

CATCH UP MATHS YEAR 6 BOOK B © PASCAL PRESS ISBN: 9781925726190

PRACTICE

Use a corner of a piece of paper to test which angles are right angles. Circle the right angles and cross the other angles.

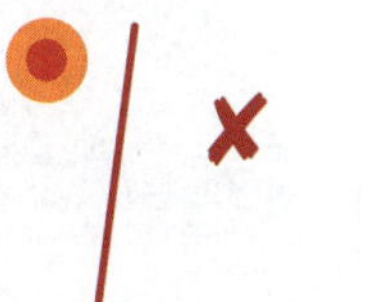

b

d

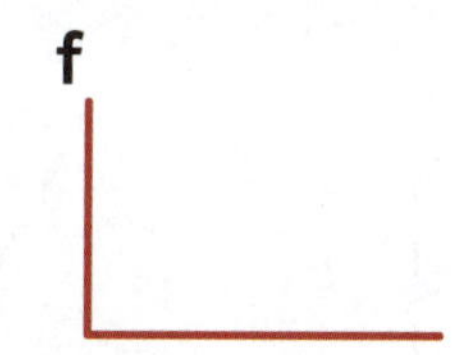

f

h

a

c

e

g

i

What is the value of x in each of these complementary angles?

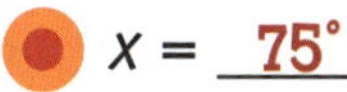

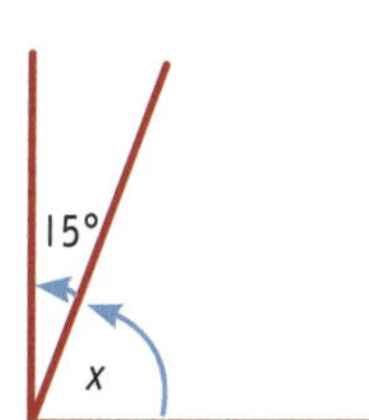

c $x =$ ____

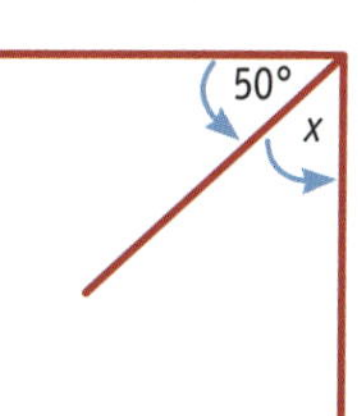

f $x =$ ____

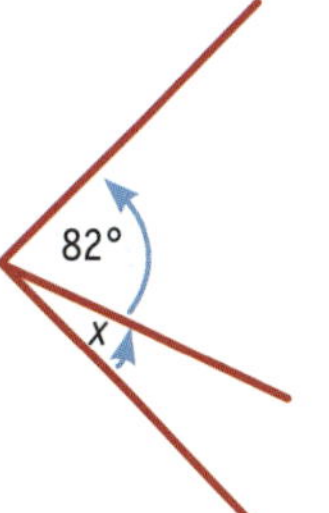

i $x =$ ____

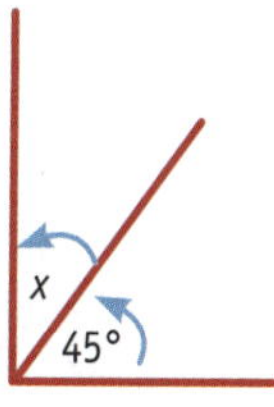

a $x =$ ____

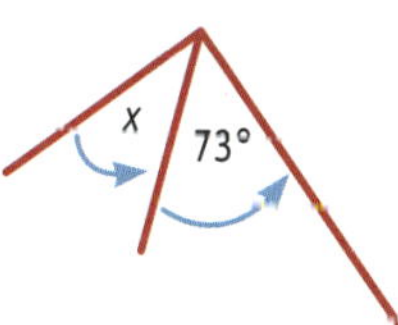

d $x =$ ____

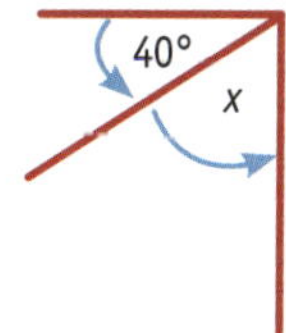

g $x =$ ____

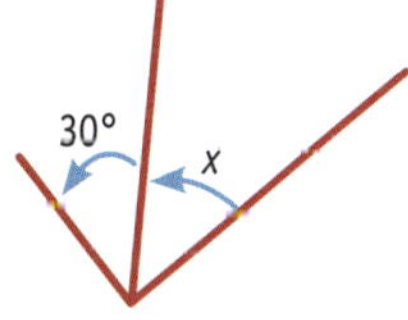

j $x =$ ____

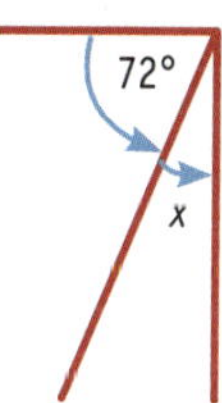

b $x =$ ____

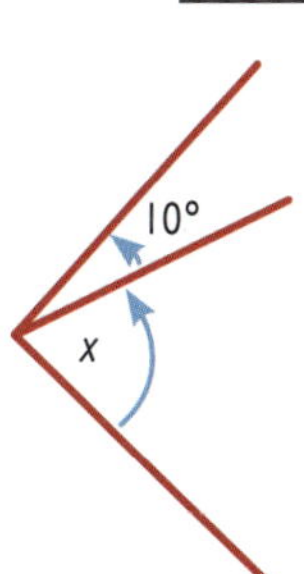

e $x =$ ____

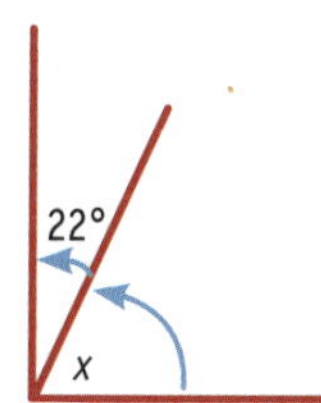

h $x =$ ____

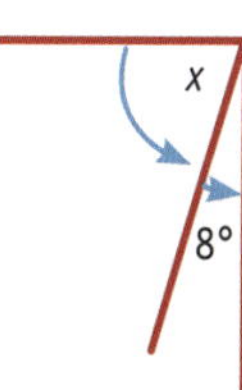

k $x =$ ____

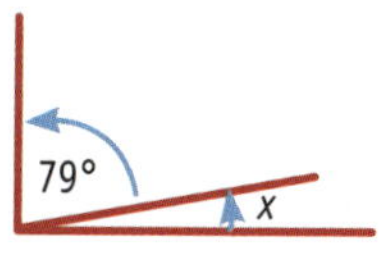

ACUTE AND OBTUSE ANGLES

Acute angles are less than 90°. Obtuse angles are more than 90°.

SCAN to watch video

Acute angles are sharp angles.

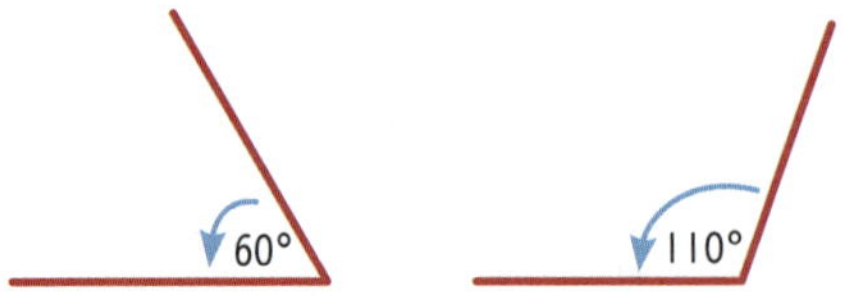

60°

110°

Obtuse angles are blunt angles.

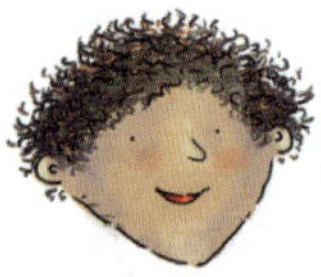

A cute little angle!

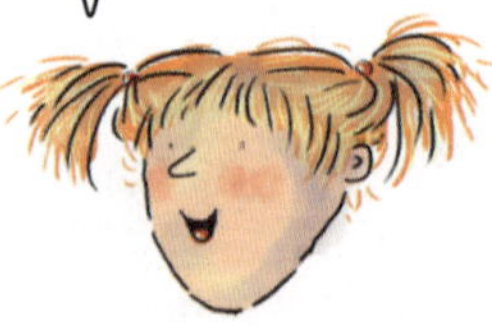

Examples: Trace over the acute angles with red and the obtuse angles with blue.

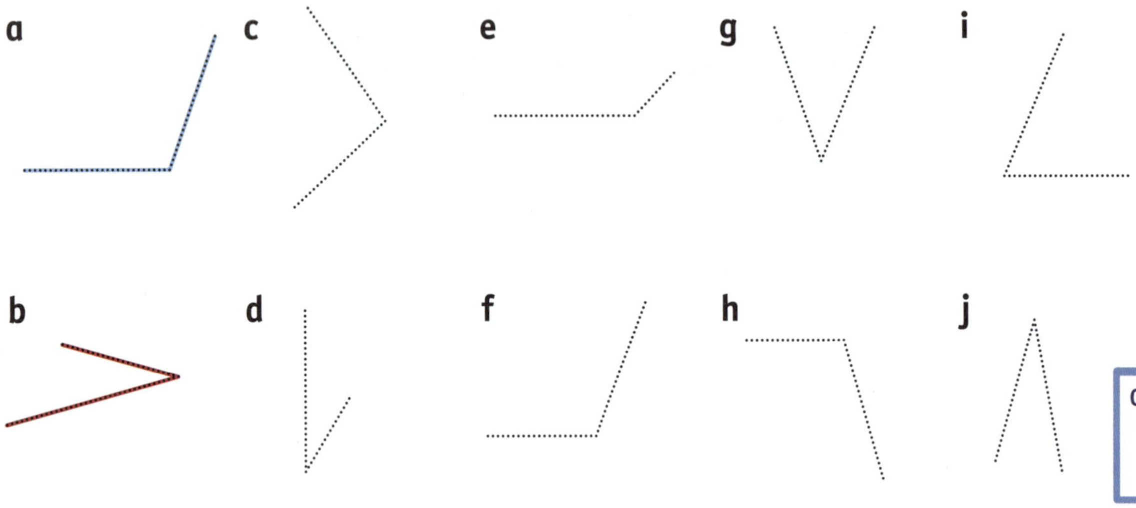

a

c

e

g

i

b

d

f

h

j

Check the answers on the video!

Your turn

Use a ruler to construct five different examples of each type of angle.

Five acute angles	Five obtuse angles

SELF CHECK Tick how you feel

Got it!	Need help...	I don't get it
☐	☐	☐

Check your answers

How many did you get correct?

PRACTICE

Use red to colour the acute angles and use blue to colour the obtuse angles. Then name the 2D shapes.

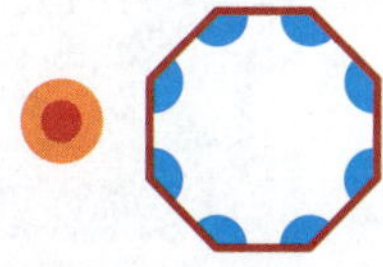 regular octagon

d ____________________

a  ____________________

e ____________________

b 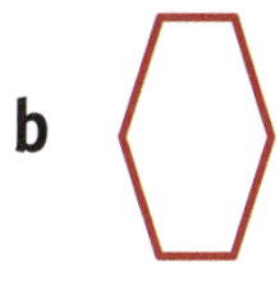 ____________________

f  ____________________

c 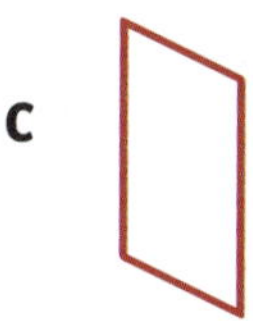 ____________________

g  ____________________

We name angles using three letters.
Put the letter at the vertex in the middle and use ∠ for the angle sign.

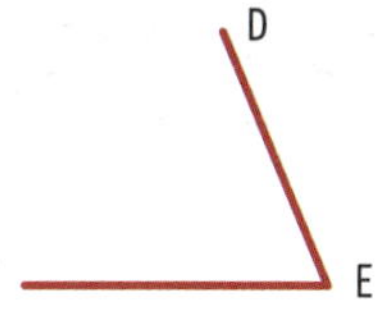

Name: ∠DEF or ∠FED

Type: Acute

a

Name: ____________

Type: ________

b

Name: ____________

Type: ________

Draw angles of the given size. Label them acute or obtuse.

45° acute (45°)	b 130° ________	d 85° ________	f 115° ________
a 75° ________	c 165° ________	e 10° ________	g 170° ________

USING A PROTRACTOR

Protractors are used to find the size of an angle.

How to measure an angle

1 Put the lower arm of the angle on the base line.

2 Put the vertex of the angle at the centre point.

3 Find 0° on the lower arm to work out if you use the inside or outside scale.

4 Read the scale.

This angle measures 75°.

Base line

Centre point

There is an inside scale and an outside scale.
The scales go from 0° to 180°.

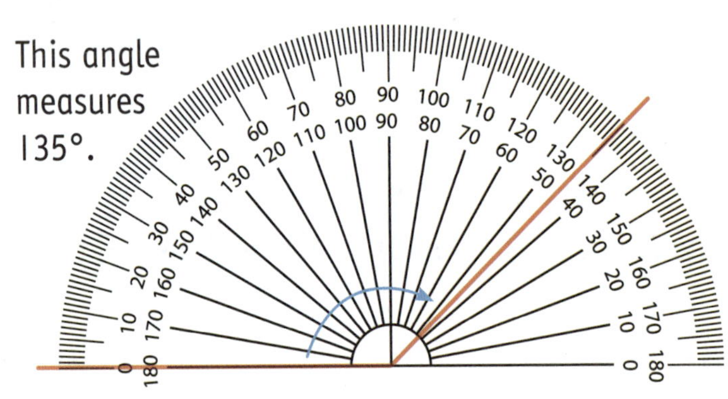

Should you use the inside scale or the outside scale?

Example:

Label the protractor at the right with these features.

a Inside numbers

b Base line

c Centre point

d Outside numbers

Check the answers on the video!

Your turn

Should you use the inside numbers (I) or the outside numbers (O) to measure these angles?

● O

a ☐

b ☐

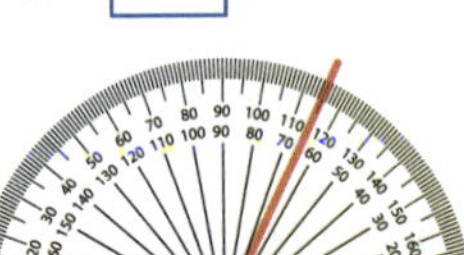

c ☐

SELF CHECK Tick how you feel		
Got it! ☐	Need help... ☐	I don't get it ☐

Check your answers
How many did you get correct? ☐

CATCH UP MATHS YEAR 6 BOOK B © PASCAL PRESS ISBN: 9781925726190

1 What is the size of each angle?

● 65°

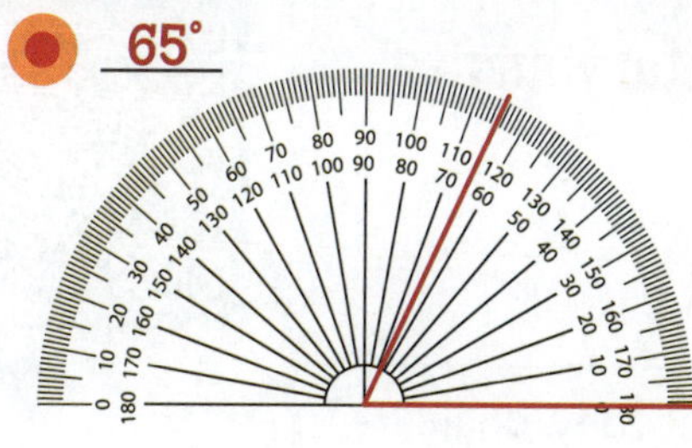

b ____

d ____

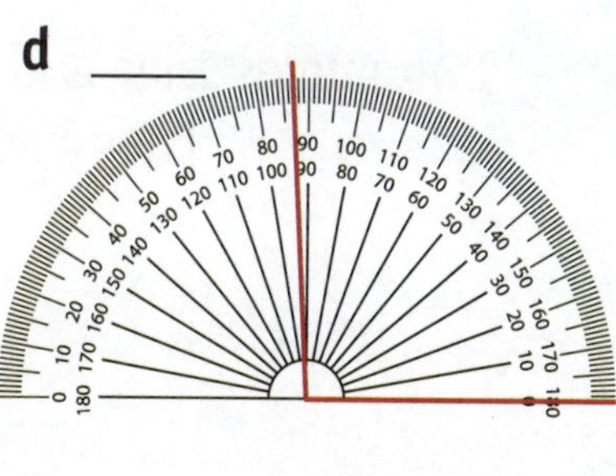

a ____

c ____

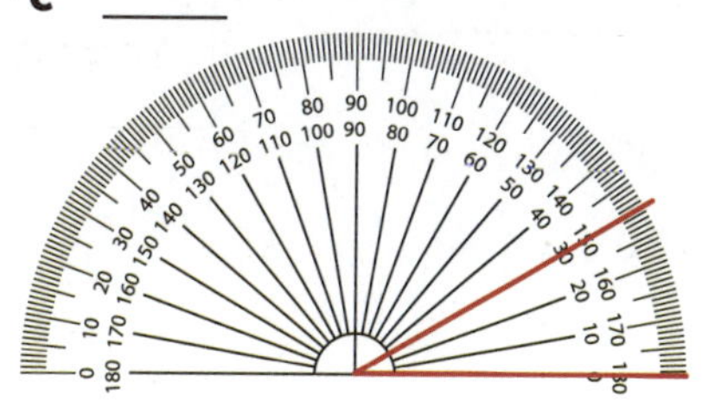

e ____

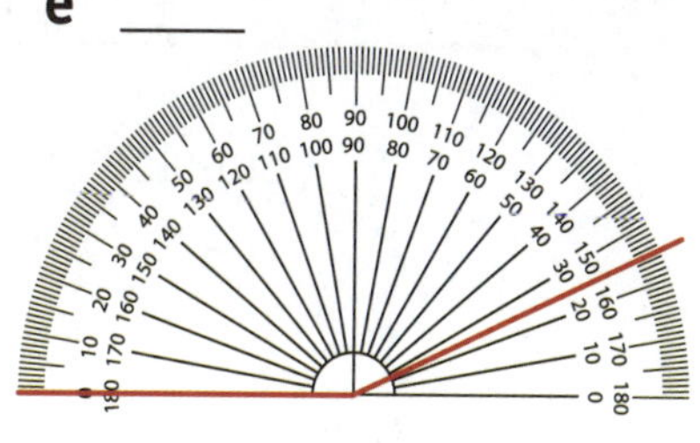

2 Use a protractor to construct angles of these sizes.

● 40°	c 77°	f 99°
a 62°	d 144°	g 182°
b 34°	e 236°	h 197°

3 Measure these angles.

● 40°

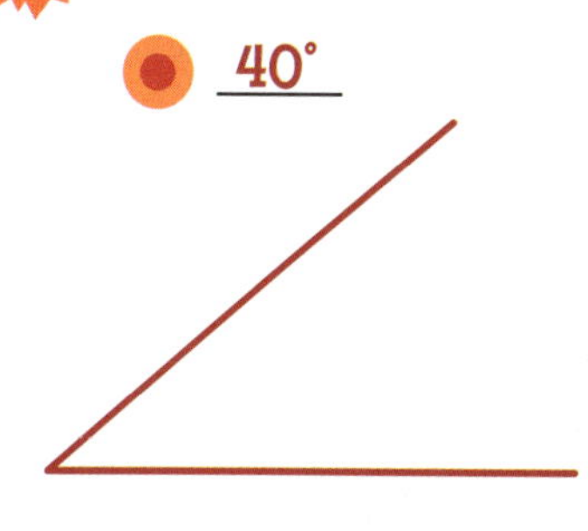

a ____

b ____

c ____

SUPPLEMENTARY ANGLES

Two angles that add up to 180° are called supplementary angles. They make a straight line, which is 180°.

105° 75°

Supplementary angles:
105° + 75° = 180°

Supplementary angles make 180° – the same size as a straight angle.

Examples:

Draw a line on each straight angle to form two supplementary angles. Label the supplementary angles *x* and *y*.

Check the answers on the video!

a *x* *y*

b

c

d

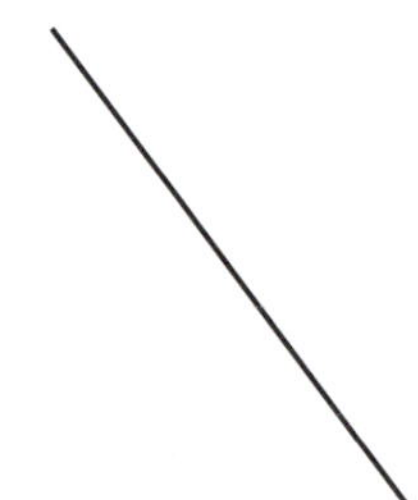

Your turn

Angles that add together to make a straight angle are supplementary angles. What is the missing supplementary angle?

● 140°

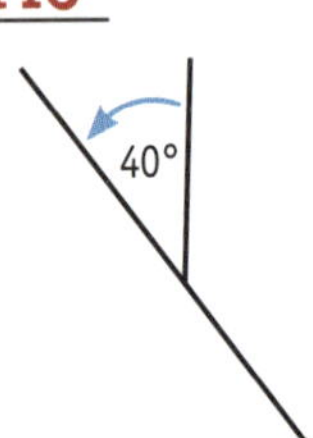

b ____

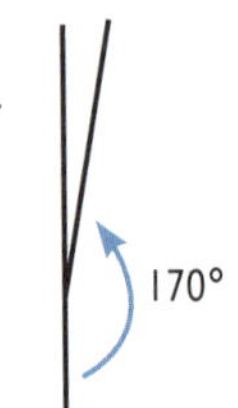

d ____

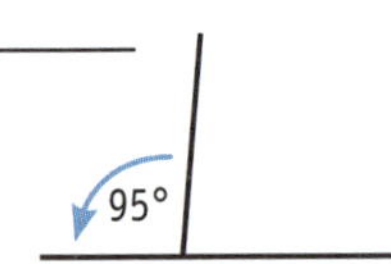

f ____

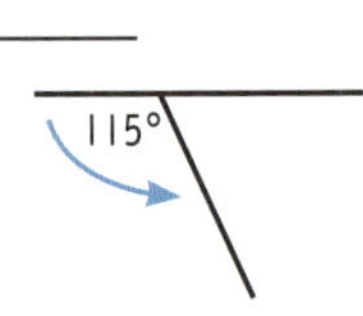

a ____

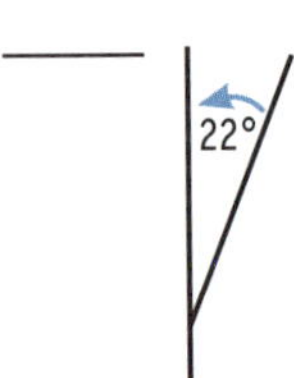

c ____

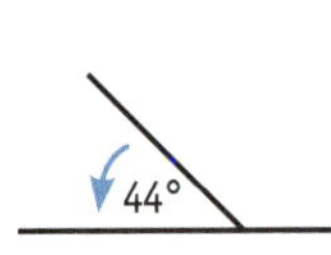

e ____

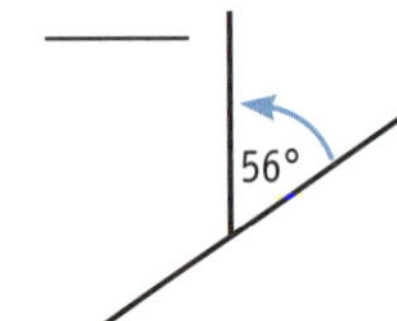

g ____

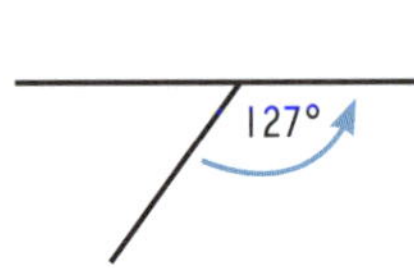

SELF CHECK Tick how you feel

Got it!	Need help...	I don't get it
		☐

Check your answers
How many did you get correct?

CATCH UP MATHS YEAR 6 BOOK B © PASCAL PRESS ISBN: 9781925726190

PRACTICE

1 What is the supplementary angle for each of these angles?

● 67° 113°	d 43° ____	h 10° ____	l 34° ____
a 21° ____	e 92° ____	i 86° ____	m 95° ____
b 103° ____	f 150° ____	j 70° ____	n 132° ____
c 171° ____	g 115° ____	k 147° ____	o 2° ____

2 Work out the missing values.

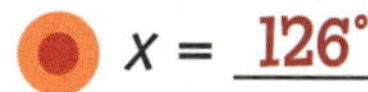

● x = 126°

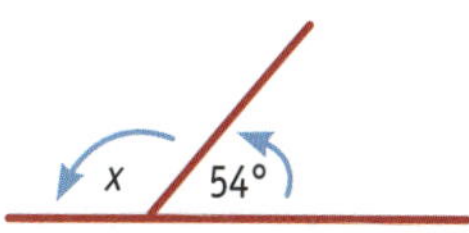

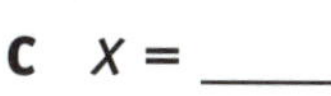

c x = ____

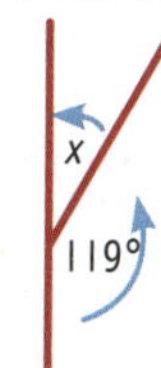

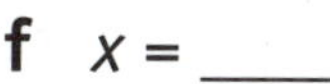

f x = ____

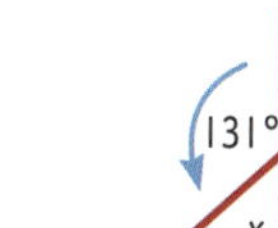

i x = ____

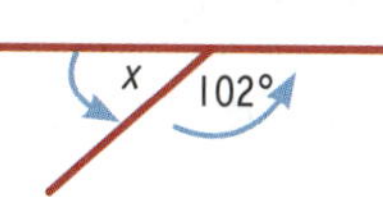

a x = ____

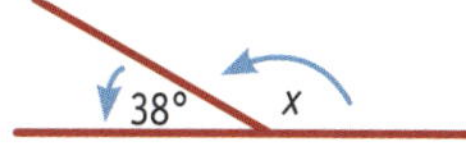

d x = ____

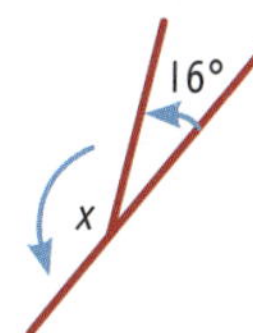

g x = ____

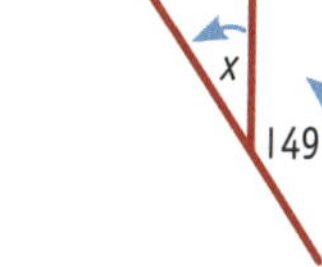

j x = ____

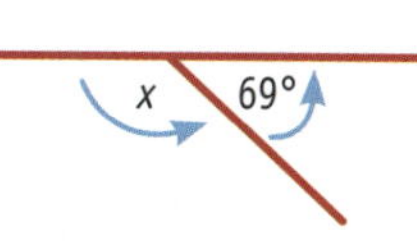

b x = ____

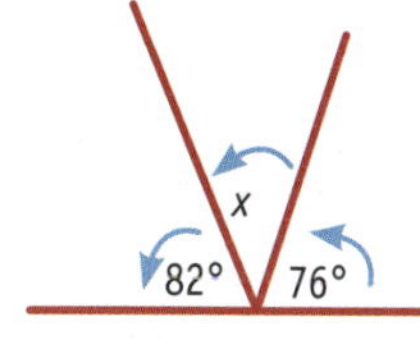

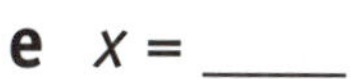

e x = ____

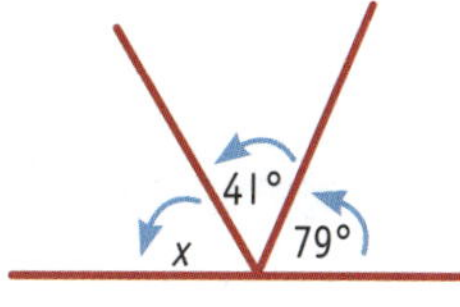

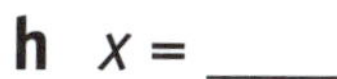

h x = ____

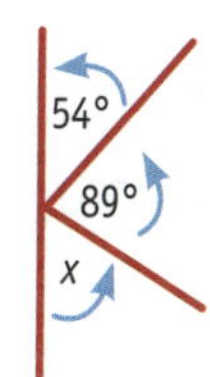

k x = ____

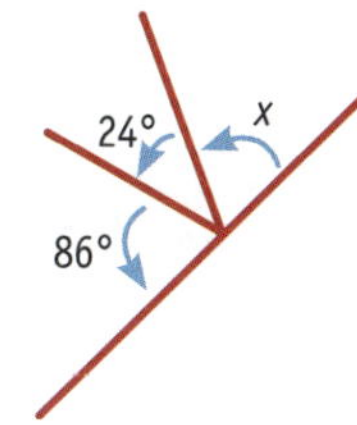

3 Use a protractor to construct these pairs of supplementary angles.

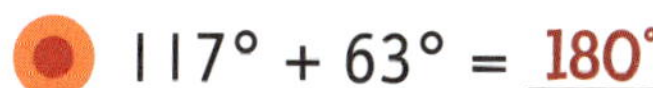

● 117° + 63° = 180°

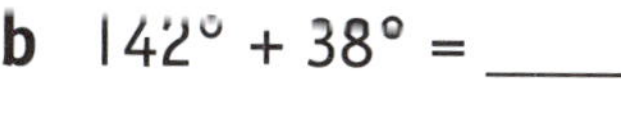

b 142° + 38° = ____

a 103° + 77° = ____

c 134° + 46° = ____

STRAIGHT, REFLEX AND REVOLUTION ANGLES

SCAN to watch video

Straight angles

Always 180°

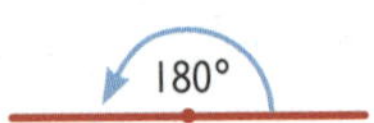

Looks like a straight line.

Reflex angles

Between 180° and 360°

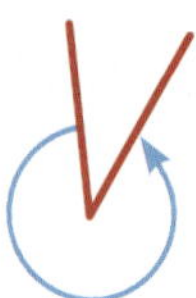

The arrow on the outside shows it is a reflex angle.

Revolution (full turn)

Always 360°

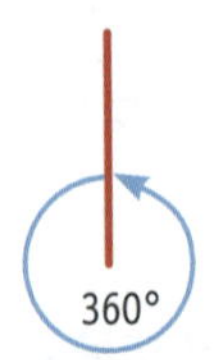

The two arms are together.

Examples:

Trace the straight angles with yellow, the reflex angles with green and the revolutions with purple.

a

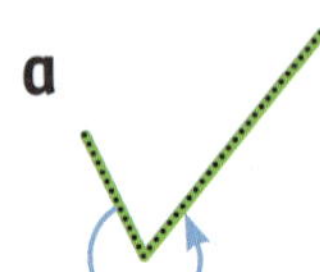

c

e

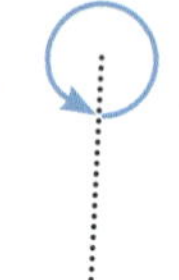

g

i

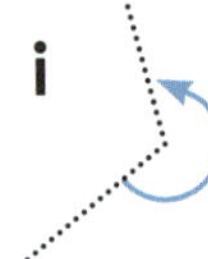

b

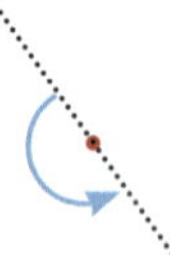

d

f

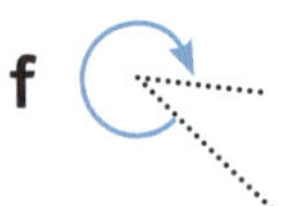

h

j

Check the answers on the video!

Your turn

Use a ruler to construct five different examples of each type of angle.

Straight angles	Reflex angles	Revolution angles
	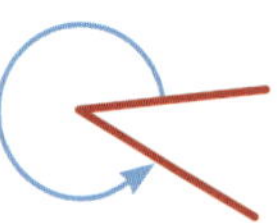	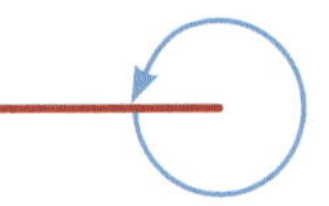

SELF CHECK Tick how you feel

Got it!	Need help...	I don't get it
☐	☐	☐

Check your answers

How many did you get correct? ☐

CATCH UP MATHS YEAR 6 BOOK B © PASCAL PRESS ISBN: 9781925726190

1 Label each angle as straight, reflex or revolution.

● revolution

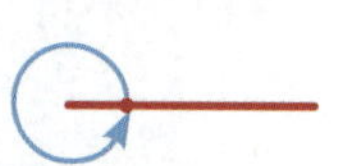

b ____________

d ____________

f ____________

a ____________

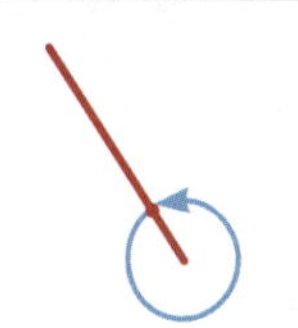

c ____________

e ____________

g ____________

2 Write the letter of the angle that matches the angle type.

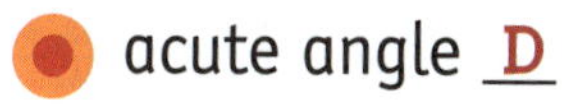

● acute angle D

a right angle __

b reflex angle __

c straight angle __

d obtuse angle __

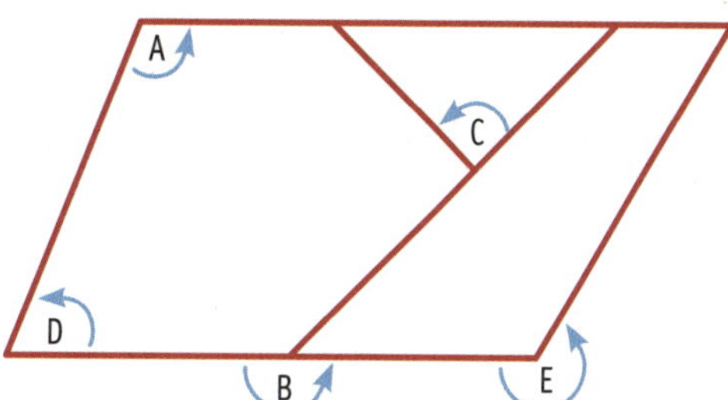

3 Find three other acute angles in the diagram above and colour them red.

4 Find two other obtuse angles in the diagram above and colour them blue.

5 Use a ruler and a protractor to draw these angles.

● A reflex angle of 184°	b A straight angle	d A reflex angle of 240°
a A reflex angle of 320°	c A revolution	e A reflex angle of 195°

  ISBN: 9781925726190

VERTICALLY OPPOSITE ANGLES

SCAN to watch video

Vertically opposite angles are equal.

When two lines cross, opposite angles are equal.

'Vertical' refers to the vertex where they cross, NOT up and down.

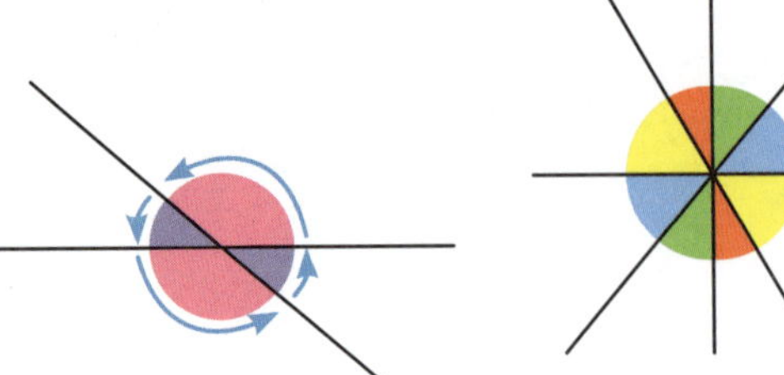

These angles add to 360°, a revolution.

Examples: Colour the angles that are equal the same colour.

a

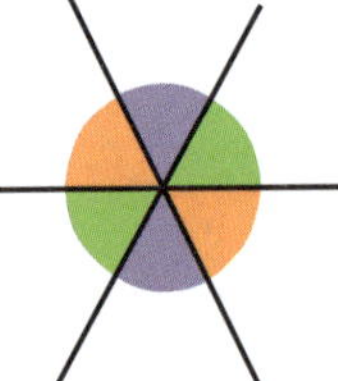

c

e

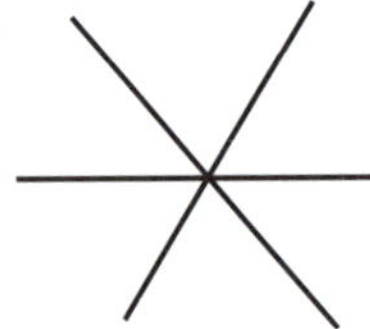

b

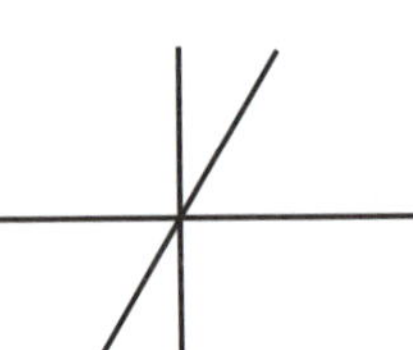

d

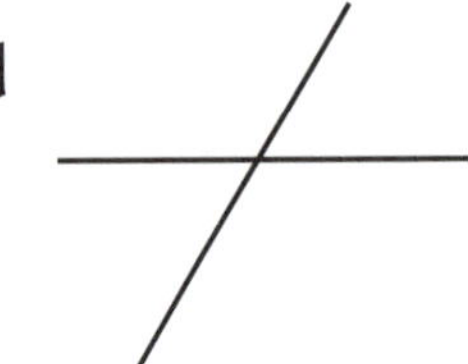

f

Check the answers on the video!

Your turn Find the value of x and y.

● $x =$ 50°, $y =$ 130°

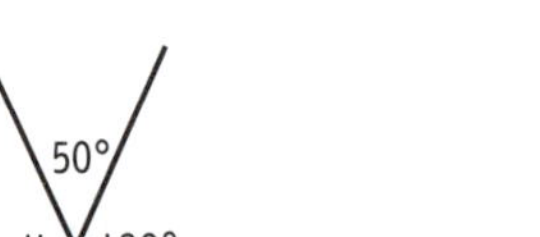

a $x =$ _____, $y =$ _____

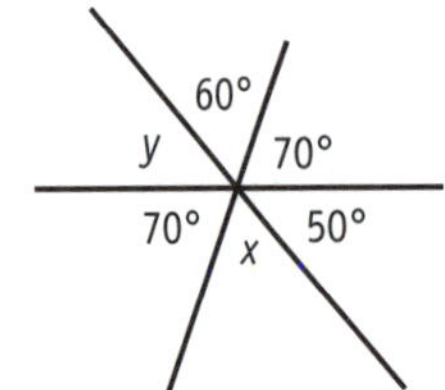

b $x =$ _____, $y =$ _____

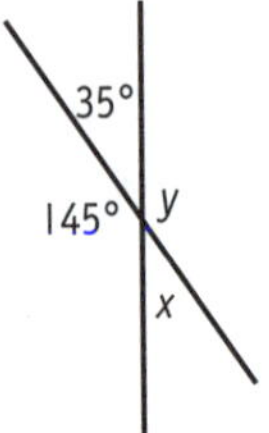

SELF CHECK Tick how you feel

Got it!	Need help...	I don't get it
 ☐	 ☐	☐

Check your answers

How many did you get correct? ☐

CATCH UP MATHS YEAR 6 BOOK B © PASCAL PRESS ISBN: 9781925726190

PRACTICE

1 Which angle is vertically opposite?

- ∠AOB ∠DOC

a ∠BOF ______

b ∠COD ______

c ∠DOE ______

d ∠AOE ______

e ∠FOC ______

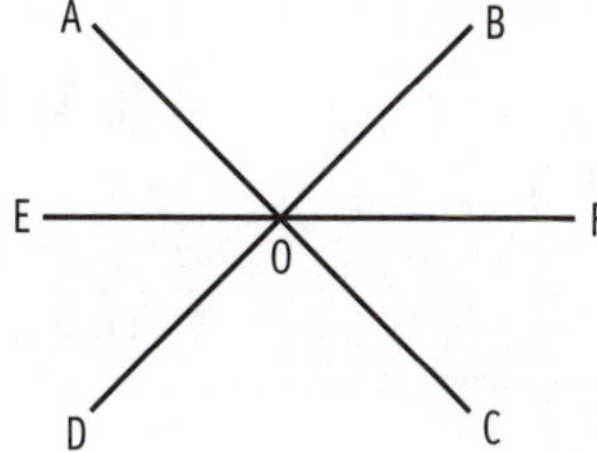

2 Colour the angle vertically opposite:

- ∠JOK blue

a ∠LOI red

b ∠LOG green

c ∠JOG orange

d ∠HOI pink

e ∠HOK grey

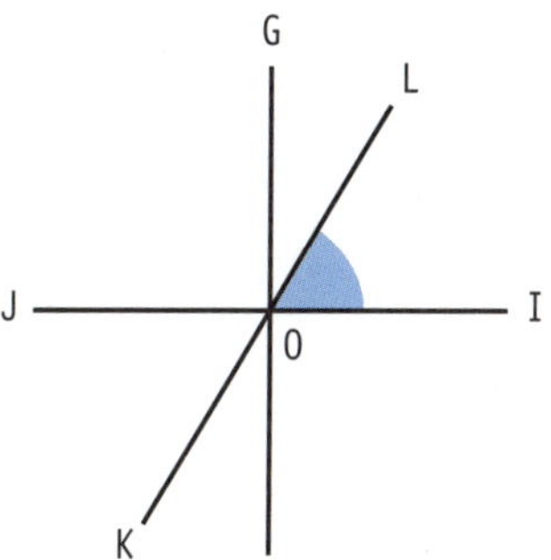

3 Find the unknown angles. Remember, the angles add to 360°.

- x = 65°, y = 115°

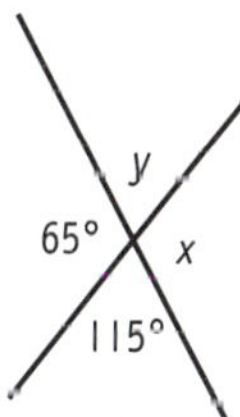

b x = ____, y = ____

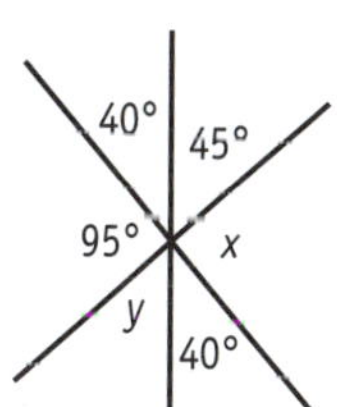

d x = ____, y = ____

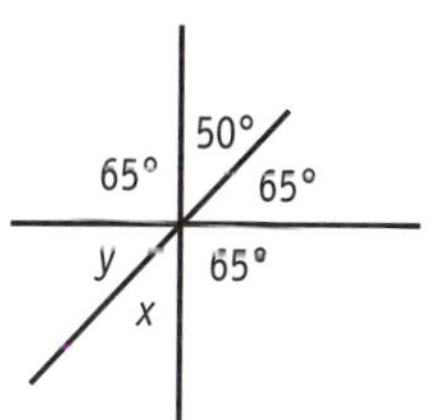

a x = ____, y = ____

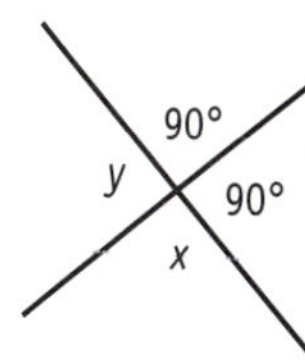

c x = ____, y = ____, z = ____

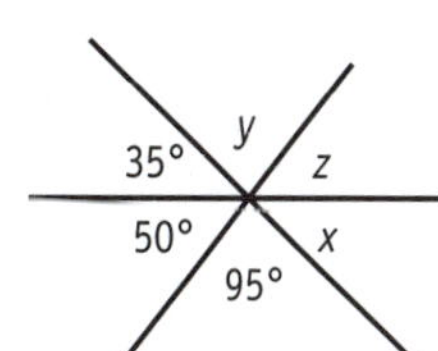

e x = ____, y = ____, z = ____

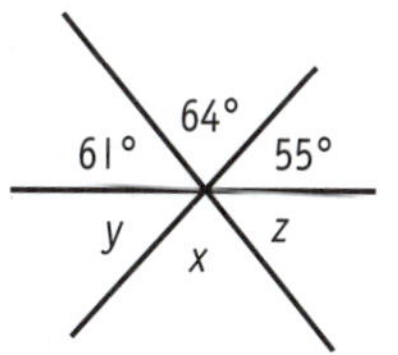

ANGLE SUM OF A TRIANGLE

The three angles in any triangle always add up to 180°.

Equilateral triangle

60° + 60° + 60°
= 180°

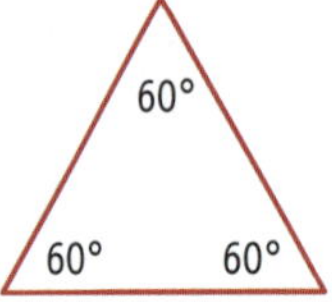

Right-angle triangle

45° + 45° + 90°
= 180°

Isosceles triangle

50° + 65° + 65°
= 180°

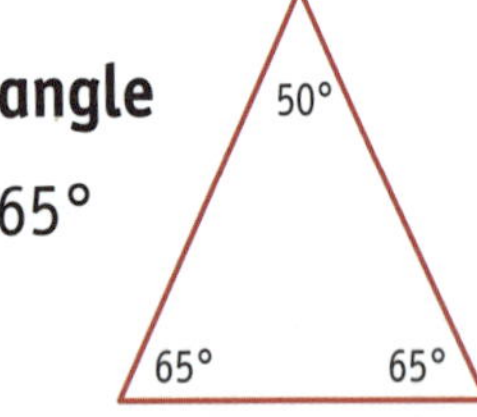

Scalene triangle

105° + 35° + 40°
= 180°

Example 1: Look at the sizes of the angles and then name the triangle.

a equilateral

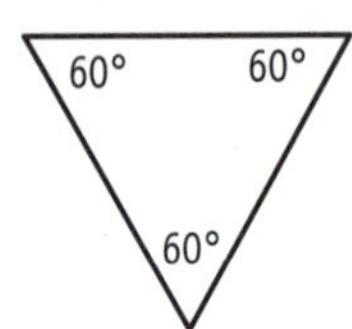

b ____________

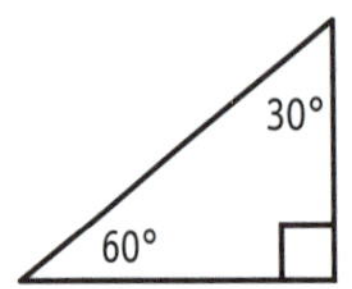

c ____________

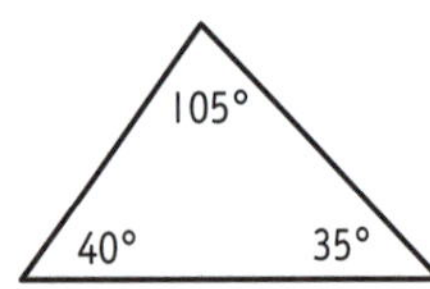

d ____________

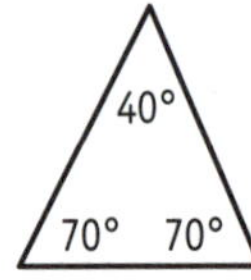

Check the answers on the video!

Example 2: Find the size of *x* and name the triangle.

a

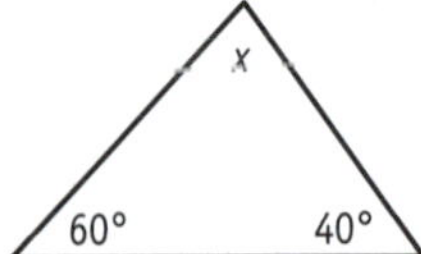

x = 80°

scalene

b

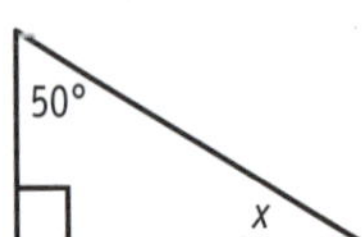

x = ____

____

c

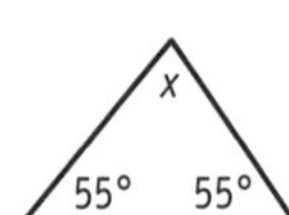

x = ____

____

d

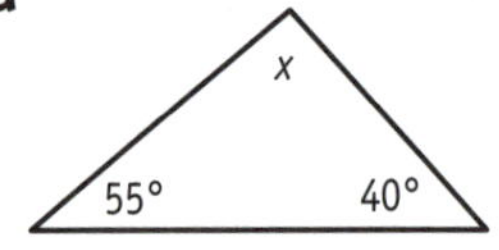

x = ____

____

Your turn Find x and name the triangle.

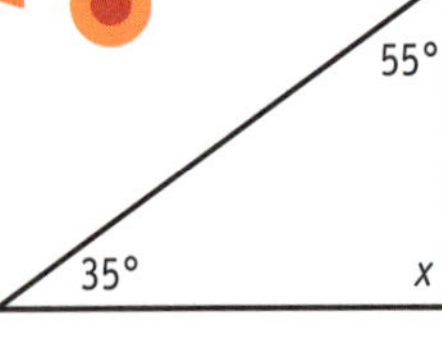

x = 90°

right angled

a

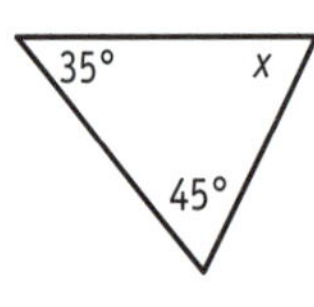

x = ____

____

b

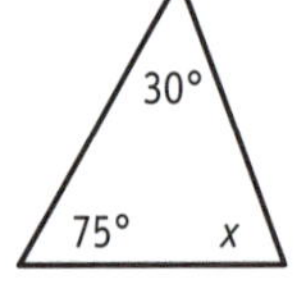

x = ____

____

c

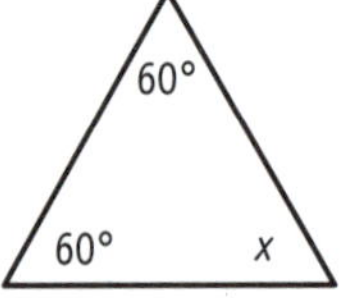

x = ____

____

SELF CHECK Tick how you feel

Got it!	Need help...	I don't get it
 ☐	 ☐	 ☐

Check your answers
How many did you get correct? ☐

 CATCH UP MATHS YEAR 6 BOOK B © PASCAL PRESS ISBN: 9781925726190

PRACTICE

1 Minh worked out the angle sizes of these triangles.
Write the total on the line inside the triangle.
Tick or cross the box to show whether Minh is correct.

● ✗
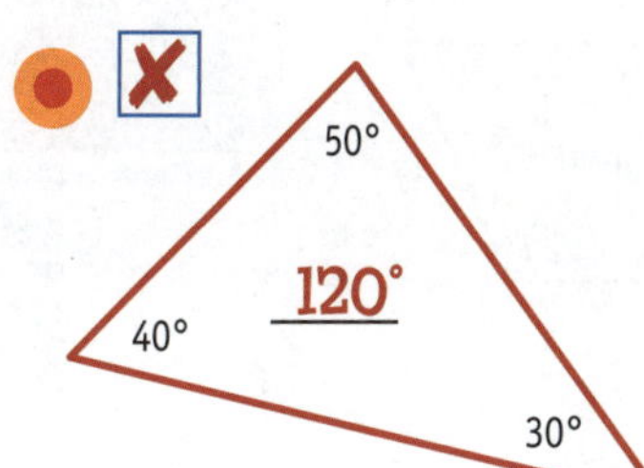

a ☐
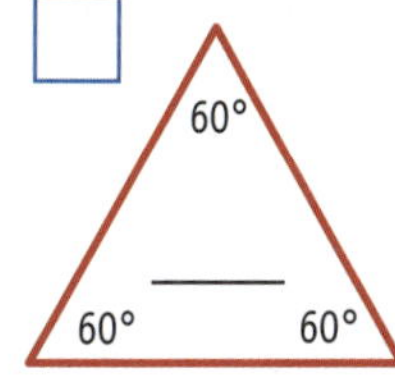

b ☐

45°
______
45°

c ☐
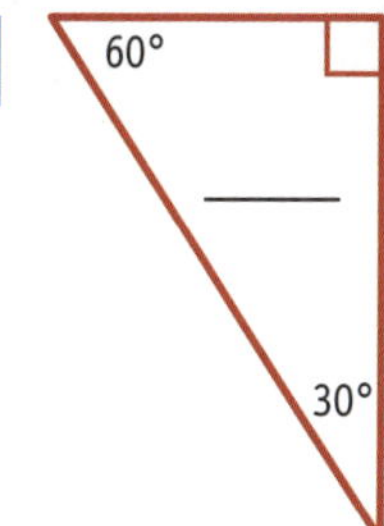

d ☐
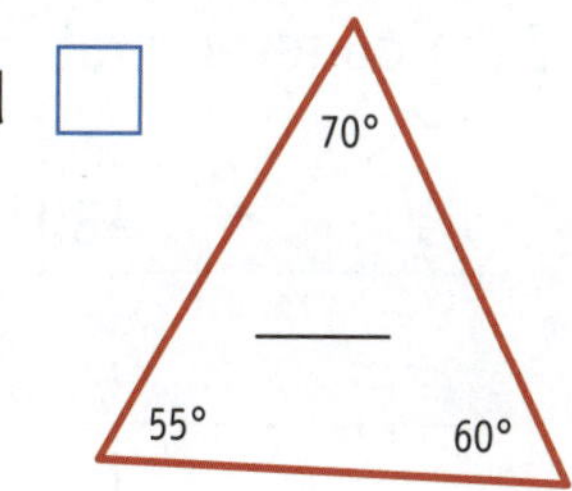

e ☐
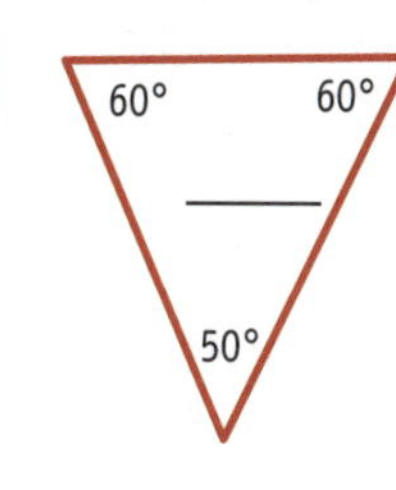

2 Find the size of the missing angle.

x = 50°

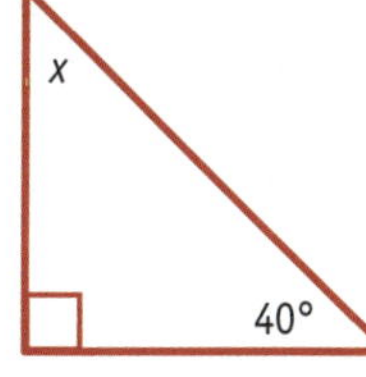

a x = ______

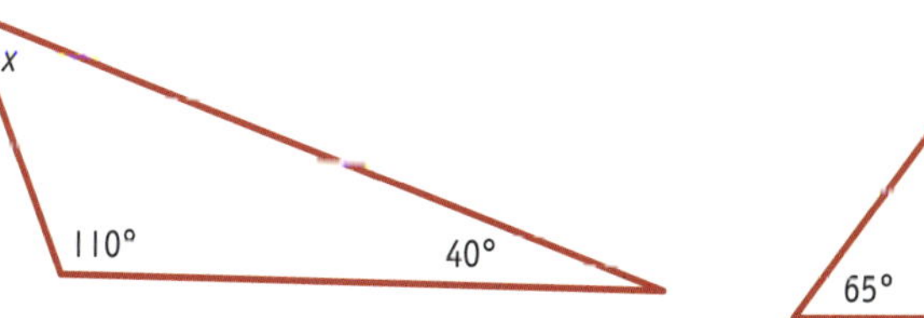

b x = ______

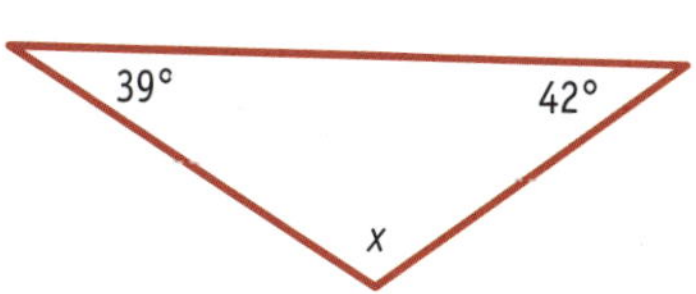

c x = ______

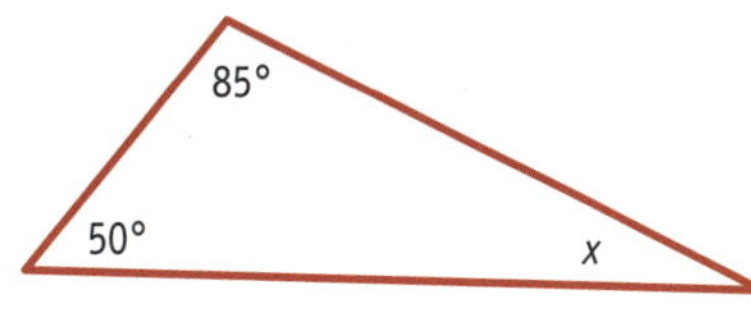

d x = ______

50°
65°
x

e x = ______

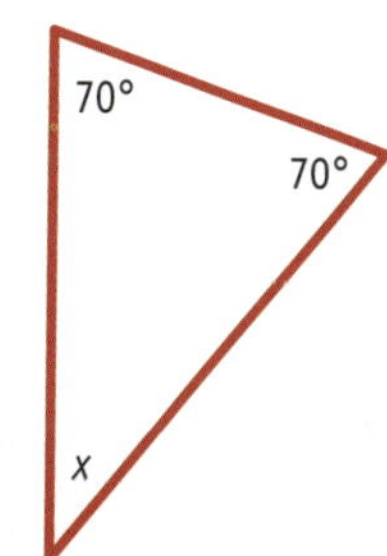

f x = ______

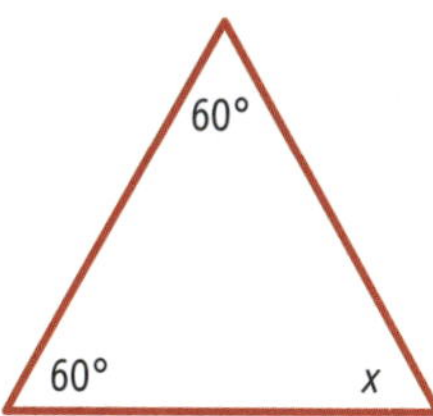

g x = ______

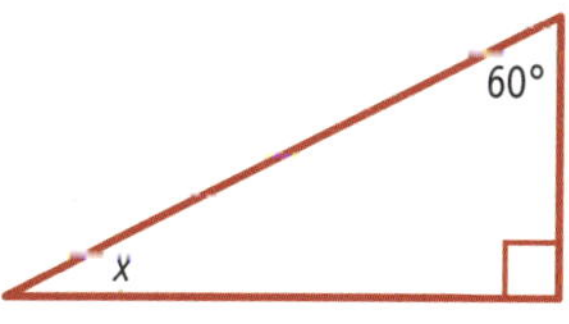

h x = ______

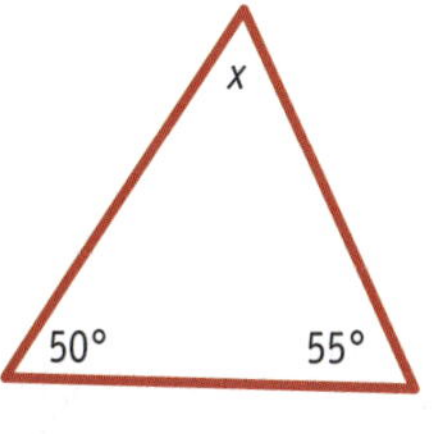

  ISBN: 9781925726190

ANGLE SUM OF QUADRILATERALS

The four angles in any quadrilateral always add to 360°.

SCAN to watch video

Each quadrilateral is made up of two triangles.

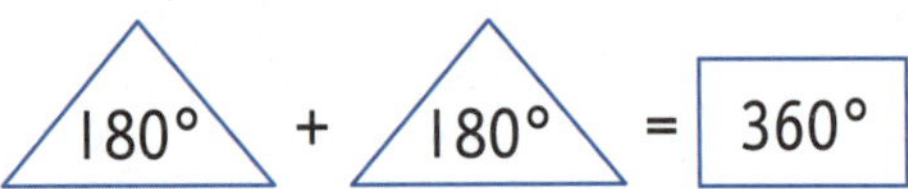

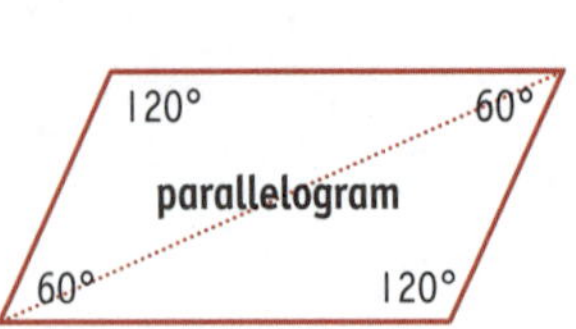

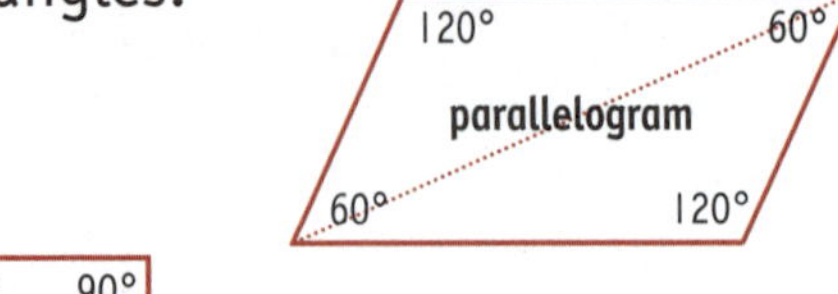

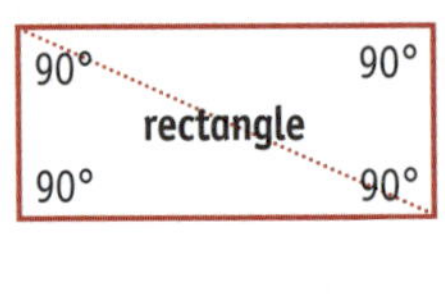

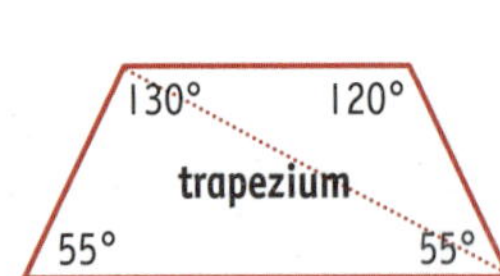

square: 90° 90° 90° 90°

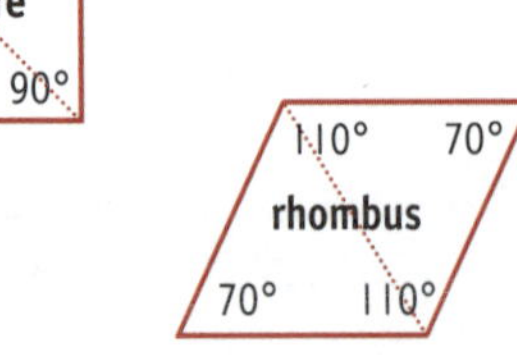

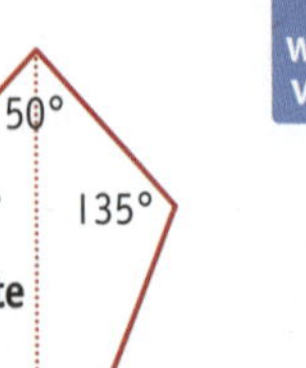

Examples: Divide each quadrilateral into two triangles.

a 

c 

e 

g 

b

d 

f 

h 

Check the answers on the video!

Your turn Find the size of the missing angle in each quadrilateral.

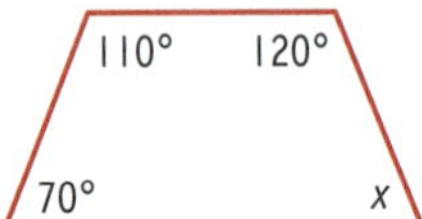

x = 360° – (110° + 120° + 70°)

= 60°

b 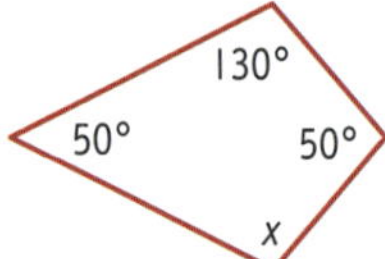

x = ____________________

= _____

a 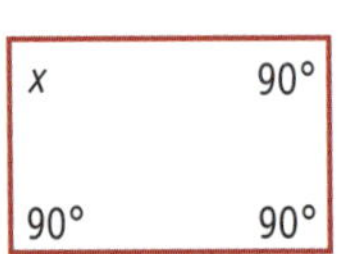

x = ____________________

= _____

c 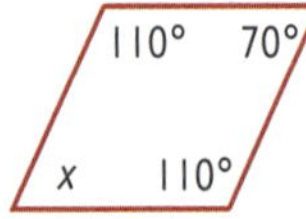

x = ____________________

= _____

SELF CHECK Tick how you feel

Got it!	Need help...	I don't get it
☐	☐	☐

Check your answers

How many did you get correct? ☐

CATCH UP MATHS YEAR 6 BOOK B © PASCAL PRESS ISBN: 9781925726190

1 Find the value of y in each quadrilateral.

● y = 360° - (175° + 55° + 50°)

= 80°

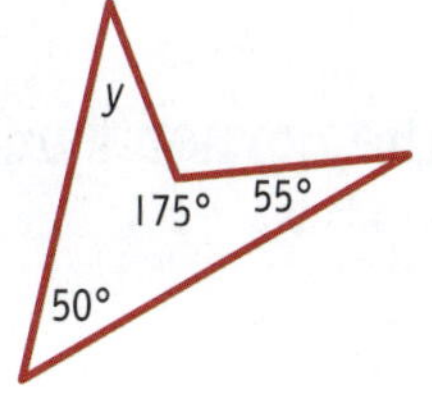

a ______________________

______

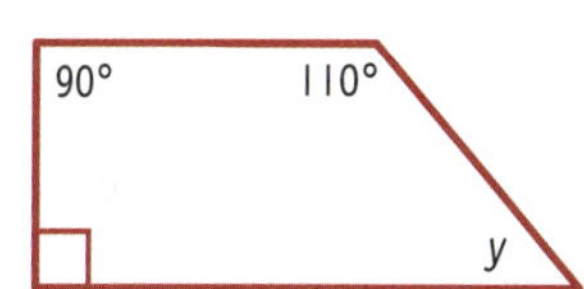

b ______________________

______

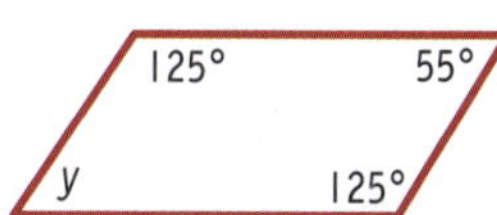

c ______________________

______

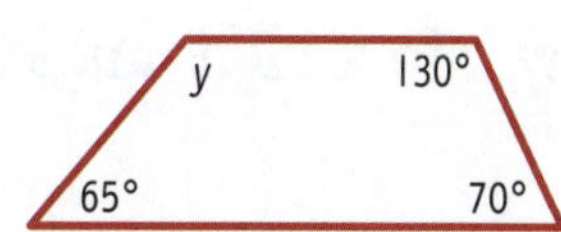

d ______________________

______

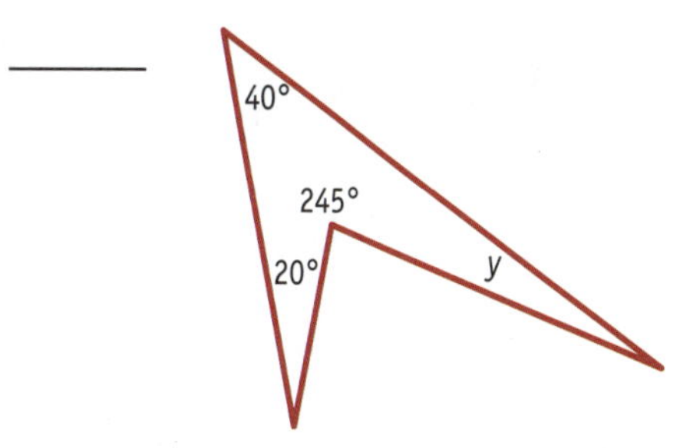

e ______________________

______

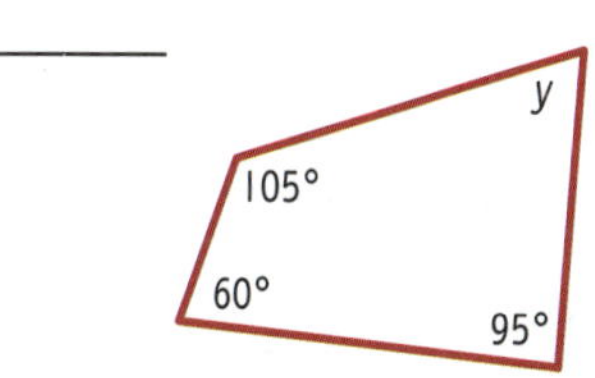

2 Measure the angles and then find the angle sum.

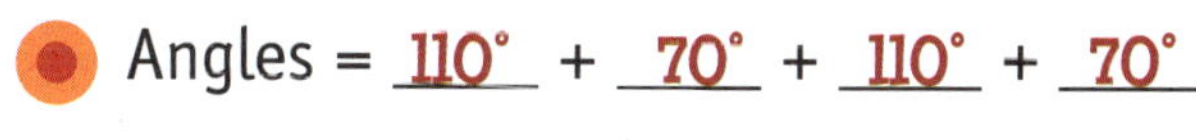

● Angles = 110° + 70° + 110° + 70°

Angle sum = 360°

a Angles = ____ + ____ + ____ + ____

Angle sum = ____

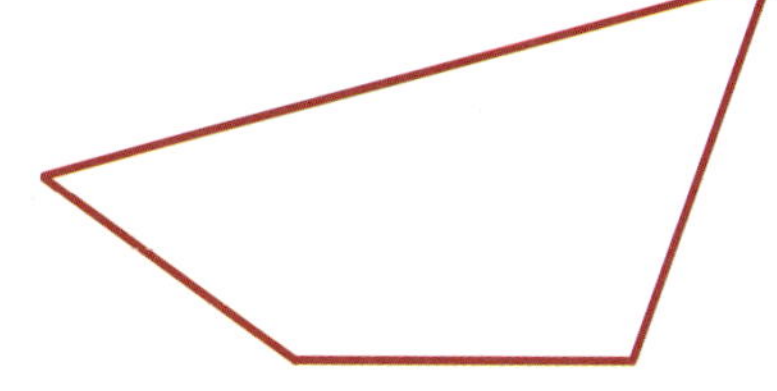

b Angles = ____ + ____ + ____ + ____

Angle sum = ____

c Angles = ____ + ____ + ____ + ____

Angle sum = ____

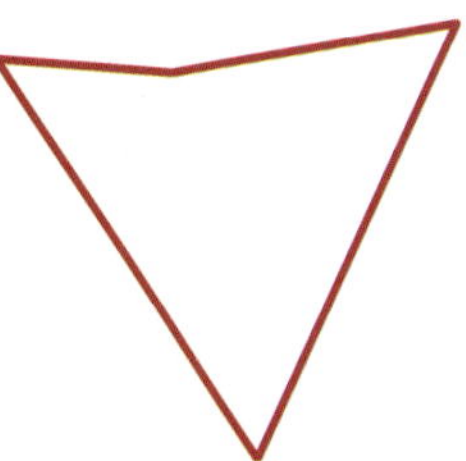

ANGLES REVIEW

1 Complete the sentence.

A right angle measures ___ degrees.

2 Turn each ray into a right angle. Mark each angle with the corner symbol.

3 What is the value of x in each of these complementary angles?

a x = ____ **b** x = ____ **c** x = ____ **d** x = ____ **e** x = ____

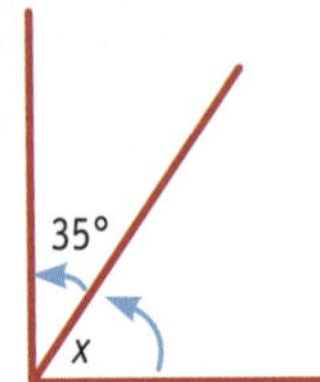

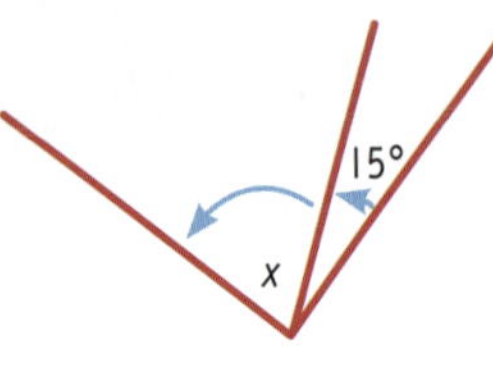

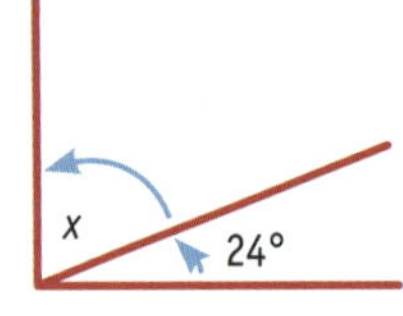

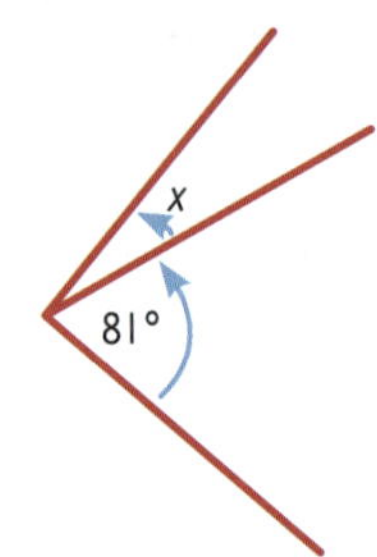

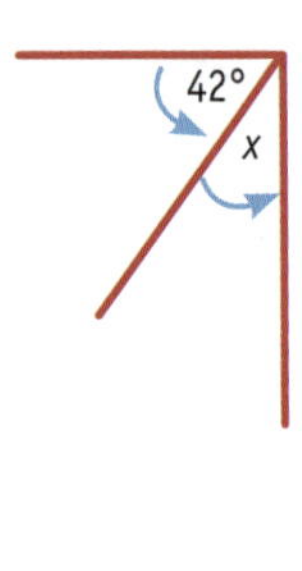

4 Complete the sentences.

a Acute angles are _______ than 90°. **b** Obtuse angles are _______ than 90°.

5 Measure the angles and write the size.
Trace the acute angles with red and the obtuse angles with blue.

a ____ **b** ____ **c** ____ **d** ____ **e** ____

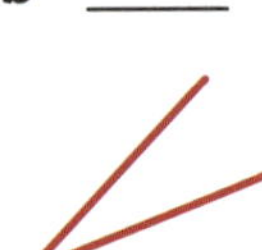

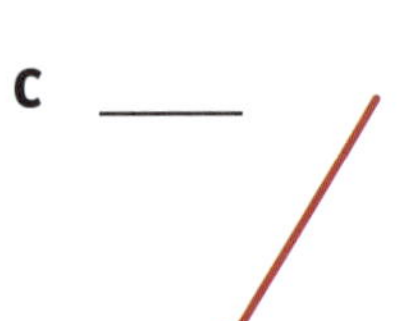

6 Draw angles of the given size. Label them acute or obtuse.

a 47° ________	**b** 126° ________	**c** 169° ________	**d** 33° ________

  ISBN: 9781925726190

7 Complete the sentences.

a Supplementary angles add up to ____.

b They make a ____________ angle.

8 What is the supplementary angle?

a	64° ____	c	101° ____	e	50° ____	g	3° ____
b	32° ____	d	167° ____	f	149° ____	h	11° ____

9 Use a protractor to measure and label the supplementary angles.

a
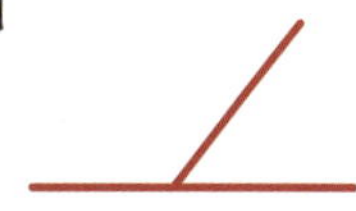

b

c

d
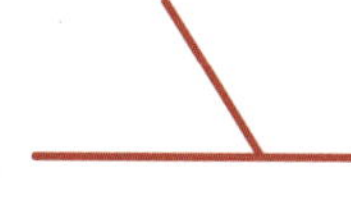

10 Construct the pairs of supplementary angles on the straight angles.

a 114° + 66°

b 33° + 147°

c 104° + 76°

d 82° + 98°

11 Complete the angle definitions.

____________ angle = ____°	____________ angle between ____° and ____°	____________ angle = ____°
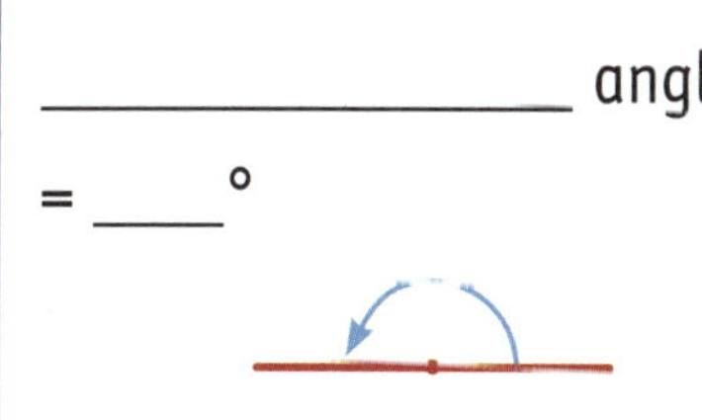	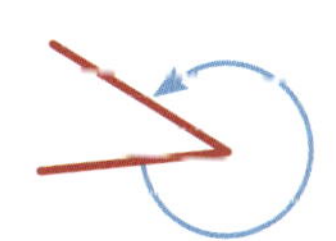	 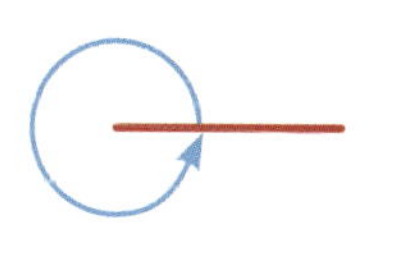

12 Draw angles of the given size.

180°	245°	360°

REVIEW

13 **Complete the sentence.**

Vertically opposite angles are __________.

14 **Find the missing values.**

a $x =$ ____, $y =$ ____

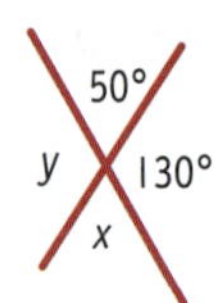

b $x =$ ____, $y =$ ____

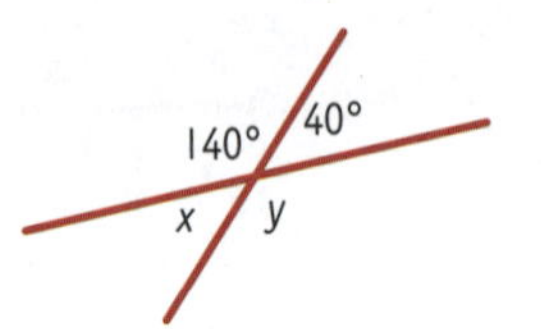

c $x =$ ____, $y =$ ____

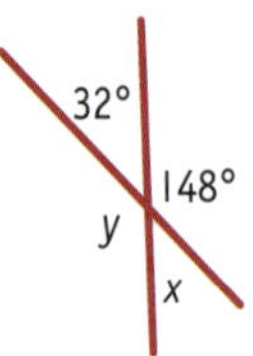

d $x =$ ____, $y =$ ____

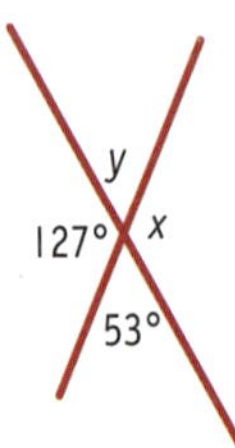

e $x =$ ____, $y =$ ____, $z =$ ____

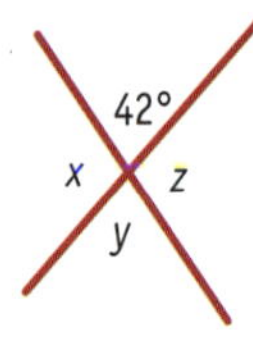

f $x =$ ____, $y =$ ____, $z =$ ____

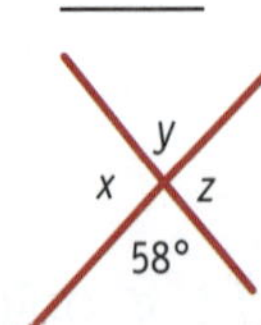

g $x =$ ____, $y =$ ____, $z =$ ____

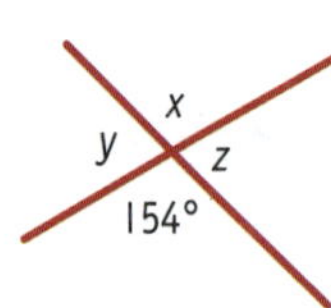

h $x =$ ____, $y =$ ____, $z =$ ____

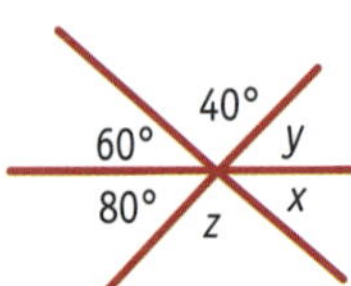

i $x =$ ____, $y =$ ____, $z =$ ____

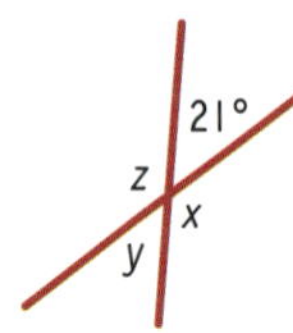

j $x =$ ____, $y =$ ____, $z =$ ____

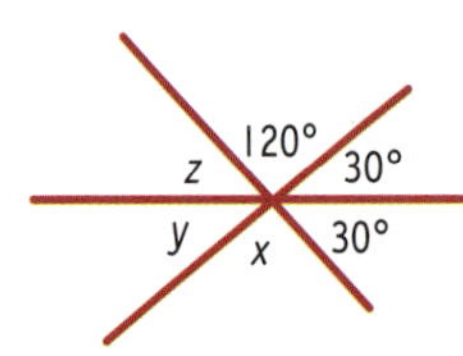

k $x =$ ____, $y =$ ____, $z =$ ____

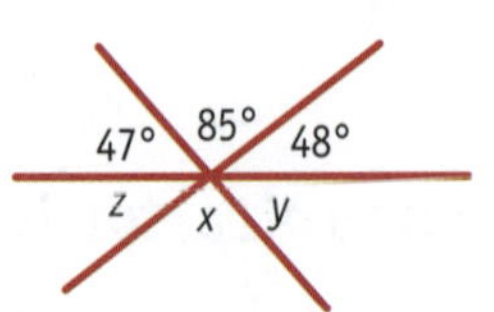

l $w =$ ____, $x =$ ____, $y =$ ____, $z =$ ____

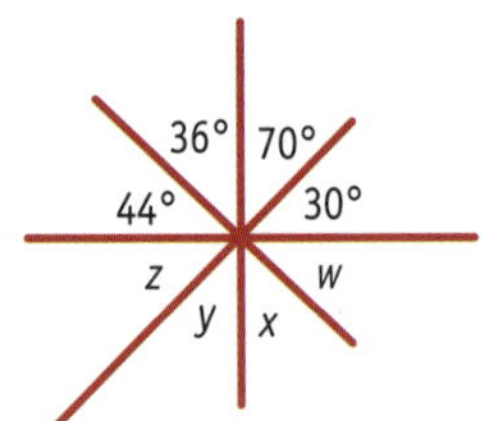

m $x =$ ____, $y =$ ____, $z =$ ____

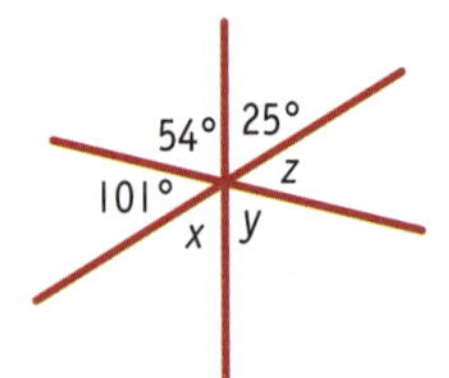

n $w =$ ____, $x =$ ____, $y =$ ____, $z =$ ____

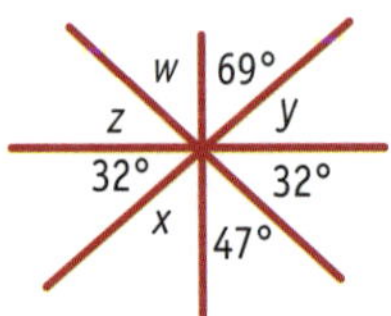

CATCH UP MATHS YEAR 6 BOOK B © PASCAL PRESS ISBN: 9781925726190

15 Complete the sentence.

The angle sum of a triangle is ______.

16 Find the size of the missing angles.

a x = _____

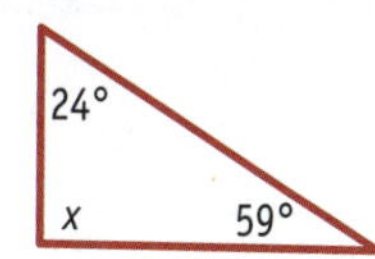

d x = _____

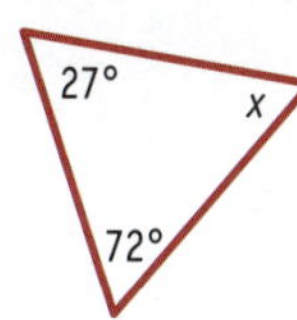

g x = _____

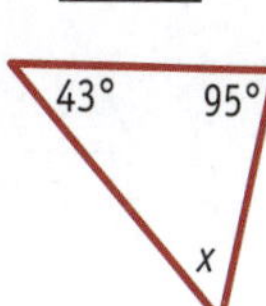

j x = _____

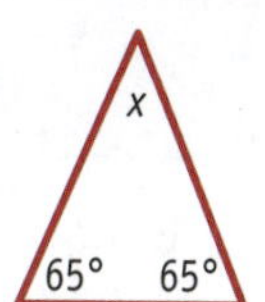

b x = _____

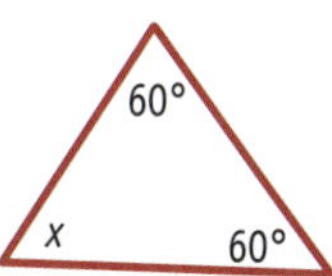

e x = _____

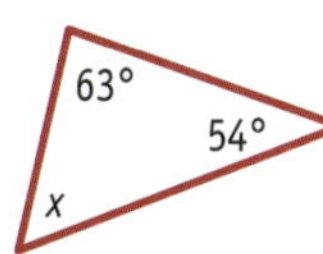

h x = _____

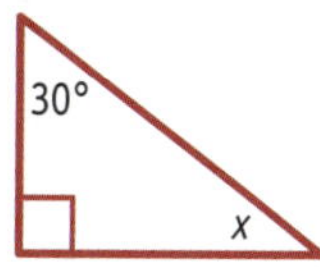

k x = _____

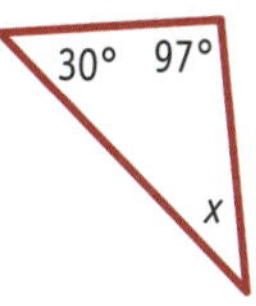

c x = _____

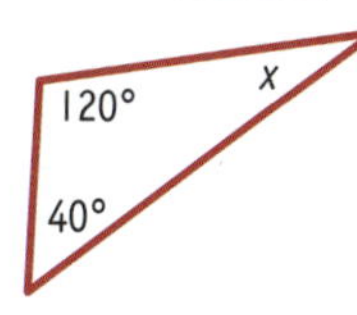

f x = _____

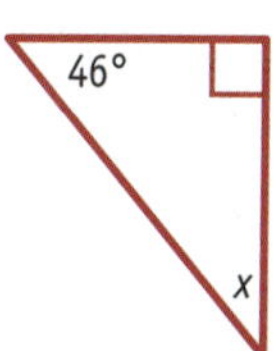

i x = _____

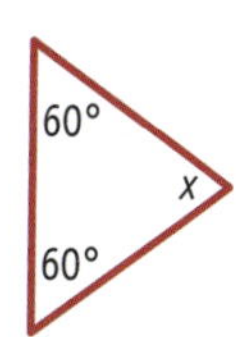

l x = _____

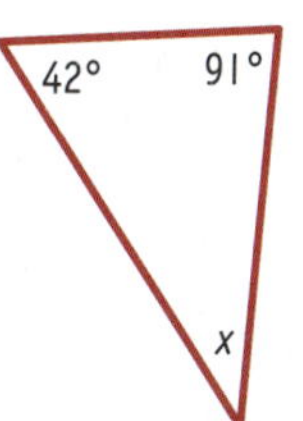

17 What type of triangle is each triangle in Question 16: equilateral, isosceles, right angled or scalene?

a ____________________ **e** ____________________ **i** ____________________

b ____________________ **f** ____________________ **j** ____________________

c ____________________ **g** ____________________ **k** ____________________

d ____________________ **h** ____________________ **l** ____________________

18 Complete the sentence.

The four angles in a quadrilateral add up to ______.

19 Find the value of x in each quadrilateral.

a x = _____

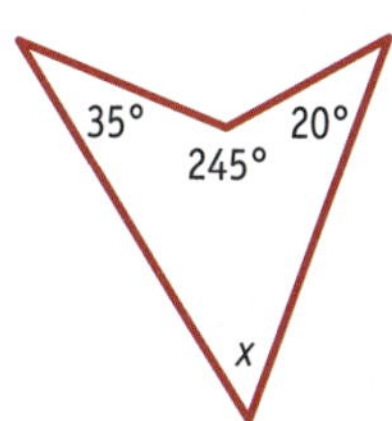

b x = _____

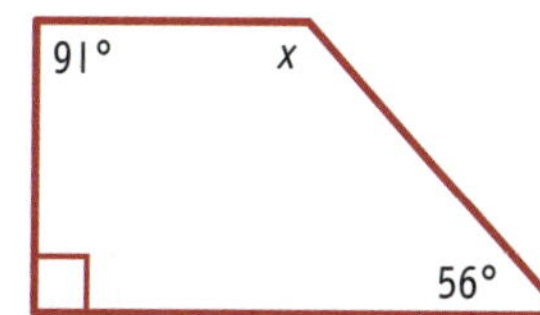

c x = _____

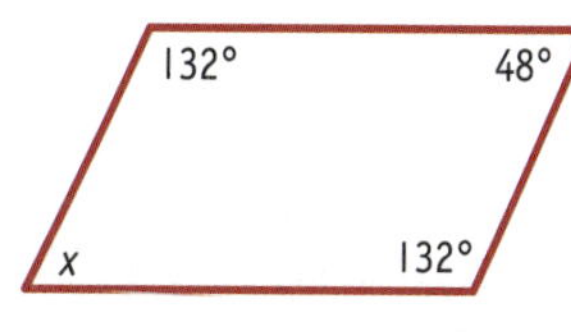

d x = _____

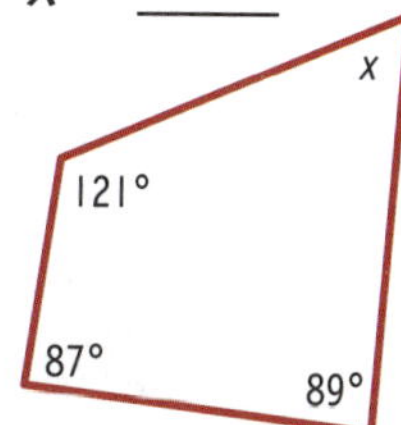

2D SHAPES

2D shapes (polygons) are flat.
They have two dimensions: length and width (height).

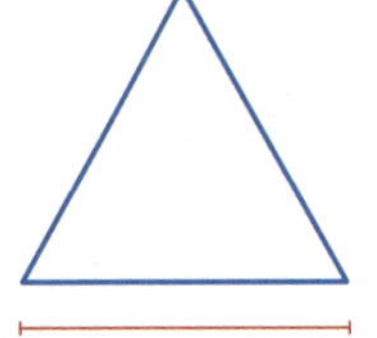

pentagon:
5 sides and
5 angles

triangle:
3 sides and
3 angles

Many shapes are named for the number of sides they have. For example, 'penta' means '5' and names the pentagon.

Regular polygons have equal sides and angles.

regular hexagon

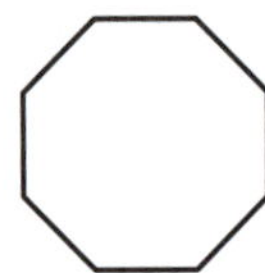
regular octagon

regular quadrilateral (square)

Irregular polygons have sides and angles of different sizes.

irregular hexagon

irregular octagon

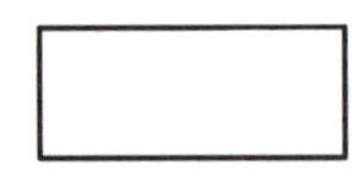
irregular quadrilateral (rectangle)

Examples:
Colour the irregular polygons blue and the regular polygons red.

a  **b**  **c**  **d** 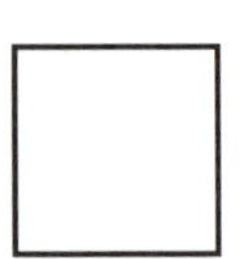 **e**  **f** 

Check the answers on the video!

Your turn

Name each polygon in the Examples above.

a irregular quadrilateral 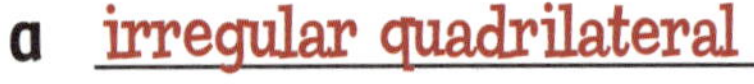

d ____________

b ____________

e ____________

c ____________

f ____________

SELF CHECK Tick how you feel

Got it!	Need help...	I don't get it
☐	☐	☐

Check your answers
How many did you get correct? ☐

CATCH UP MATHS YEAR 6 BOOK B © PASCAL PRESS ISBN: 9781925726190

PRACTICE

1 Trace the regular polygons with red and the irregular polygons with blue, and then name them.

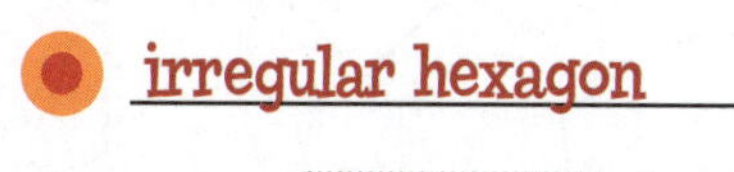

irregular hexagon

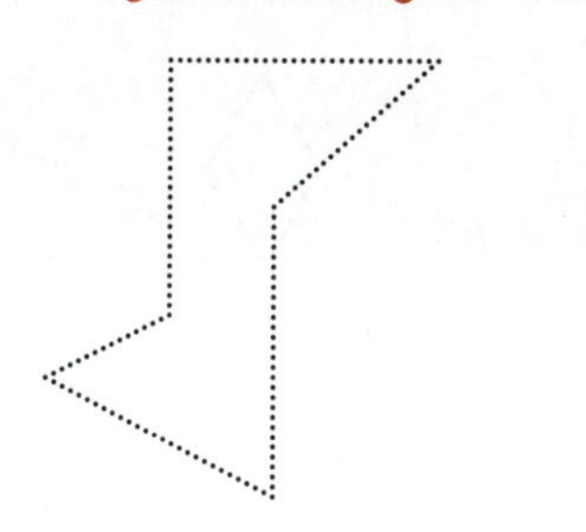

a ____________

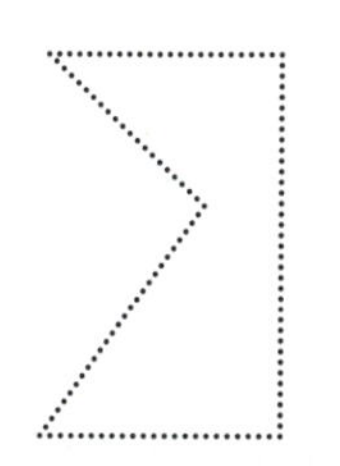

b ____________

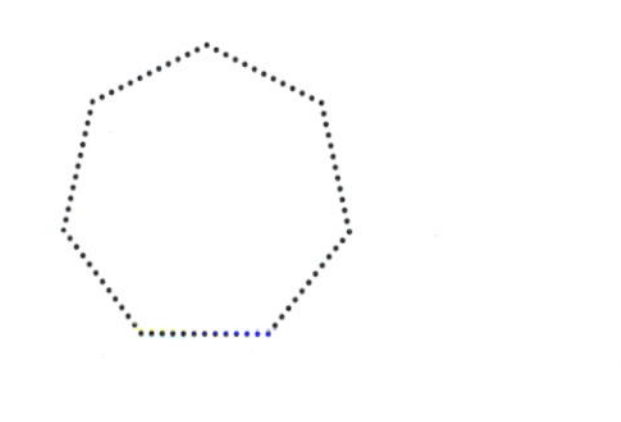

c ____________

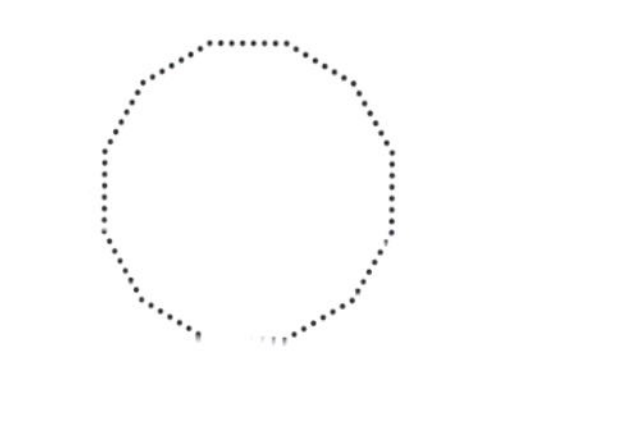

d ____________

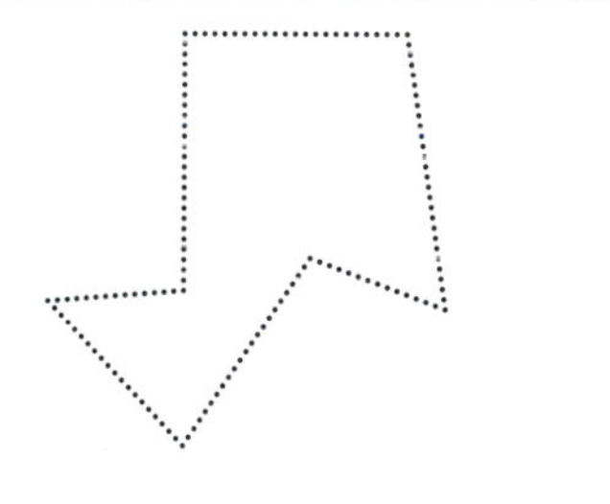

e ____________

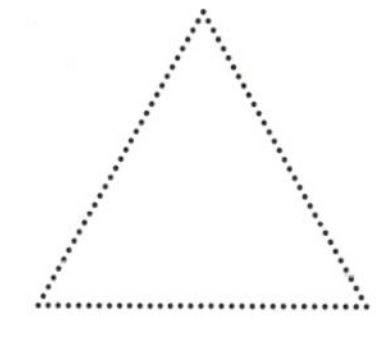

f ____________

g ____________

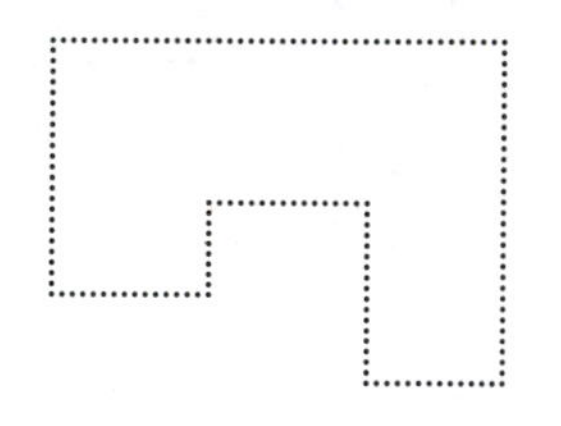

h ____________

i ____________

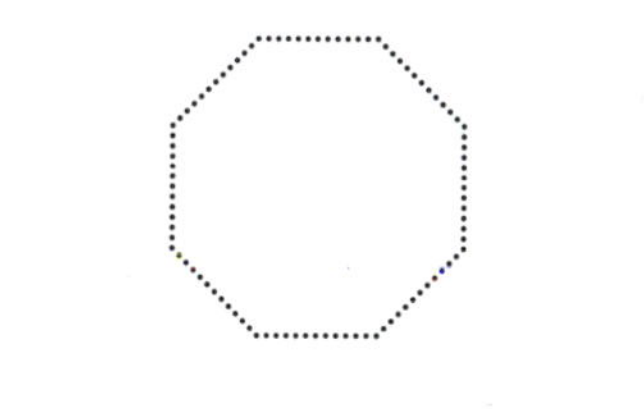

j ____________

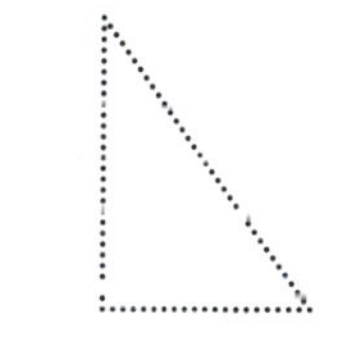

k ____________

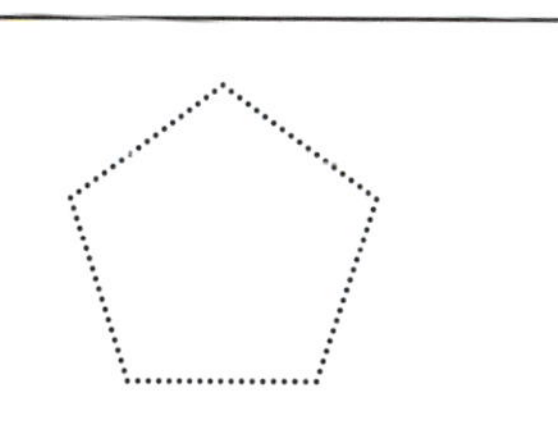

l ____________

m ____________

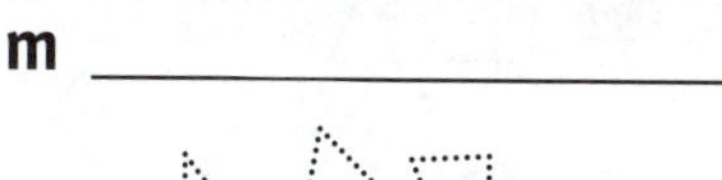

n ____________

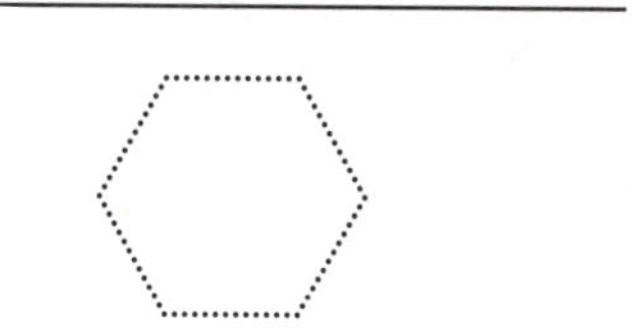

o ____________

p ____________

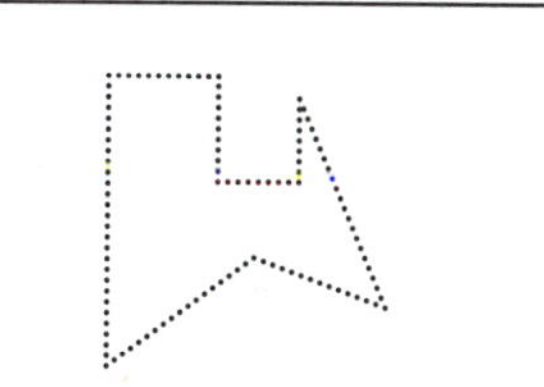

q ____________

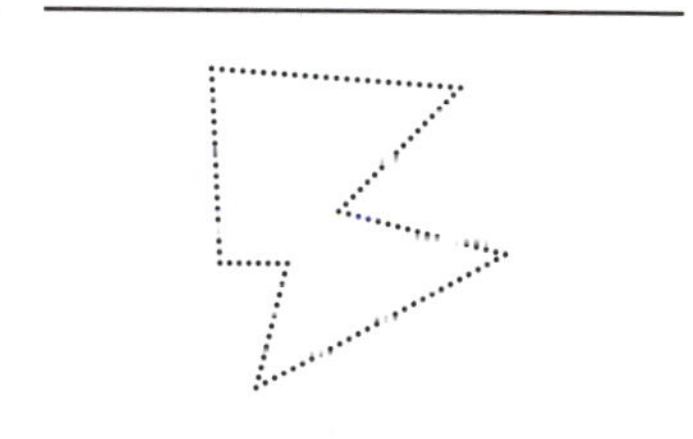

r ____________

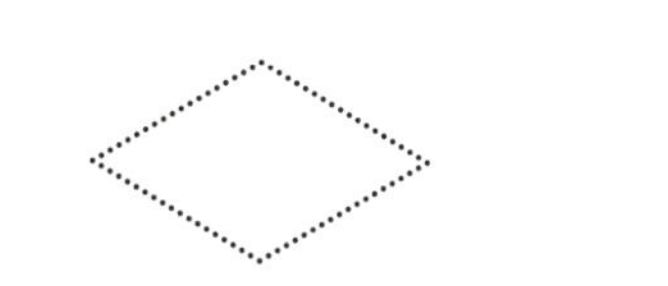

s ____________

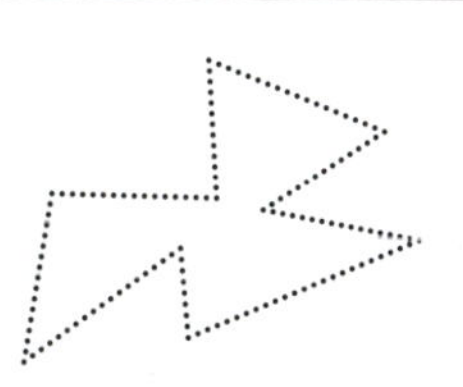

Colour the rectangles red, squares orange, rhombuses blue, parallelograms green, trapeziums yellow and kites black.

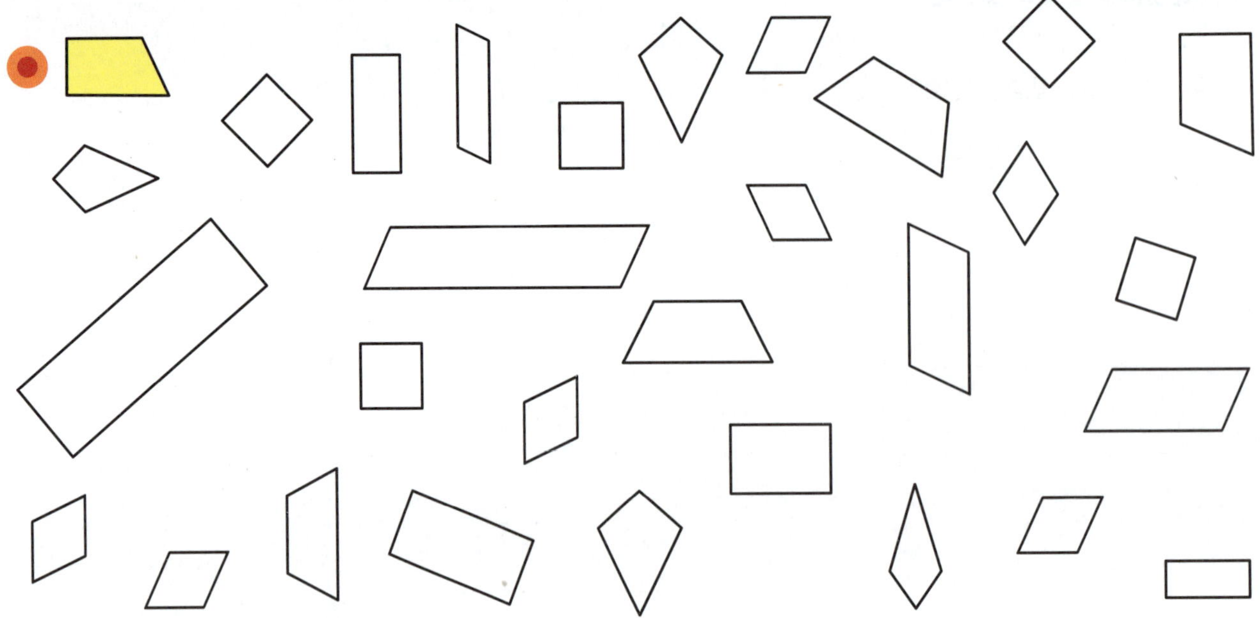

3 Trace the pentagons with purple, the hexagons with brown and the heptagons with black.

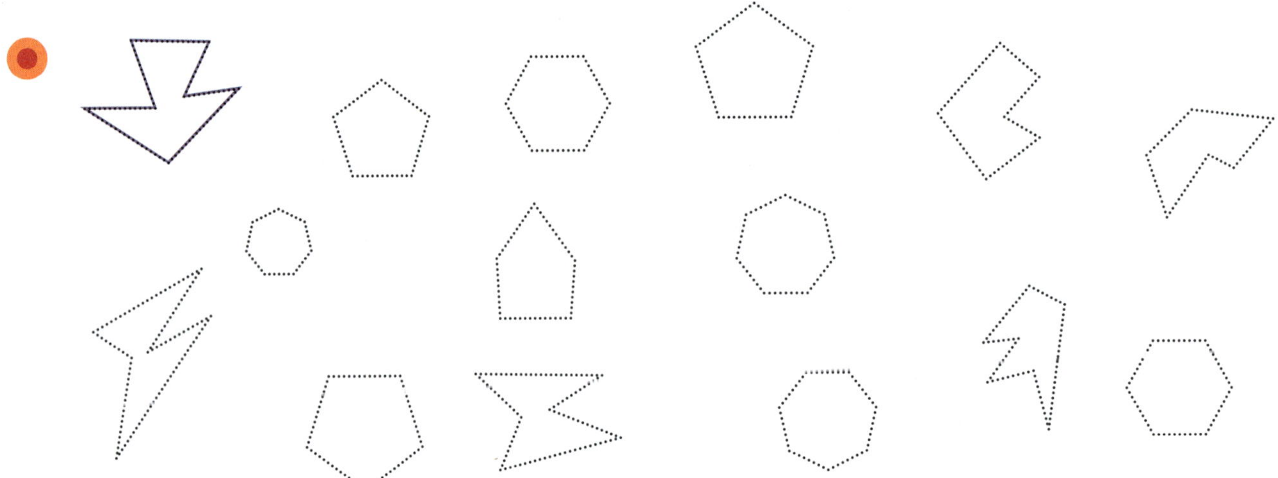

Draw one example of each shape.

	Regular	Irregular
	pentagon	
a	nonagon	

	Regular	Irregular
b	quadrilateral	
c	octagon	

	Regular	Irregular
d	triangle	
e	hexagon	

CATCH UP MATHS YEAR 6 BOOK B © PASCAL PRESS ISBN: 9781925726190

Draw two examples of each shape.

Rectangle	Trapezium	Irregular Heptagon	Kite	Irregular Dodecagon
Rhombus	Irregular Octagon	Parallelogram	Irregular Decagon	Triangle

Complete the chart.

	Polygon	Name	Letters that help identify shape	No. angles	No. sides	No. vertices
●		quadrilateral	quad	4	4	4
a						
b						
c						
d						
e						
f						
g						

  ISBN: 9781925726190

TYPES OF LINES

Many shapes have lines that are vertical, horizontal, parallel and perpendicular.

Vertical lines go up and down.

Horizontal lines go left to right.

Parallel lines never meet and are an equal distance apart.

Perpendicular lines meet at right angles.

Many shapes are drawn using these types of lines.

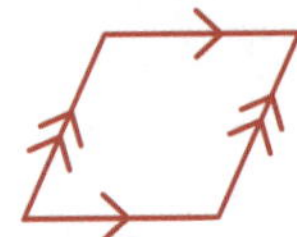

A rhombus has two sets of parallel lines.

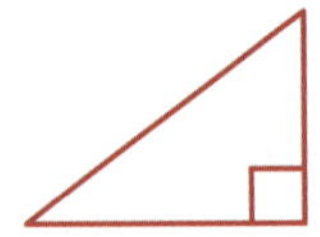

A right-angled triangle has two lines that meet at right angles.

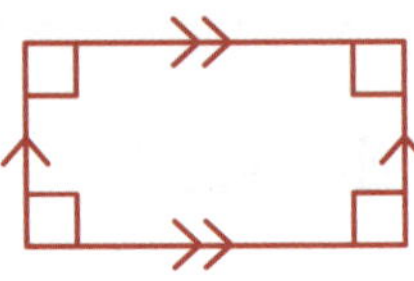

A rectangle has two sets of parallel lines and all the lines meet at right angles.

Examples: Label the types of lines.

a horizontal **b** ____________ **c** ____________ **d** ____________

Check the answers on the video!

Your turn Draw three examples of each type of line.

Horizontal	Vertical	Parallel	Perpendicular

SELF CHECK Tick how you feel

Got it!	Need help...	I don't get it
 ☐	 ☐	 ☐

Check your answers
How many did you get correct? ☐

CATCH UP MATHS YEAR 6 BOOK B © PASCAL PRESS ISBN: 9781925726190

PRACTICE

1 Trace the vertical lines with blue, the horizontal lines with orange, parallel lines with green and perpendicular lines with purple.

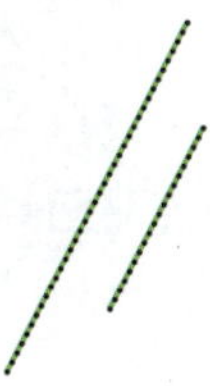

e

j

f

k

a

g

l

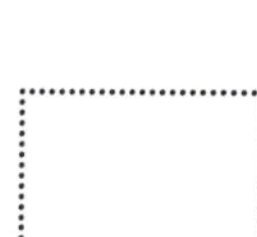

b

c

h

m

d

i

n

2 Circle the lines that match the line type.

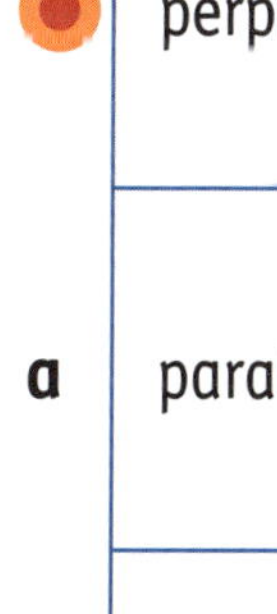

	Line type	Lines
	perpendicular	
a	parallel	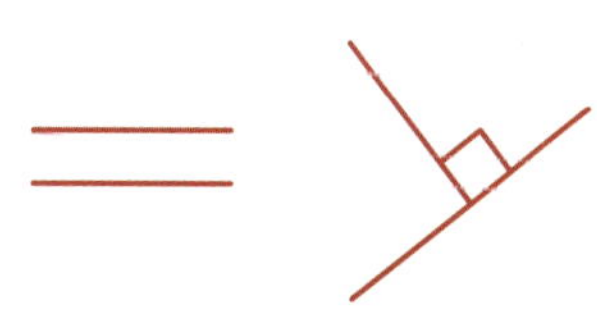 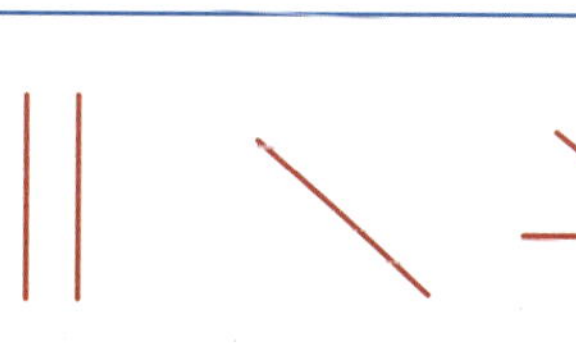 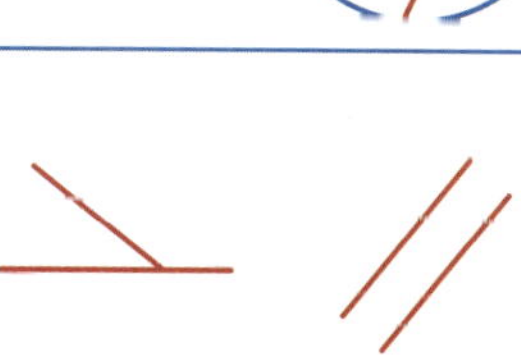
b	vertical	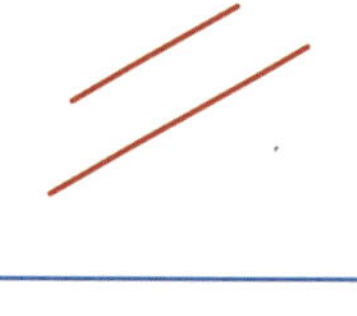
c	perpendicular	 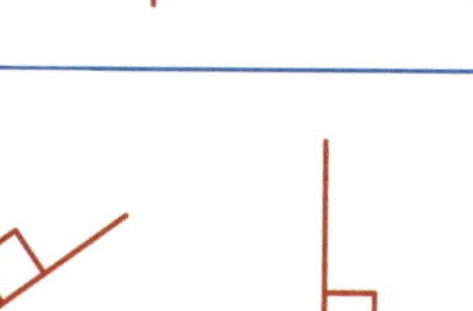

CIRCLES

A circle is a 2D shape. It has one curved edge and no corners.

A semicircle is half a circle. It has one curved edge and one straight side.

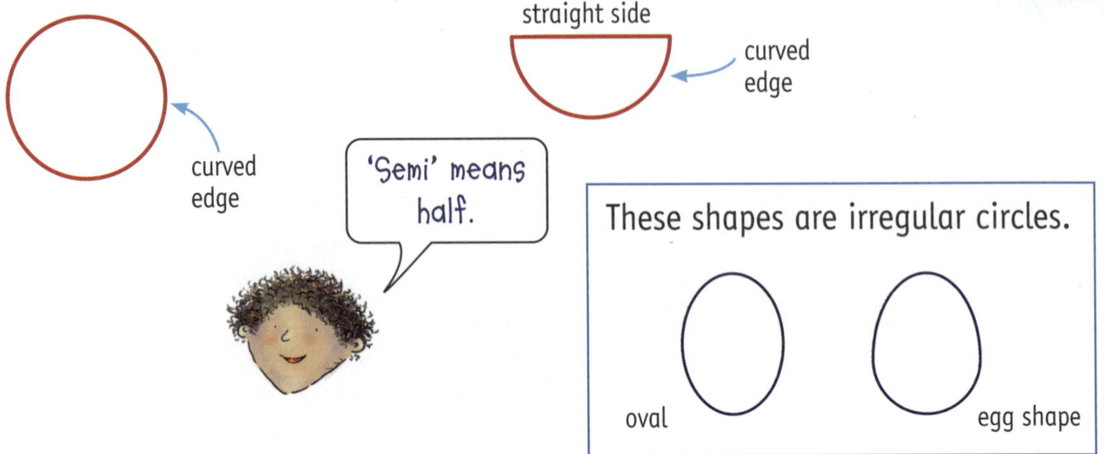

These shapes are irregular circles.

oval

egg shape

Examples: Colour the regular circles red and the irregular circles blue.

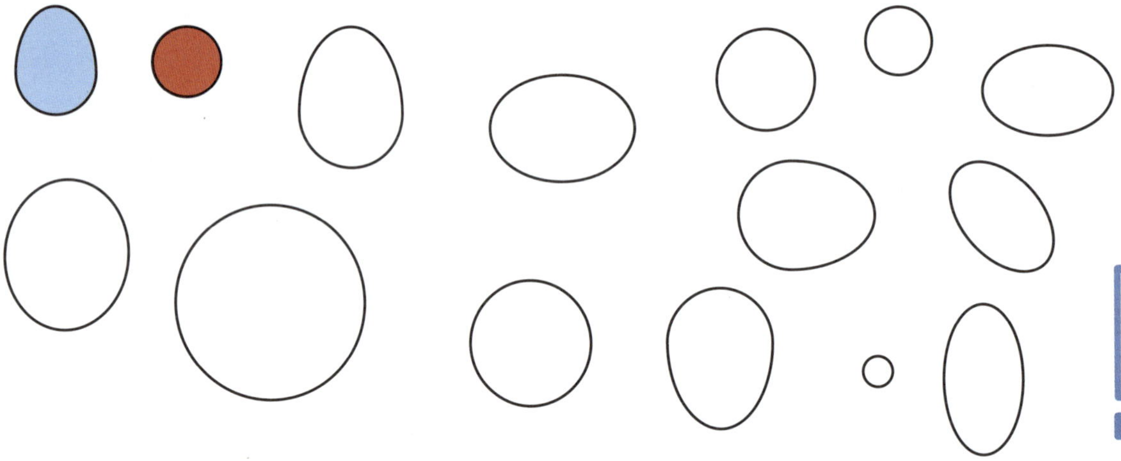

Check the answers on the video!

Your turn

Use green to trace each shape and then label it.

circle

a ____________

b ____________

c ____________

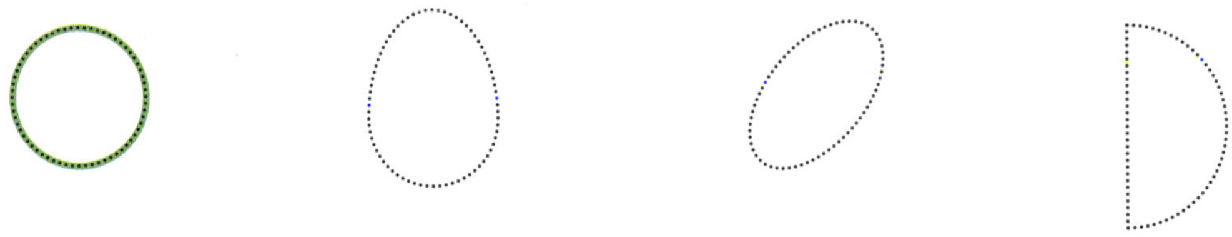

SELF CHECK Tick how you feel

Got it!	Need help...	I don't get it
☐	☐	☐

Check your answers

How many did you get correct?

ISBN: 9781925726190

PRACTICE

1 Mark the centre of each circle with ×.

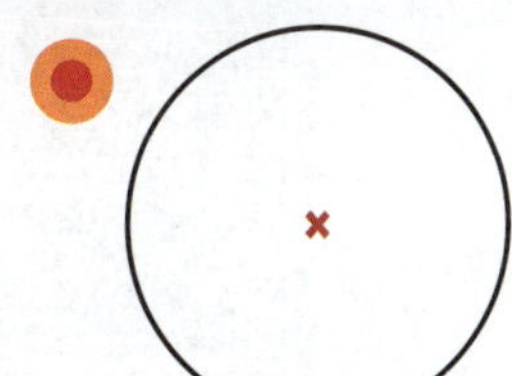

a

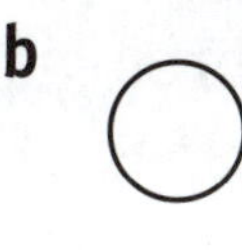

b

c

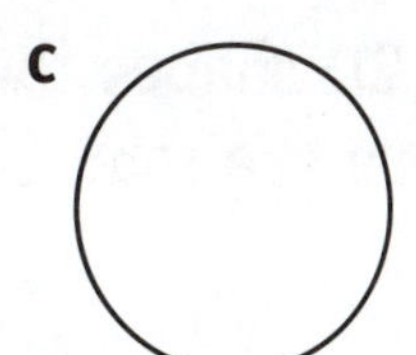

d

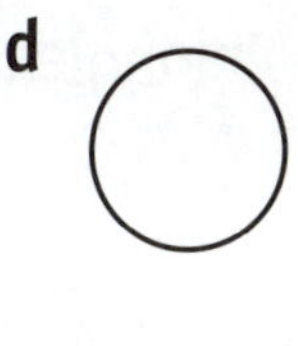

2 Use the labelled circle to complete the definitions.

Ci r c u m f e r e n c e: the outside edge (curved edge) of a circle

a C _ _ _ _ _: the middle of a circle

b A _ _: a section of the circumference

c S _ _ _ _ _: the area made by two radii and an arc

d D _ _ _ _ _ _ _: a straight line from one side of a circle to the other that goes through the centre

e C _ _ _ _: a straight line joining any two parts of the circumference

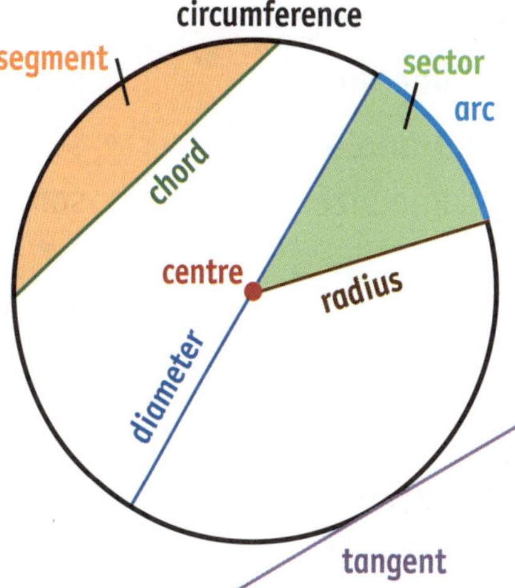

f S _ _ _ _ _ _: the area between the circumference and a chord

g R _ _ _ _ _: a straight line from the centre of a circle to the circumference

h T _ _ _ _ _ _: a straight line that touches the circumference at a single point

3 Mark the part on the circle.

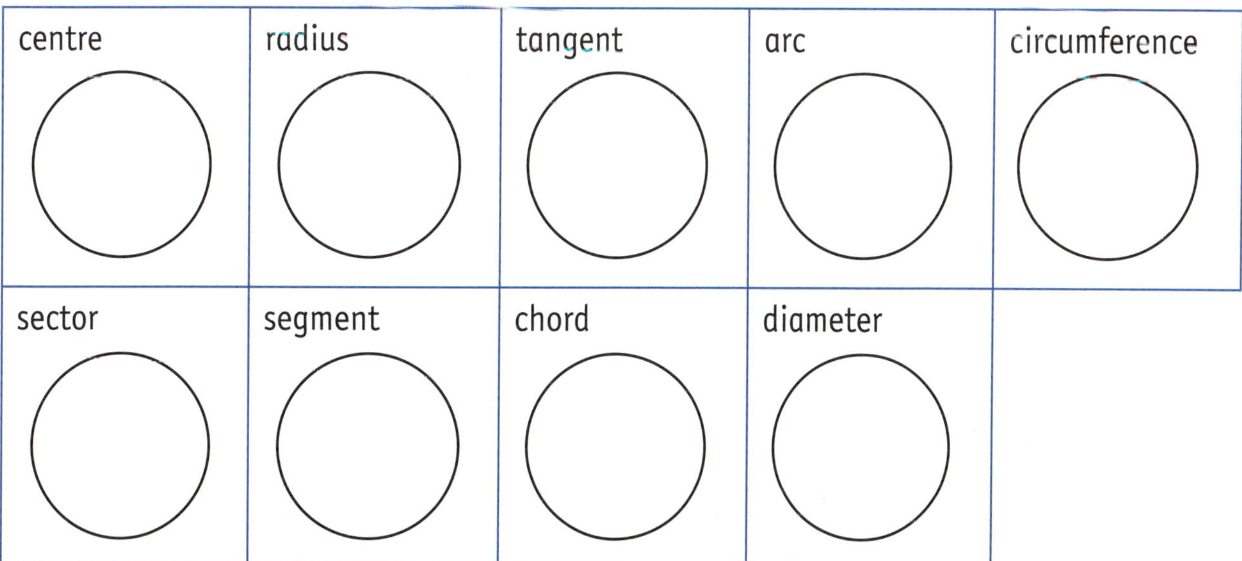

TRIANGLES

Triangles are 2D shapes that have three sides and three angles. There are four different types of triangles.

Equilateral
- Regular triangle
- All sides the same length
- All angles the same size

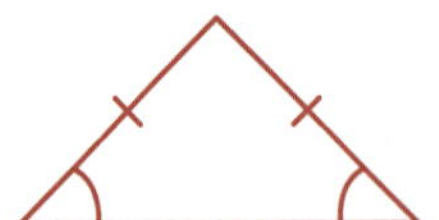

Isosceles
- Irregular triangle
- Two sides the same length
- Two angles the same size

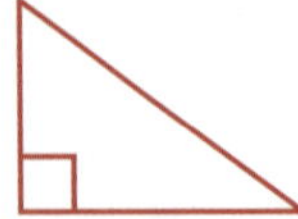

Right angled
- Irregular triangle
- One right angle

Scalene
- Irregular triangle
- All sides different lengths
- All angles different sizes

Examples: Colour the regular triangles red and the irregular triangles blue.

a 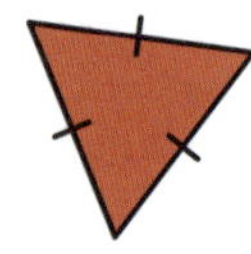

c 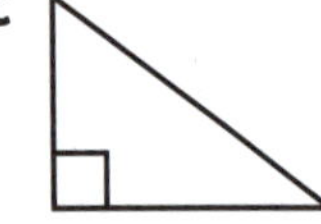

e 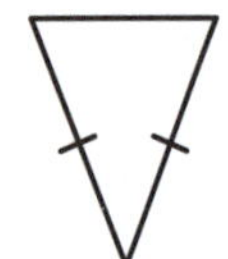

g 

i 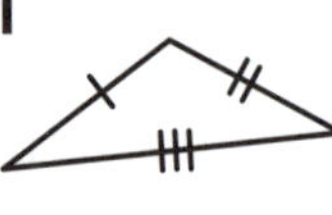

b 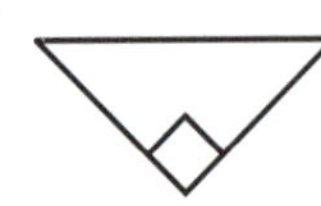

d 

f 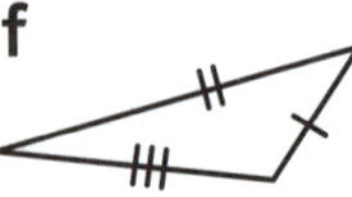

h 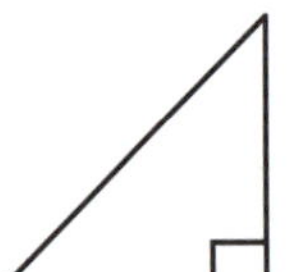

j

Check the answers on the video!

Name each of the triangles in the Examples above.

a equilateral

b ______

c ______

d ______

e ______

f ______

g ______

h ______

i ______

j ______

Check your answers
How many did you get correct?

CATCH UP MATHS YEAR 6 BOOK B © PASCAL PRESS ISBN: 9781925726190

PRACTICE

1 Describe and draw each triangle.

a Isosceles triangle	c Right-angled triangle
______________ ______________ ______________ ______________ ______________ ______________	______________ ______________ ______________ ______________ ______________ ______________
b Equilateral triangle	**d** Scalene triangle
______________ ______________ ______________ ______________ ______________ ______________	______________ ______________ ______________ ______________ ______________ ______________

2 Measure the sides of the triangles and look for any markings to help you label them.

 equilateral

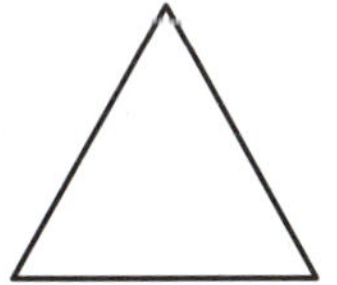

b 

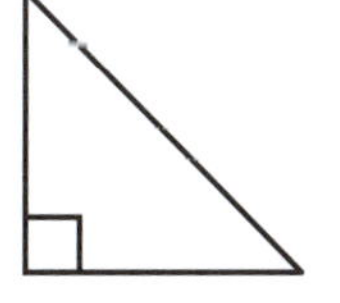

d 

f 

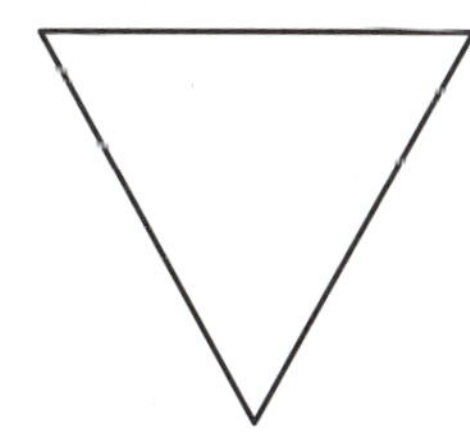

a ______________

c ______________

e 

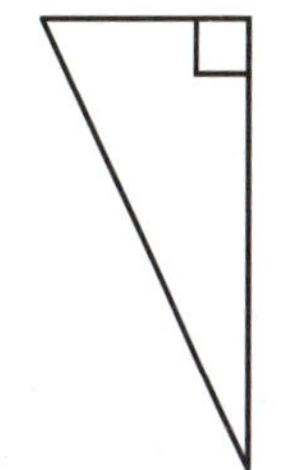

g ______________

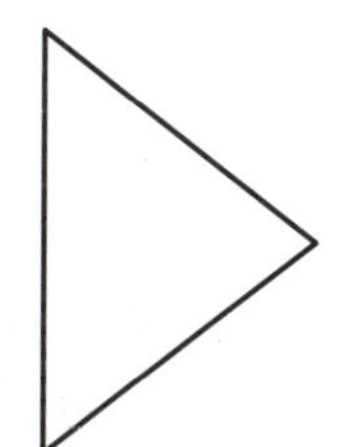

TRANSFORMATIONS

When a shape is changed or moved into a different position, it is called a transformation.

Flip (Reflection)

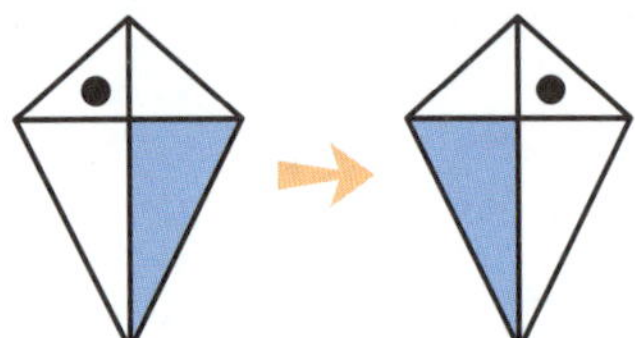

The kite is lifted and flipped over.

Slide (Translation)

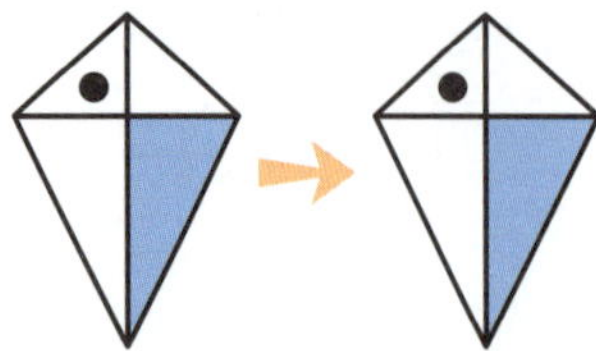

The kite is not lifted or turned. Its direction does not change.

Turn (Rotation)

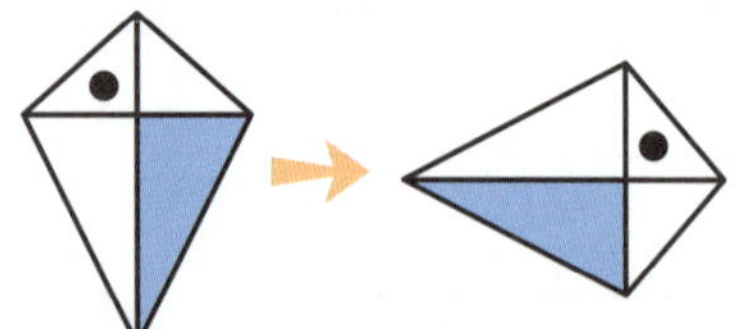

This is a quarter turn. The kite is turned clockwise around a point.

Examples:

Match the words that mean the same by colouring them the same colour.

a	Flip	Rotation
b	Slide	Reflection
c	Turn	Translation

Label each transformation as a reflection, rotation or translation.

●       reflection

a        __________

b        __________

c    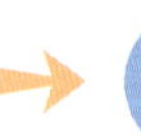          __________

SELF CHECK Tick how you feel

Got it!	Need help...	I don't get it
 ☐	 ☐	☐

Check your answers

How many did you get correct?

CATCH UP MATHS YEAR 6 BOOK B © PASCAL PRESS ISBN: 9781925726190

PRACTICE

1 Label each transformation as a reflection, rotation or translation.

- ● → → → → reflection
- **a** → → → → ______
- **b** → → → → ______
- **c** 5 → 5 → 5 → 5 → 5 ______
- **d** R → R → R → R → R ______

2 Reflect the shapes to continue the pattern.

a     ______ ______ ______ ______

b    ______ ______ ______ ______

c  ______ ______ ______ ______

d ______ ______ ______ ______

3 Rotate the shapes to continue the pattern.

a     ______ ______ ______ ______

b      ______ ______ ______ ______

c     ______ ______ ______ ______

4 Translate the shapes to continue the pattern.

a     

b     ______ ______ ______ ______

c      ______ ______ ______ ______

TESSELLATIONS

A tessellation is a pattern of 2D shapes. The shapes cover a flat surface with no gaps or overlaps.

SCAN to watch video

These squares tessellate:

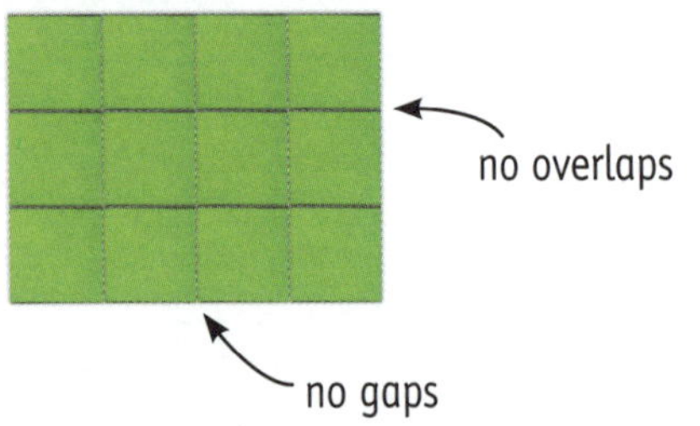

You can also use a combination of shapes to make a tessellation.

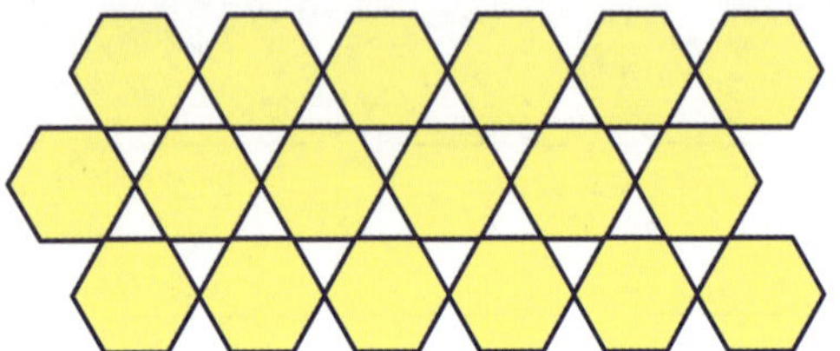

These shapes are not tessellating:

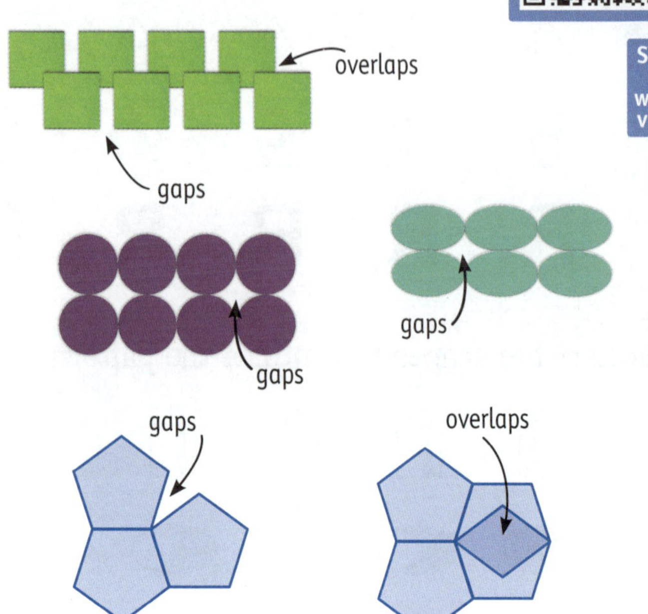

Examples: Tick the tessellations.

a ✔

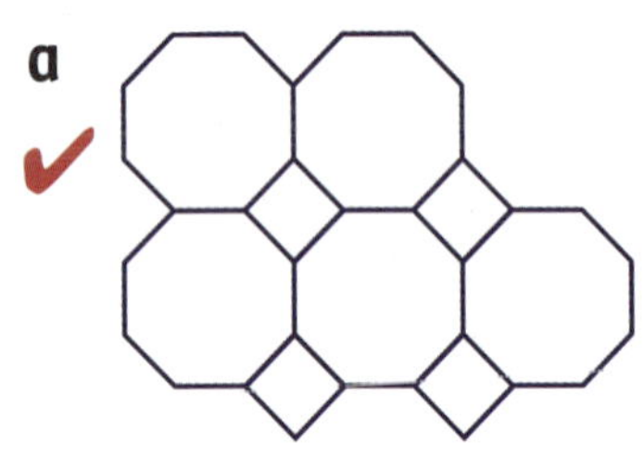

b

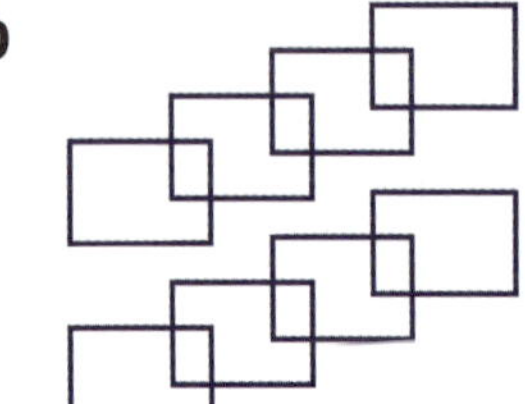

c

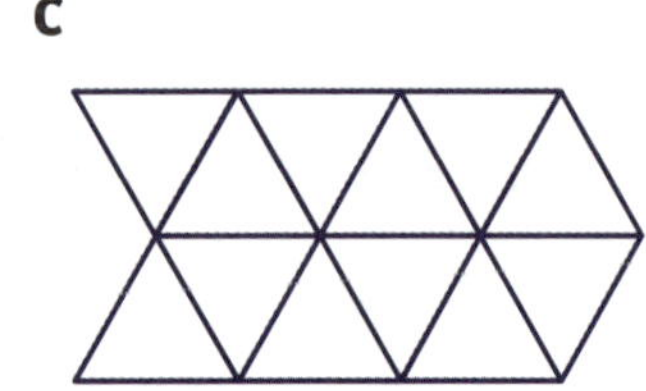

Check the answers on the video!

Trace the shapes to continue these real-life tessellations.

● Pattern on a rug

a Bricks on a house

b Chocolate block

SELF CHECK Tick how you feel

Got it!	Need help...	I don't get it
☐	☐	☐

Check your answers

How many did you get correct? ☐

CATCH UP MATHS YEAR 6 BOOK B © PASCAL PRESS ISBN: 9781925726190

PRACTICE

1 Tick the shapes that will tessellate on their own.
Cross the shapes that will not tessellate.

b 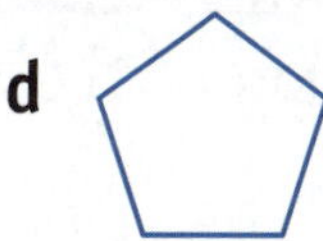

d

f 

a 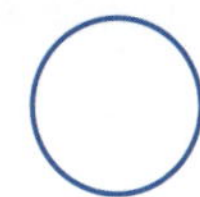

c 

e

g 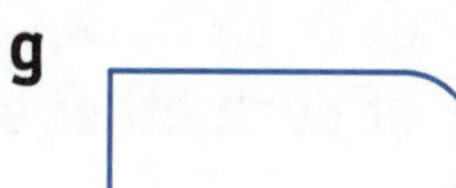

2 Continue these tessellations.

a 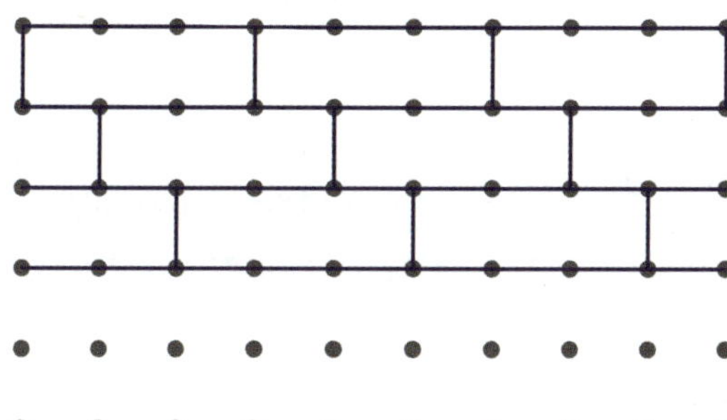

b 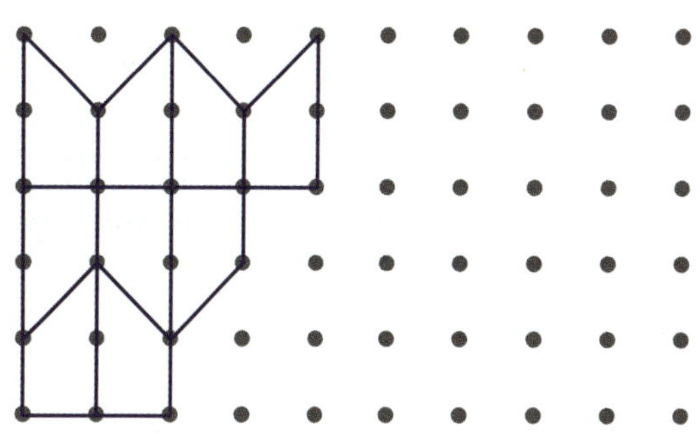

3 Make your own tessellations with these shapes.
You can rotate, reflect or slide the shapes.

a 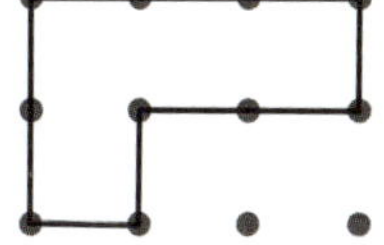

b 

c 

SYMMETRY

A shape is symmetrical when one half of the shape is an exact reflection of the other half.

The fox face is symmetrical as both halves fit on top of each other exactly.

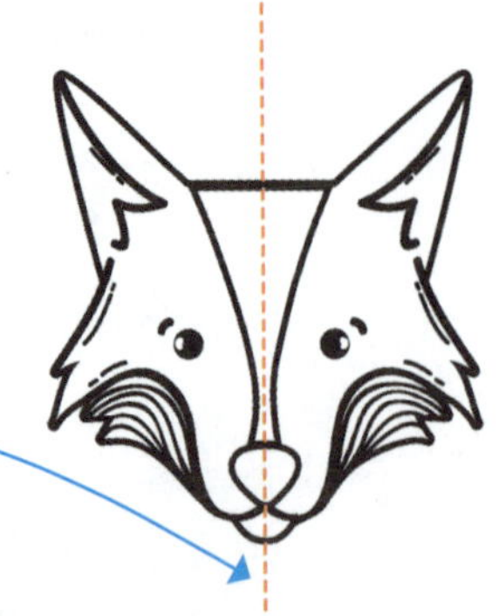

A line of symmetry is always dotted.

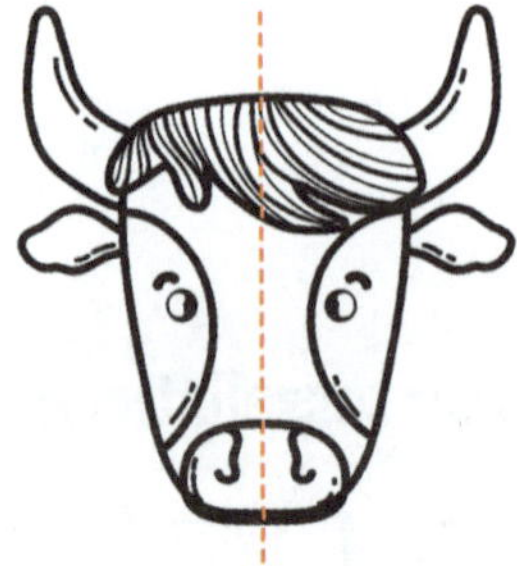

The cow face is not symmetrical because the two sides are not the same.

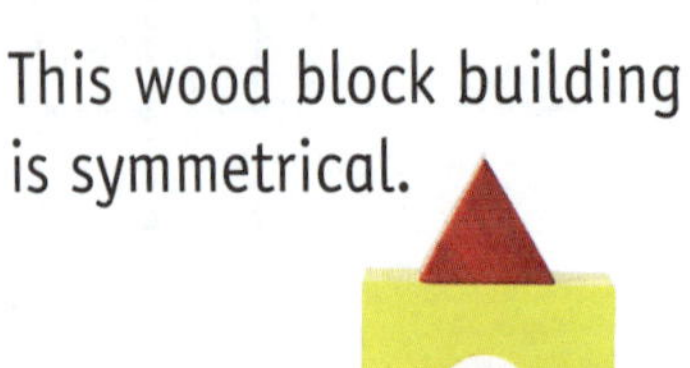

This square has four lines of symmetry.

This wood block building is symmetrical.

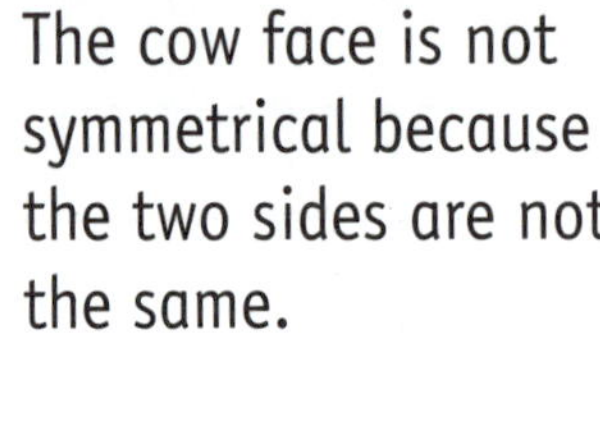

Examples: Circle the symmetrical shapes.

a  b 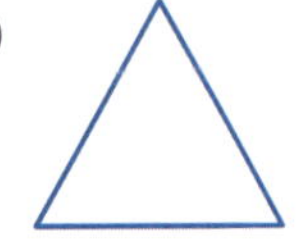 c  d 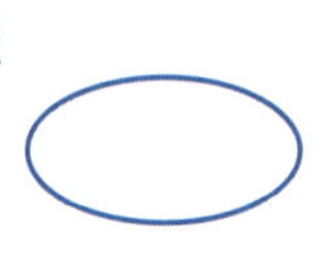 e 

Check the answers on the video!

Draw the line(s) of symmetry on the shapes below.

a 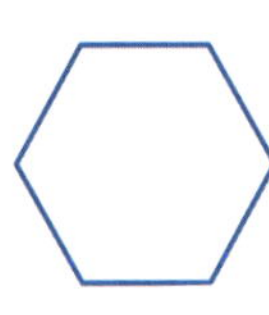 b  c 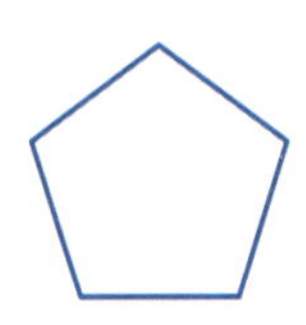 d 

SELF CHECK Tick how you feel

Got it!	Need help...	I don't get it

Check your answers
How many did you get correct?

CATCH UP MATHS YEAR 6 BOOK B © PASCAL PRESS ISBN: 9781925726190

PRACTICE

1 Draw five letters that are symmetrical and draw in the line(s) of symmetry.

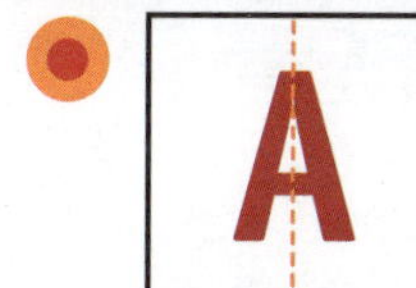

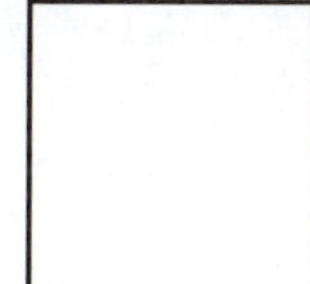
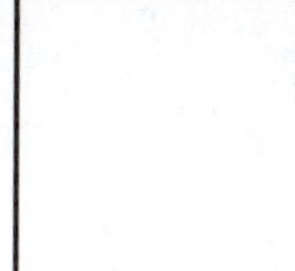

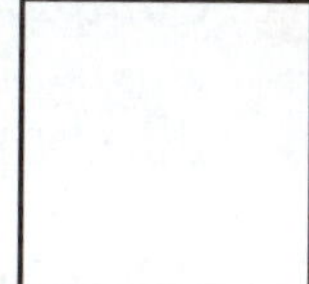

2 Complete the other half of the picture along the axis of symmetry.

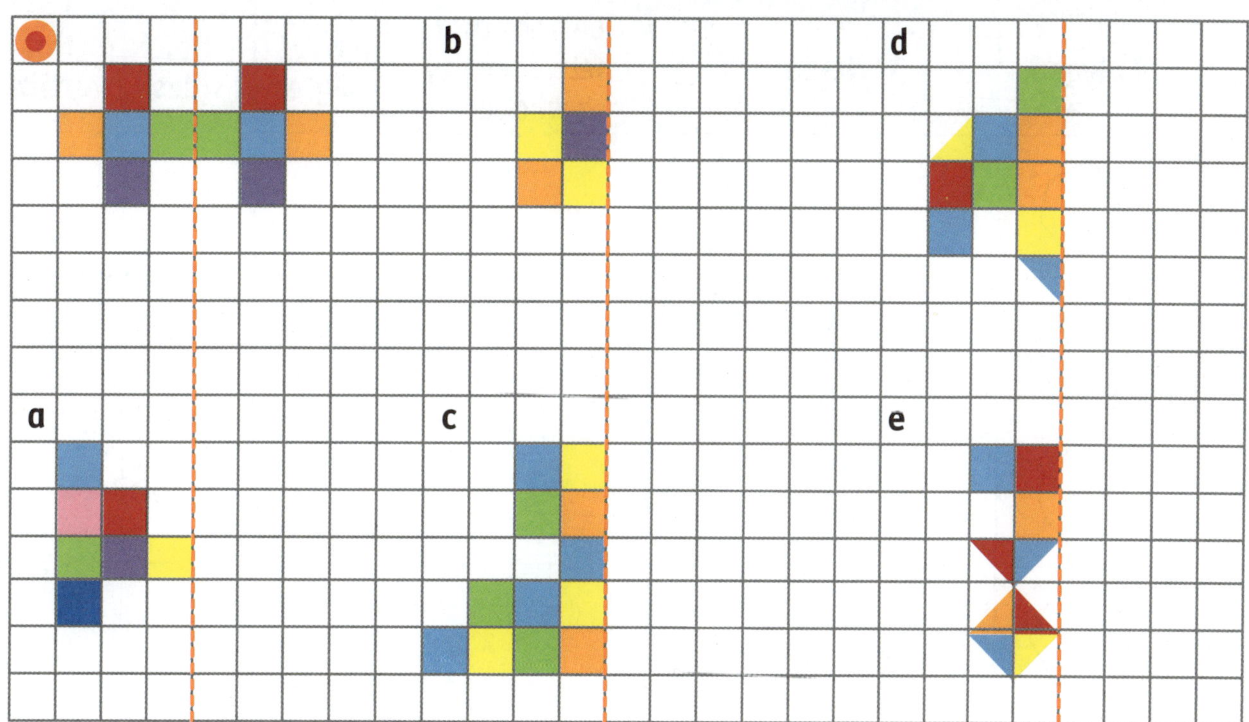

3 Draw the mirror image to complete the right side of the butterfly.

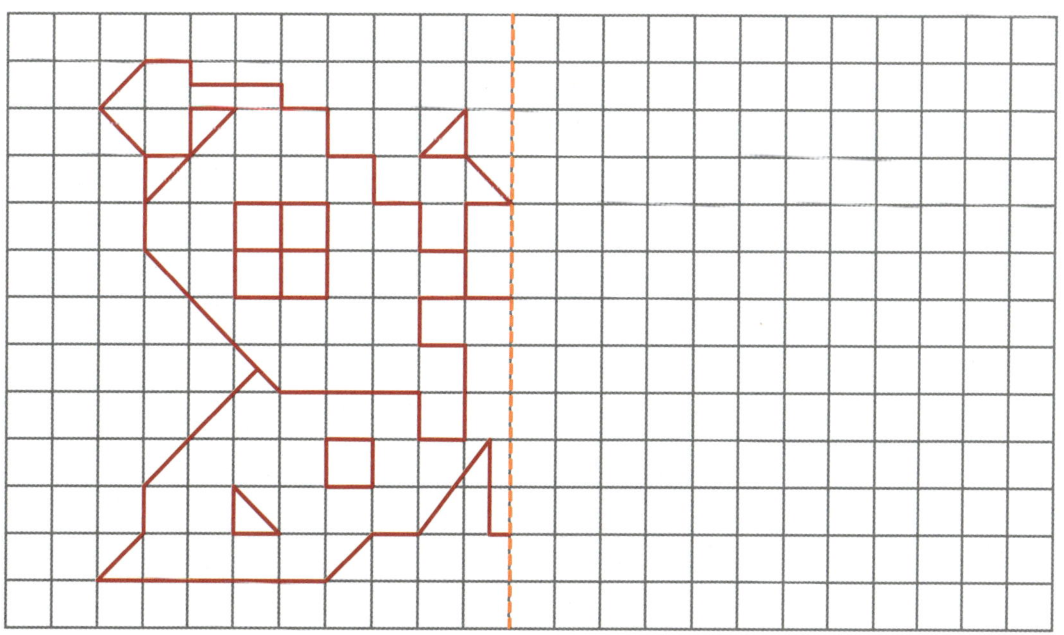

2D SHAPES REVIEW

1 Complete the sentences.

2D shapes have two _______________: a width and a ________.

2D shapes are _______.

2 Complete the table.

	Polygon	Name	Letters that help identify shape	No. angles	No. sides	No. vertices
a						
b						
c						
d						
e						
f						
g						
h						
i						

CATCH UP MATHS YEAR 6 BOOK B © PASCAL PRESS ISBN: 9781925726190

3 Draw the 2D shapes to complete the chart.

a

Regular	Irregular
triangle	

b

quadrilateral	

c

Regular	Irregular
pentagon	

d

hexagon	

e

Regular	Irregular
octagon	

f

nonagon	

4 Draw three examples of each type of line.

Perpendicular	Parallel	Horizontal	Vertical

5 Draw five regular 2D shapes. Trace the perpendicular lines with purple, parallel lines green, vertical lines blue and horizontal lines orange.

REVIEW

Label the circle with the terms.

- circumference
- diameter
- chord
- segment
- radius
- sector
- arc
- tangent
- centre

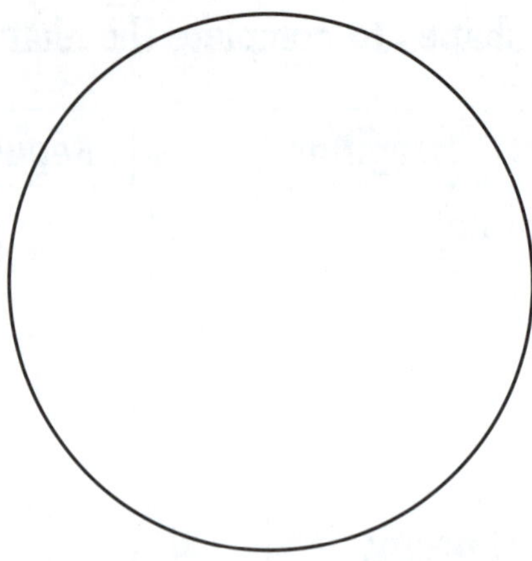

Colour the regular circles red and the irregular circles blue. Then label each shape.

a  b  c  d

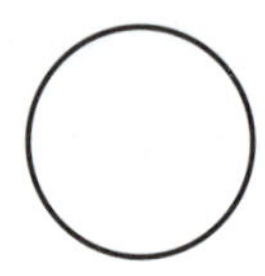

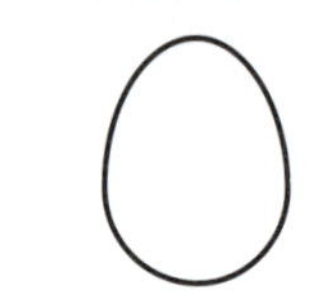

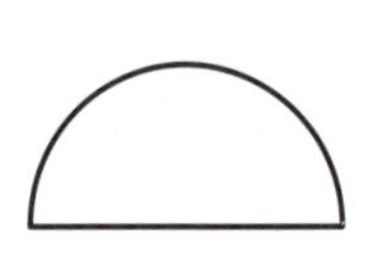

Label these triangles.

a  b  c  d

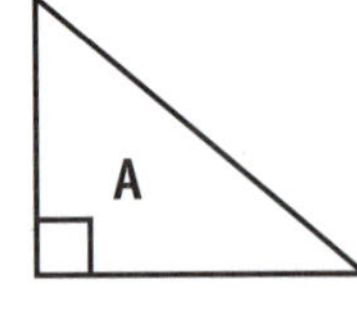

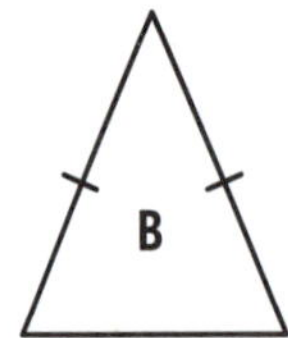

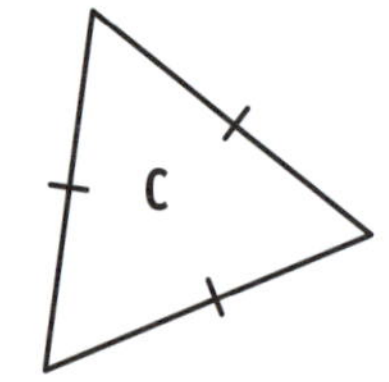

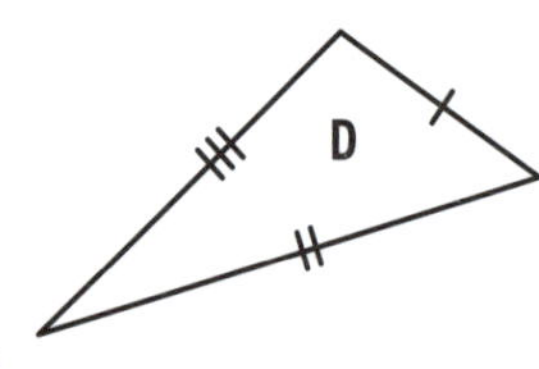

Write a description of each triangle in Question 8.

A ______________________________

____

B ______________________________

____

C ______________________________

____

D ______________________________

____

CATCH UP MATHS YEAR 6 BOOK B © PASCAL PRESS ISBN: 9781925726190

10 Measure the sides and then label each triangle.

a ____________

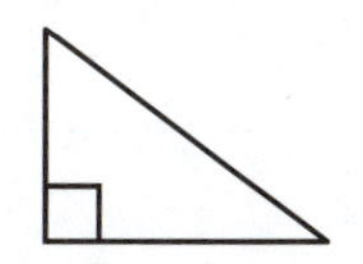

c ____________

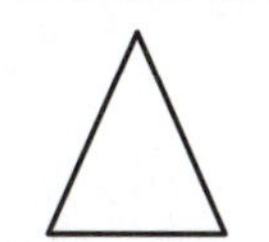

e ____________

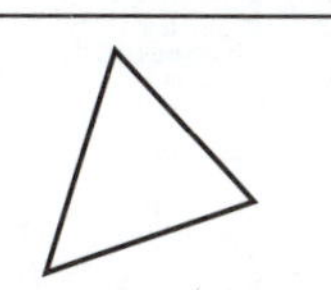

g ____________

b ____________

d ____________

f ____________

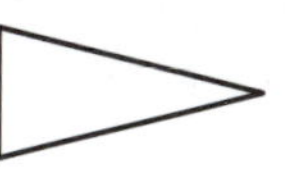

h ____________

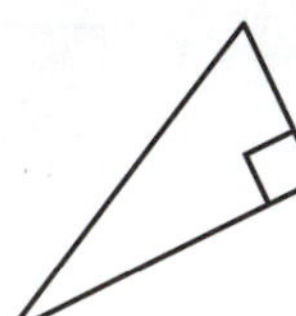

11 Label the transformations as flip, slide or turn.

a ____________

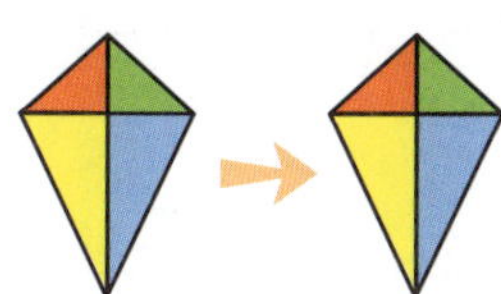

b ____________

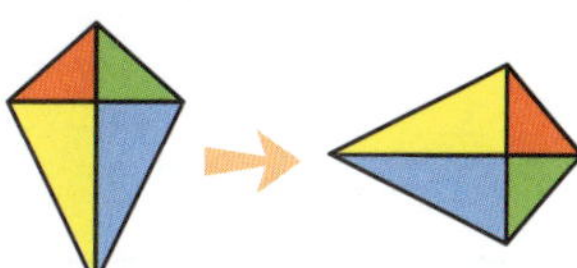

c ____________

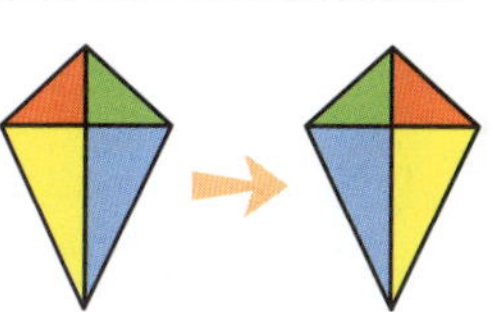

12 Write another word for each of these transformations.

a flip ________________

b slide ________________

c turn ________________

13 Write the term for each definition.

a A ________________ is lifted and turned over.

b A ________________ is a clockwise turn around a point.

c A ________________ is not lifted or turned, so its direction doesn't change.

14 Describe each transformation as a reflection, translation or rotation.

a 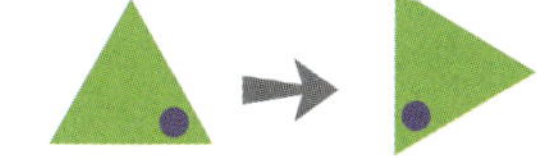 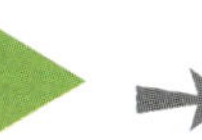  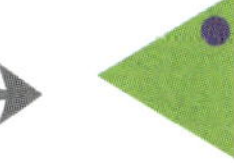 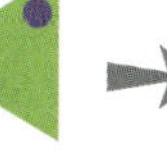   ________________

b    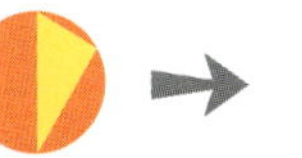  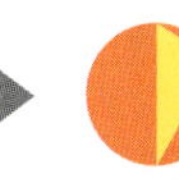  ________________

c   ________________

d      ________________

e         ________________

REVIEW

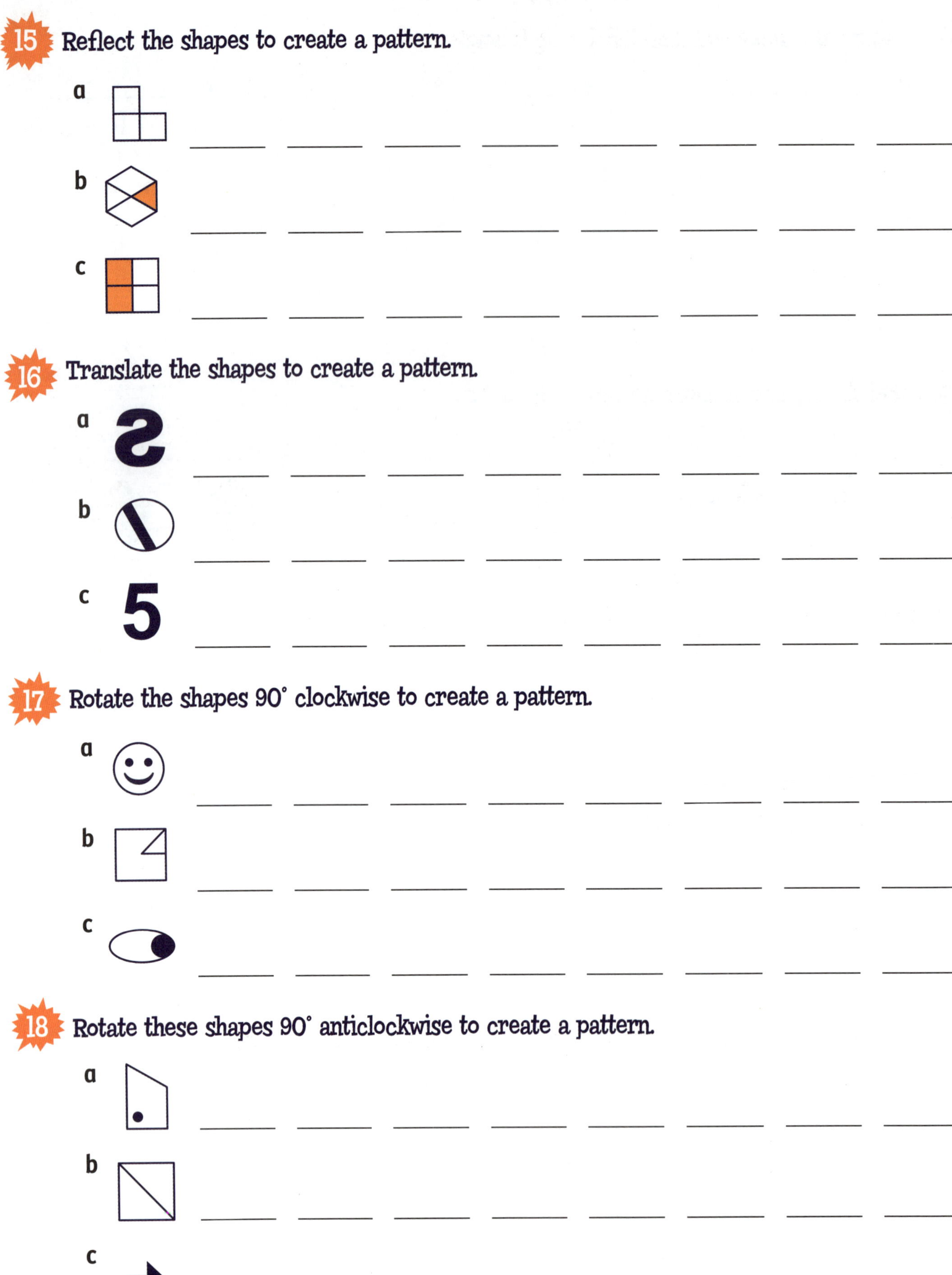

15 Reflect the shapes to create a pattern.

a

b

c

16 Translate the shapes to create a pattern.

a

b

c

17 Rotate the shapes 90° clockwise to create a pattern.

a

b

c

18 Rotate these shapes 90° anticlockwise to create a pattern.

a

b

c

CATCH UP MATHS YEAR 6 BOOK B © PASCAL PRESS ISBN: 9781925726190

19 Name five shapes that tessellate on their own.

___

20 Does a circle tessellate on its own? ___

21 Name three things in our environment that tessellate.

___

22 Make your own tessellations with these shapes.

a

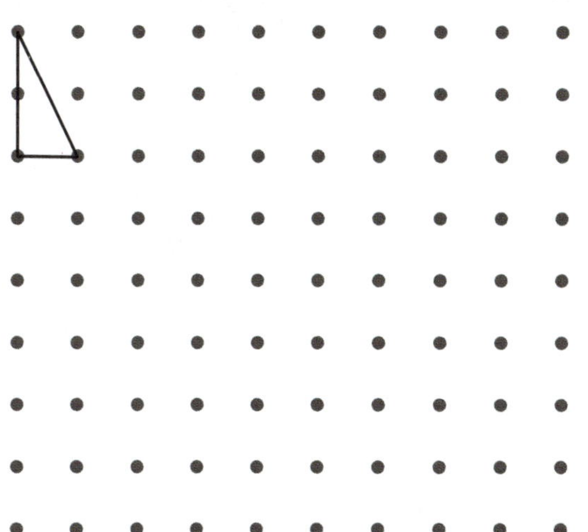

b

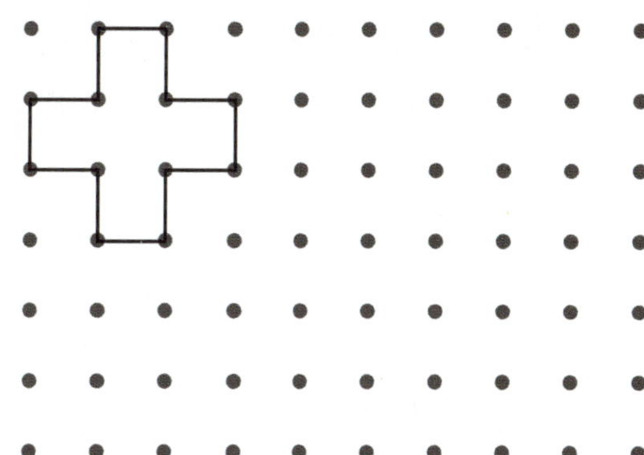

23 Tick the symmetrical shapes and cross the non-symmetrical shapes.

a 

c 

e 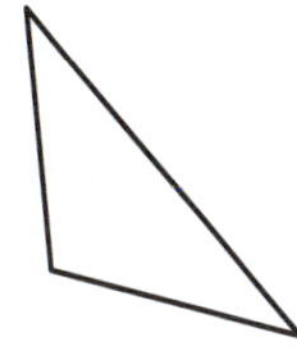

g 

b 

d 

f 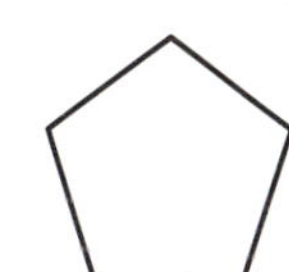

h 

24 Draw all the lines of symmetry on the shapes you ticked in Question 19.

25 Complete the picture along the axis of symmetry to make it symmetrical.

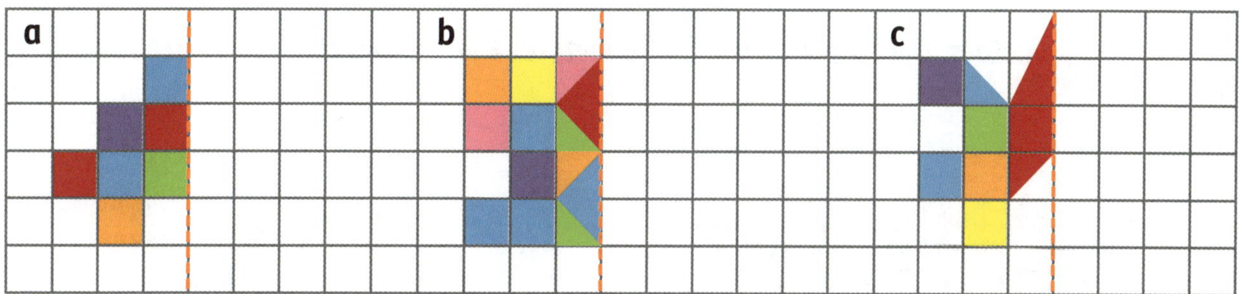

PRISMS AND PYRAMIDS

A **prism** is a three-dimensional object – it has height, width (length) and depth. Prisms have one pair of parallel bases. The shape of the bases names the prism. Rectangles join the bases.

Triangular prism

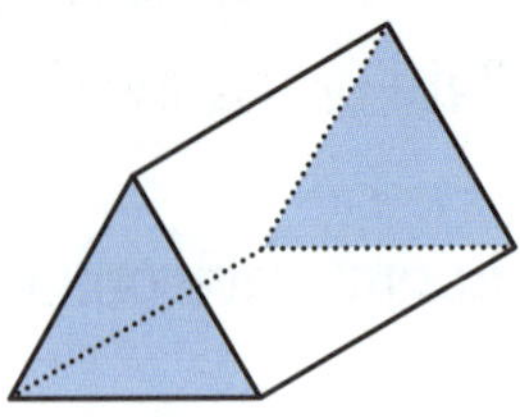

SCAN to watch video

A **pyramid** is a three-dimensional object – it has height, width (length) and depth. The shape of the base names the pyramid. The other faces are triangles and they meet at a point called the apex.

Triangular pyramid

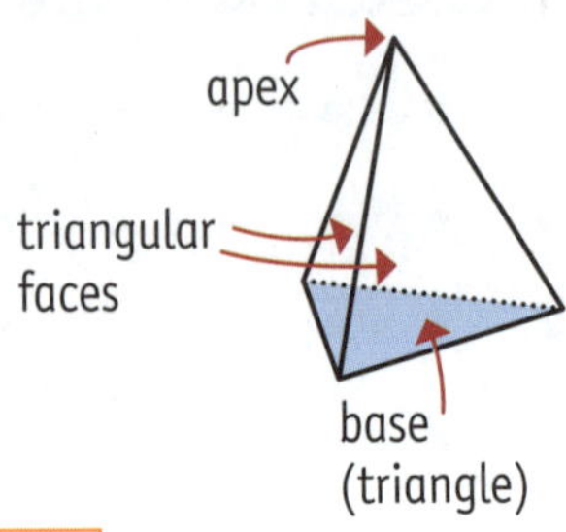

Three 3D objects have curved surfaces:

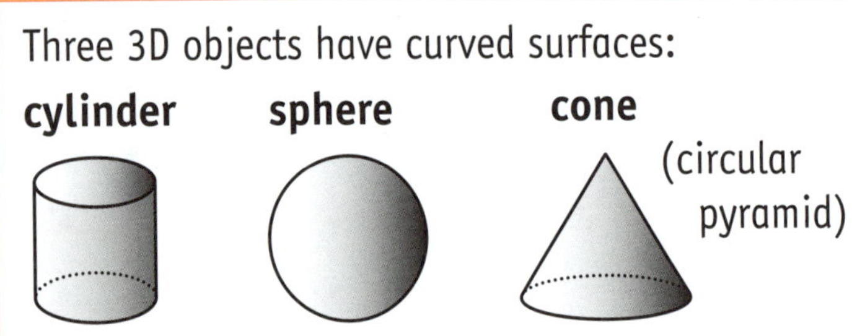

Examples: Complete the sketches and labels.

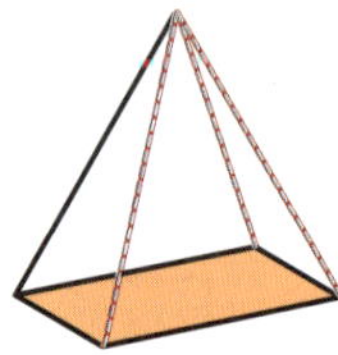

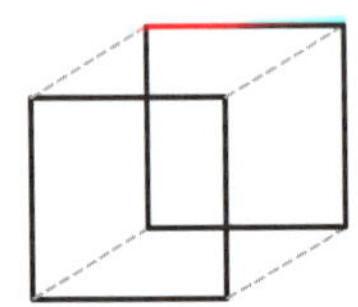

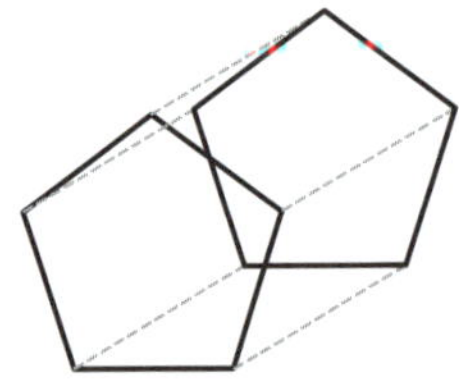

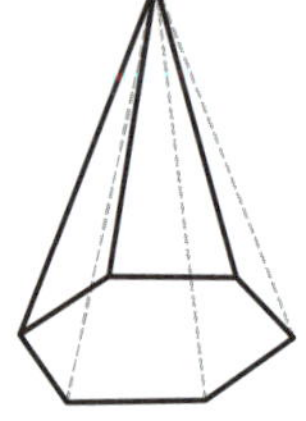

a rectangular pyramid

b c_____

c pentagonal __________

d hexagonal __________

Check the answers on the video!

Colour the bases of the 3D objects in the Examples.

SELF CHECK Tick how you feel

Got it!	Need help...	I don't get it
☐	☐	☐

Check your answers
How many did you get correct? ☐

CATCH UP MATHS YEAR 6 BOOK B © PASCAL PRESS ISBN: 9781925726190

PRACTICE

1 Join the matching labels and 3D objects.

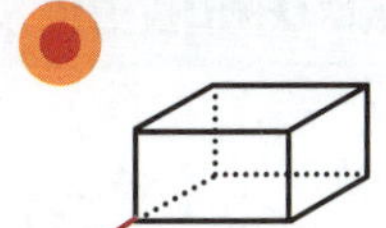

a

b 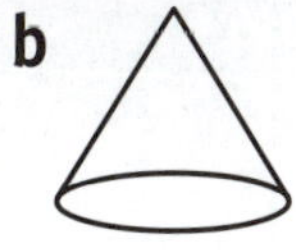

c 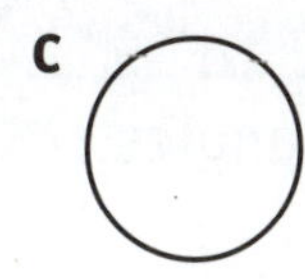

d 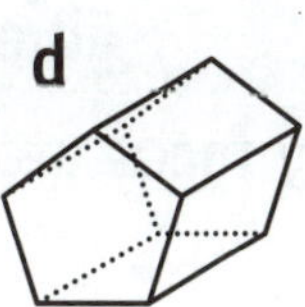

e 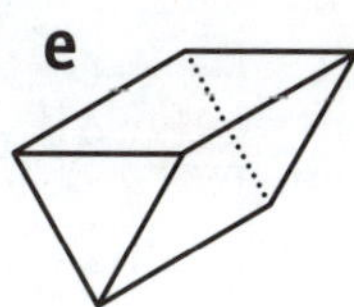

cone	pentagonal pyramid	cube	octagonal prism
rectangular prism	triangular prism	cylinder	octagonal pyramid
hexagonal prism	sphere	pentagonal prism	triangular pyramid

f 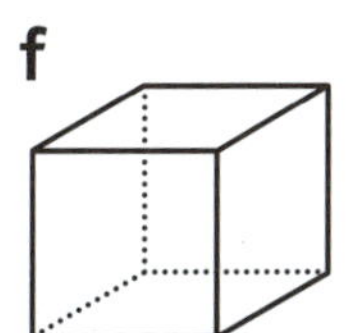

g 

h 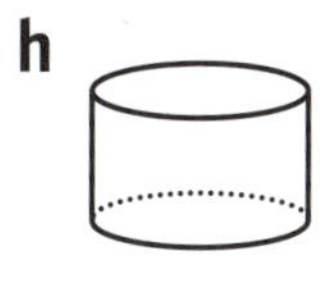

i 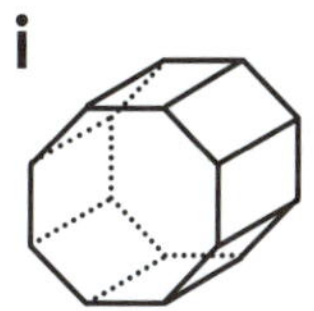

j 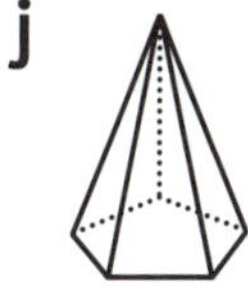

k 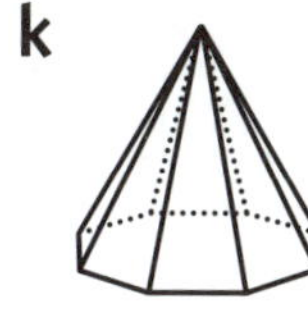

Use the 3D objects below to answer the following questions.

A 

B 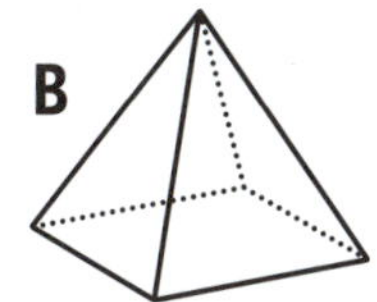

C 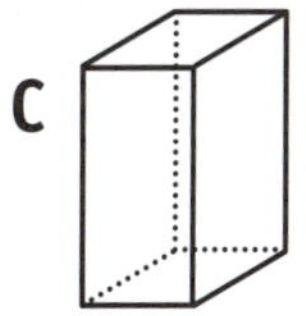

D 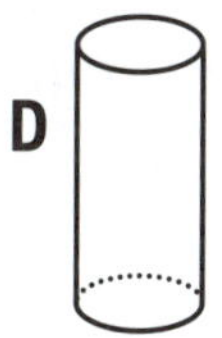

E 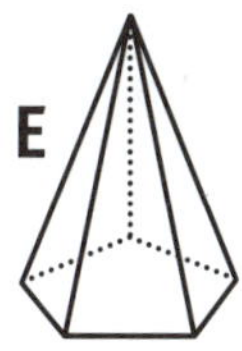

2 Use blue to circle the prisms and red to circle the pyramids.

3 Name each 3D object.

- Object A hexagonal prism
- a Object B ____________
- b Object C ____________
- c Object D ____________
- d Object E ____________

4 Use green to colour the base or bases of each 3D object.

5 Draw a yellow dot on the apex of each pyramid.

6 Draw each face or surface of the 3D objects.

Object A	Object B	Object C	Object D	Object E

Which solid am I?

Clues	Name	Diagram
All of my faces are rectangles.	rectangular prism	
a I have six rectangular sides and two hexagonal bases.		
b I have an apex and a circular base.		
c All my 4 sides are triangles.		
d I have one rectangular face and all my other faces are triangles.		
e I have two circles for bases, and one curved surface.		
f I have two octagonal bases and all my other sides are rectangles.		
g I have one round, curved surface.		

 CATCH UP MATHS YEAR 6 BOOK B © PASCAL PRESS ISBN: 9781925726190

8 Which of these 3D objects have curved surfaces and can roll? Colour them.

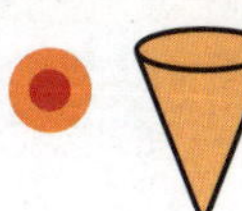

b 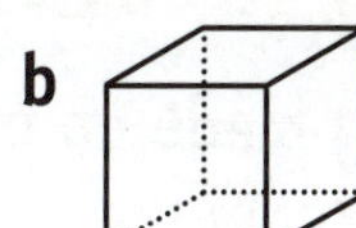

d 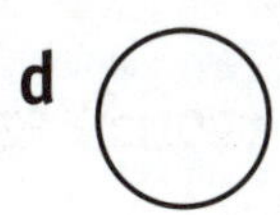

f 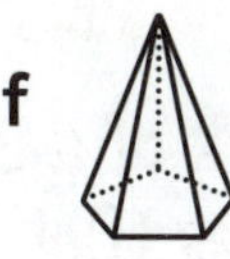

h 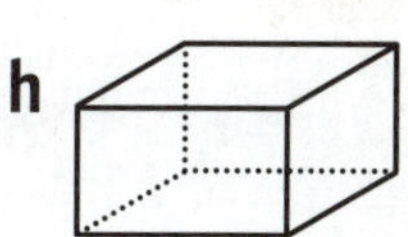

a

c 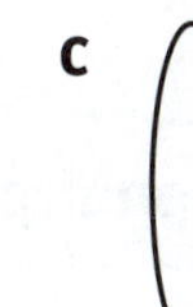

e 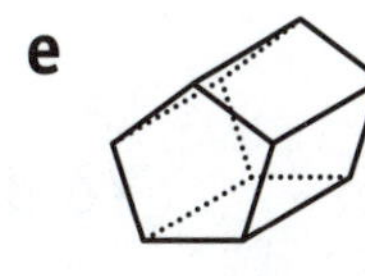

g 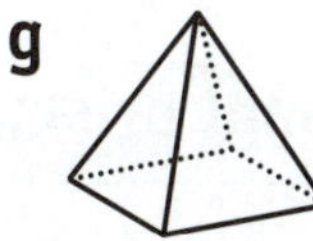

i 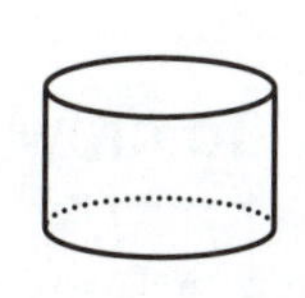

9 Name and describe each 3D object.

Diagram	Name	Description
●	rectangular pyramid	1 rectangular base 4 triangular faces
a		
b		
c		
d		
e		

DRAWING 3D OBJECTS

You can think of a 3D object (solid) as being made up of 2D shapes (flat shapes) that are joined together.

To draw a hexagonal prism

- Draw two hexagons for the bases.
- Make them the same size, and overlap them.

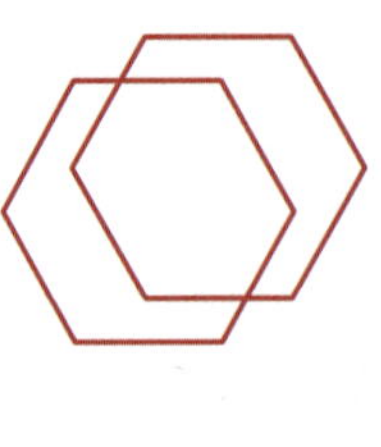

- Draw straight lines to join the matching corners of the bases.

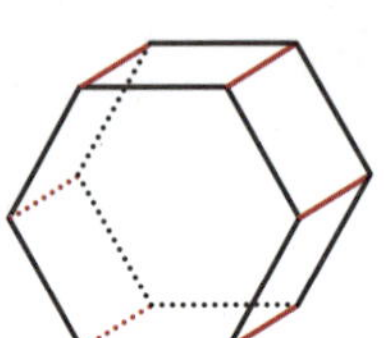

To draw a hexagonal pyramid

- Draw the hexagon for the base.
- Draw a dot for the apex.

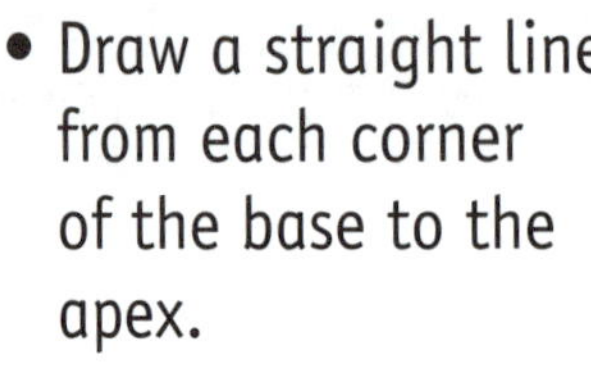

- Draw a straight line from each corner of the base to the apex.

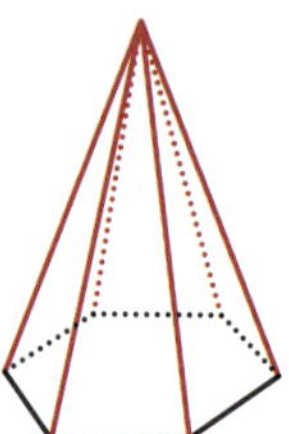

When you draw 3D objects, you can use dotted lines to show the edges that cannot be seen.

Examples: Finish the drawings by joining the corners on the prisms and joining each corner to the apex on the pyramids.

a

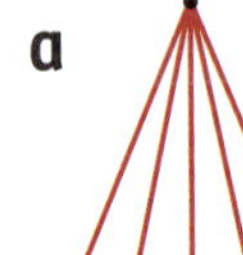

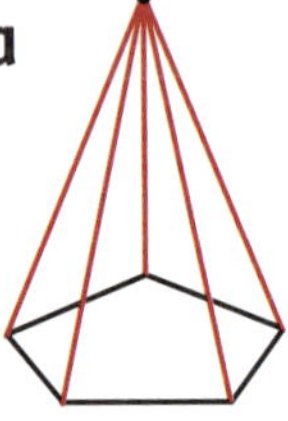

b

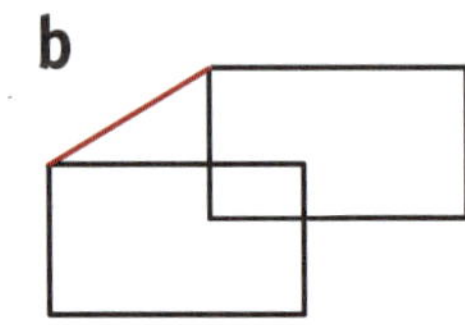

c

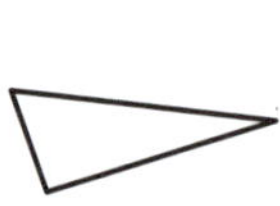

d

Check the answers on the video!

Complete your own drawing of each object in the Examples.

a

b

c

d

SELF CHECK Tick how you feel

Got it!	Need help...	I don't get it
☐	☐	☐

Check your answers

How many did you get correct?

  ISBN: 9781925726190

PRACTICE

Draw the 3D objects on the dot paper.
Try using dotted lines for the edges that cannot be seen.

- cube
- c hexagonal pyramid
- f rectangular pyramid
- a square pyramid
- d pentagonal prism
- g octagonal pyramid
- b hexagonal prism
- e triangular prism
- h triangular pyramid

When drawing curved surfaces, shading is used to make the drawing look curved.
Shade the objects to show the curves.

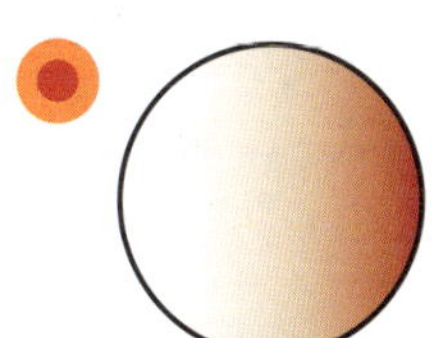

a 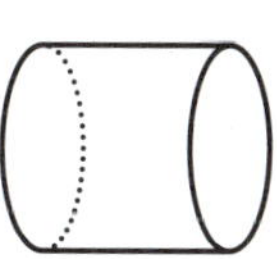

b 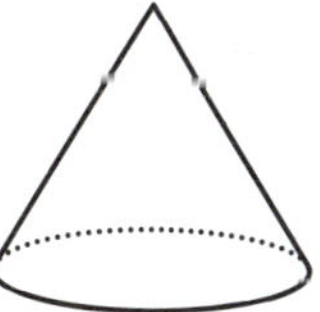

c 

Draw three examples of each curved 3D object.

Cones	Spheres	Cylinders

FACES, EDGES AND CORNERS

Face: A flat surface on a 3D object; a 2D shape

Edge: The line made where two faces join

Corner (Vertex): The point where faces on a 3D object meet

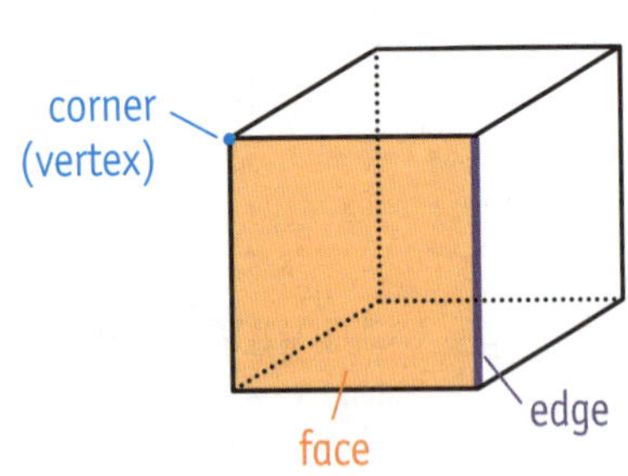

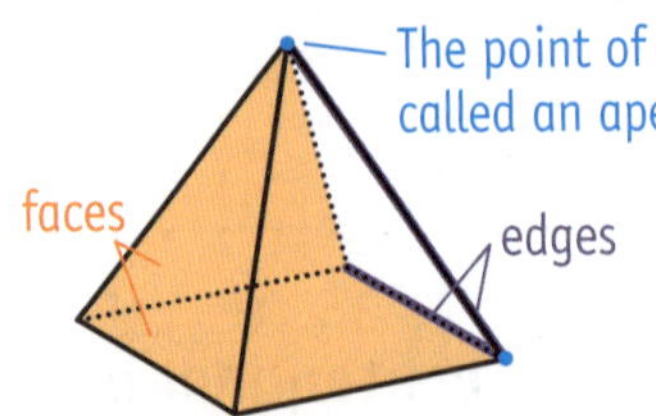

Bases are faces!

Examples: Draw a blue dot on each vertex, colour the faces orange and trace the edges with purple. Then label each 3D object.

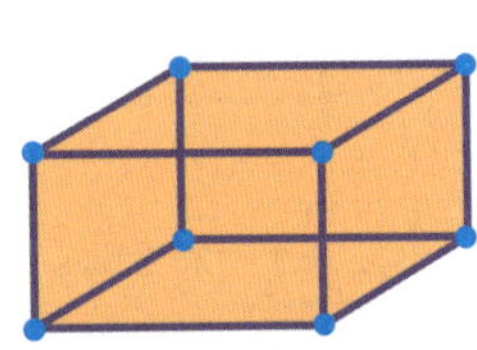

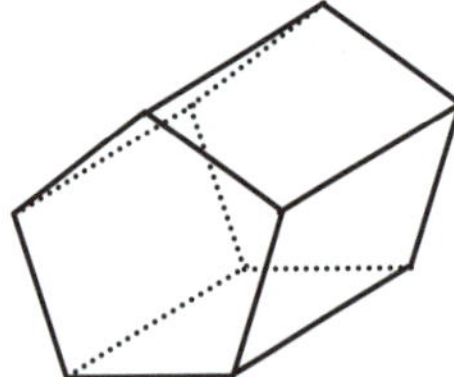

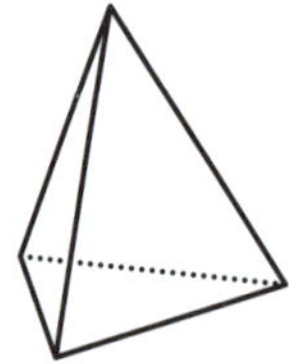

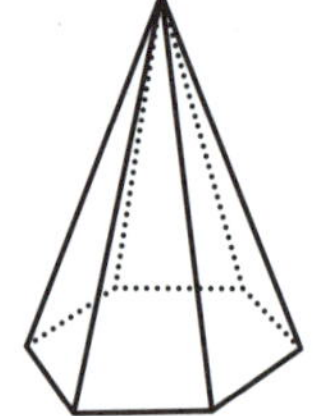

a rectangular prism

b ______________

c ______________

d ______________

Check the answers on the video!

Count each face, edge and vertex on the 3D objects in the Examples and fill in the table.

	Number of faces	Number of edges	Number of vertices
a	6	12	8
b			
c			
d			

SELF CHECK Tick how you feel

Got it!	Need help...	I don't get it
☐	☐	☐

Check your answers

How many did you get correct? ☐

CATCH UP MATHS YEAR 6 BOOK B © PASCAL PRESS ISBN: 9781925726190

PRACTICE

Use dotted lines to show the edges that cannot be seen and then name each object.

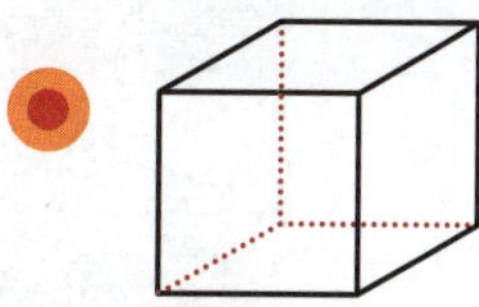

cube

b 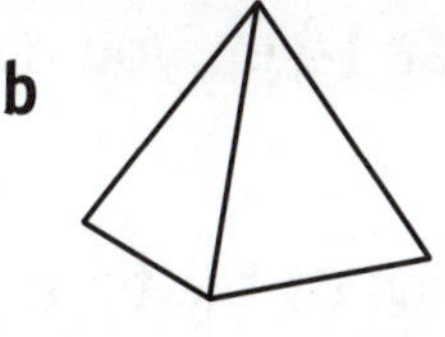

d 

f 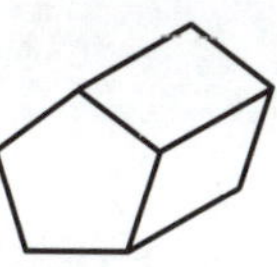

a 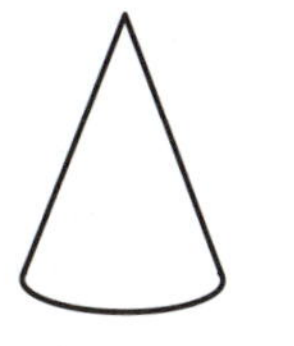

c 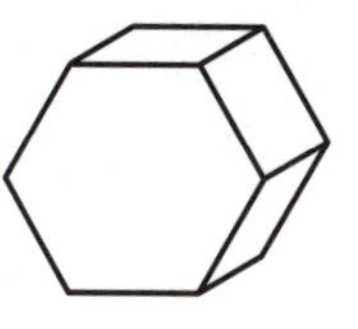

e 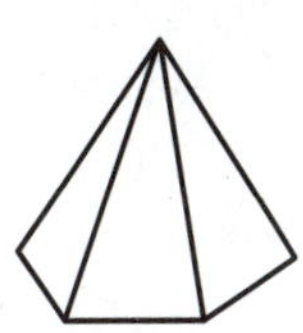

g 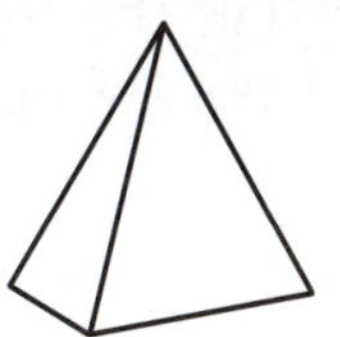

2 Complete the table.

	Name of prism	Diagram	Number of vertices	Number of edges	Number of faces
	cube		8	12	6
a					
b	hexagonal prism				
c	cone				
d					
e	square pyramid				
f					

CROSS-SECTIONS

A cross-section is the shape you see when you cut through a 3D object.

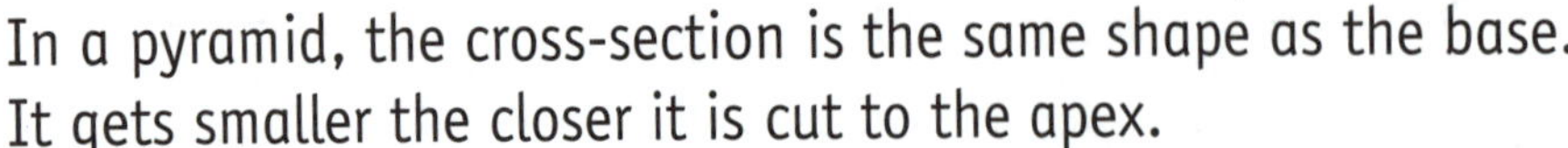

In a pyramid, the cross-section is the same shape as the base. It gets smaller the closer it is cut to the apex.

Cut the cone along the orange line.

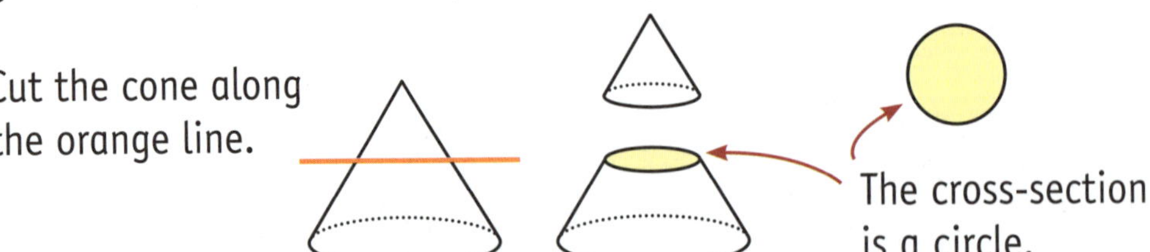

In a prism, the cross-section is the same shape and size as the base.

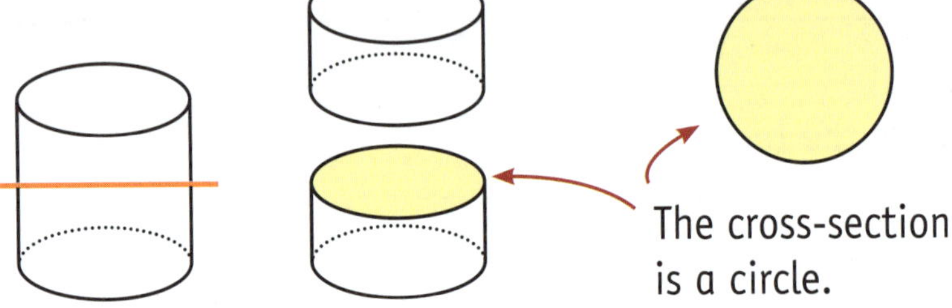

Examples: Circle the cross-section you would see if you cut the 3D objects along the orange line.

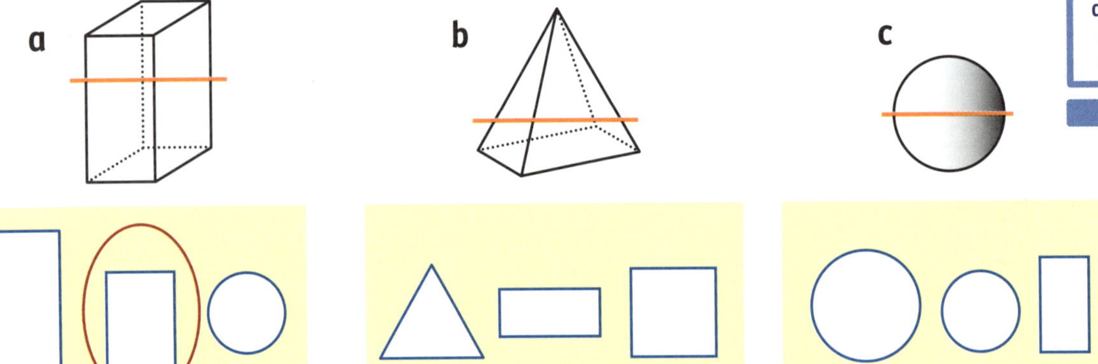

Check the answers on the video!

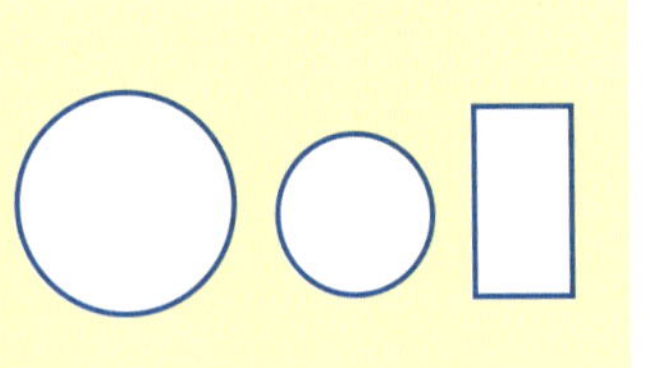

Using the 3D objects from the Examples, draw the cross-section you would see if you cut them vertically.

a

b

c

SELF CHECK Tick how you feel

Got it!	Need help...	I don't get it
☐	☐	☐

Check your answers

How many did you get correct? ☐

CATCH UP MATHS YEAR 6 BOOK B © PASCAL PRESS ISBN: 9781925726190

1 Draw a line on the 3D object to show where the cross-section was made.

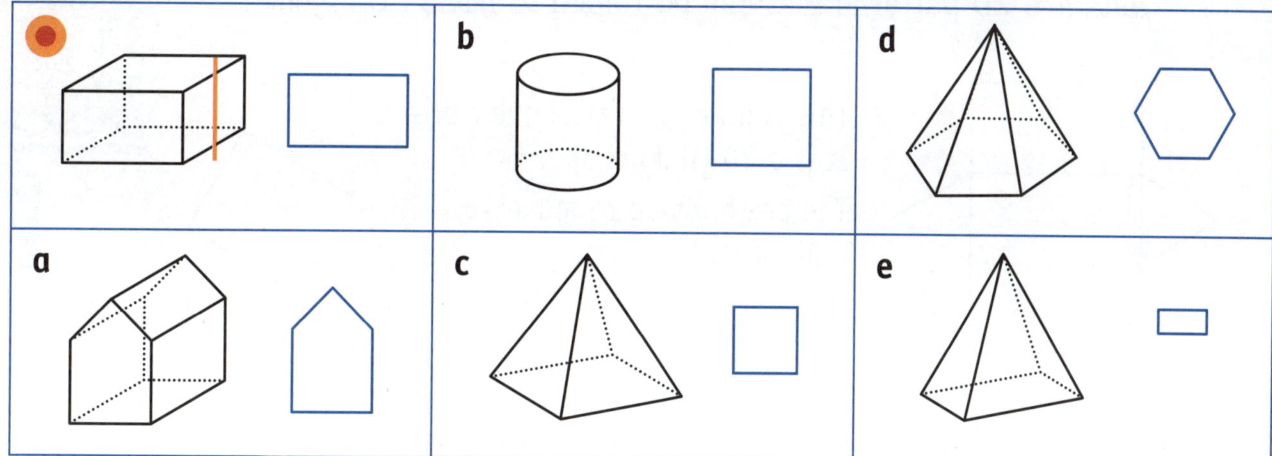

2 Name the shape of the cross-section made by the orange line.

a 

b 

c 

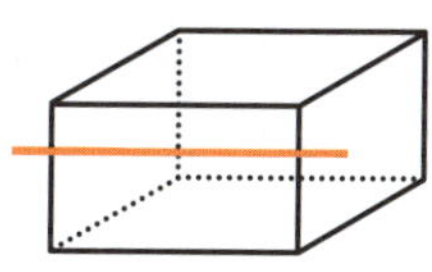

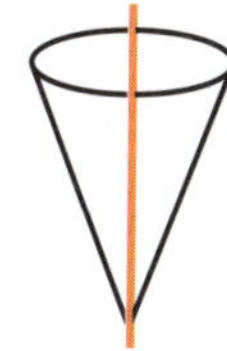

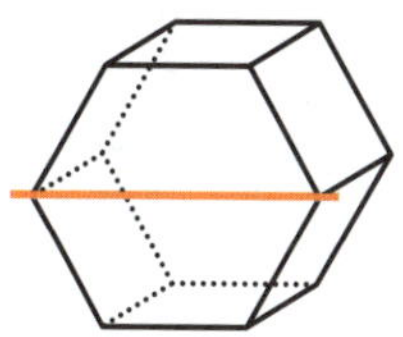

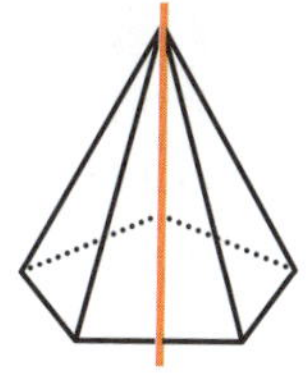

3 Draw the cross-section made by the orange line for each 3D object in Question 2.

●	a	b	c

4 Draw the horizontal and vertical cross-sections.

	●	a	b	c
Horizontal cross-section				
Vertical cross-section				

NETS

Nets are 2D patterns that can be folded to make 3D objects.

SCAN to watch video

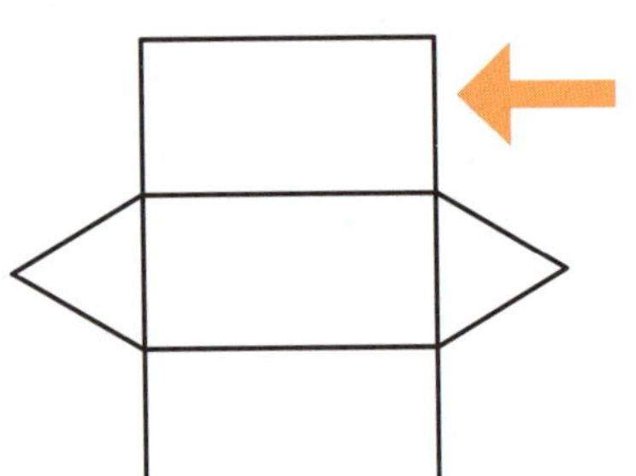

This is a net of a triangular prism.
It is a 2D (flat) shape.
The net folds up to make a 3D object.

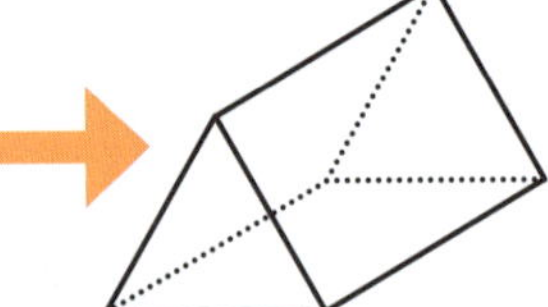

Examples: Colour the nets green and the 3D objects blue.
Then join the matching nets and 3D objects.

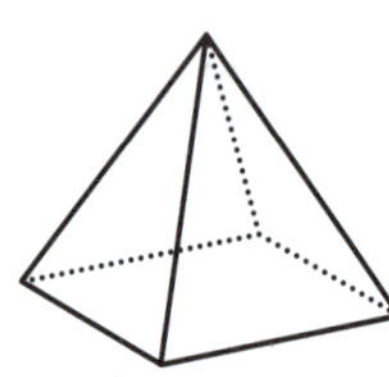

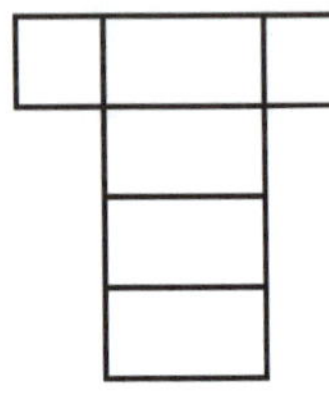

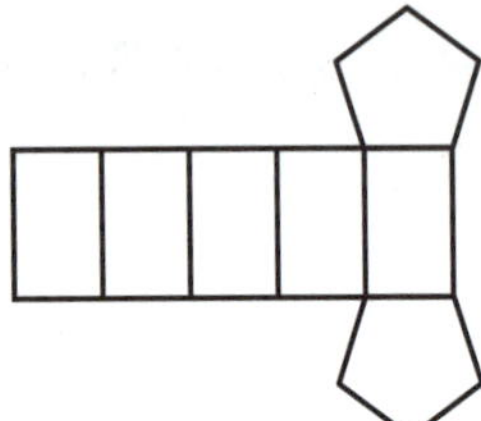

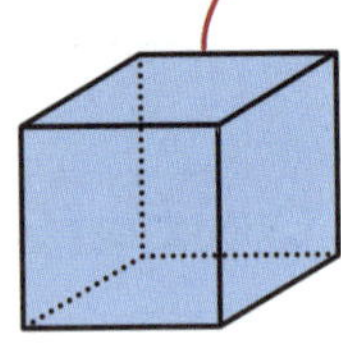

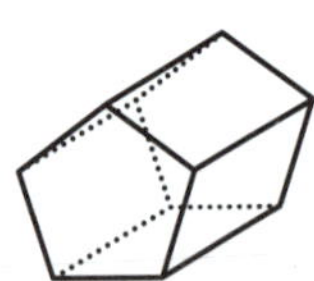

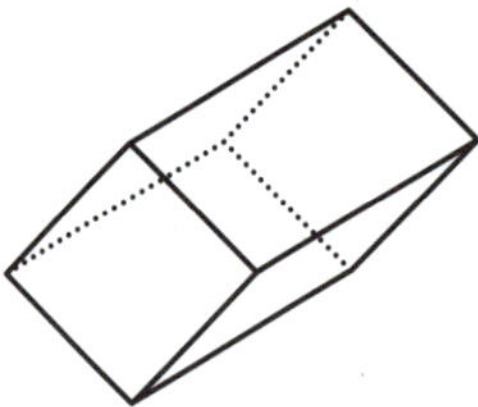

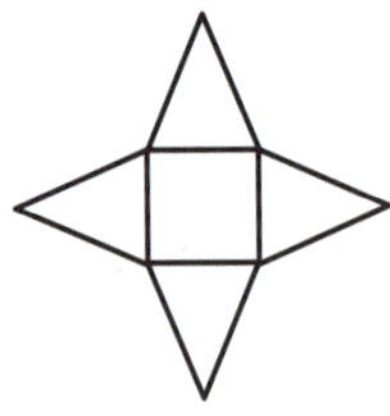

Check the answers on the video!

Your turn

Draw a net for each 3D object.

a

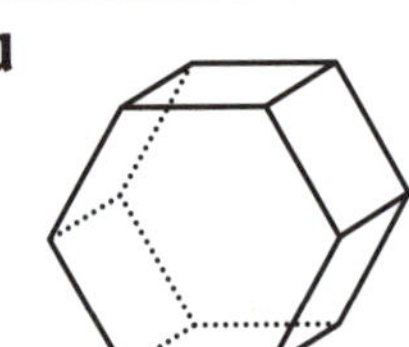

b

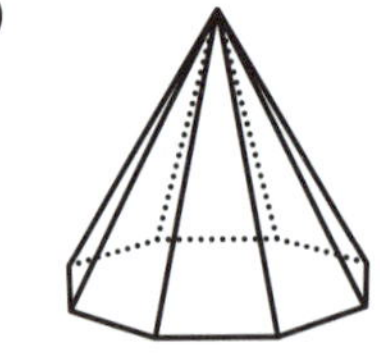

c

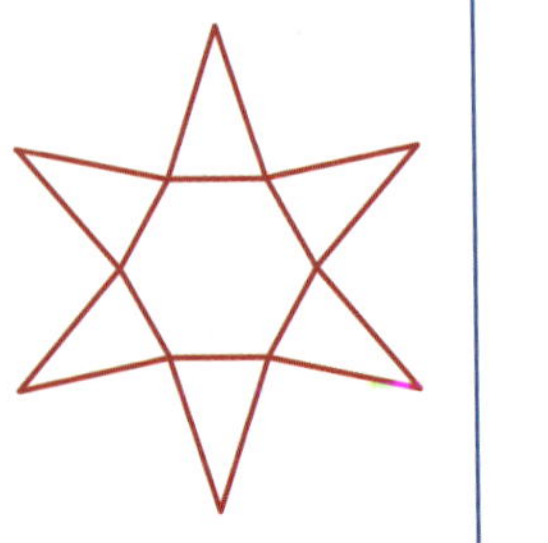

SELF CHECK Tick how you feel

Got it!	Need help...	I don't get it
☐	☐	☐

Check your answers
How many did you get correct? ☐

CATCH UP MATHS YEAR 6 BOOK B © PASCAL PRESS ISBN: 9781925726190

PRACTICE

Join the two pieces that make up a complete net.

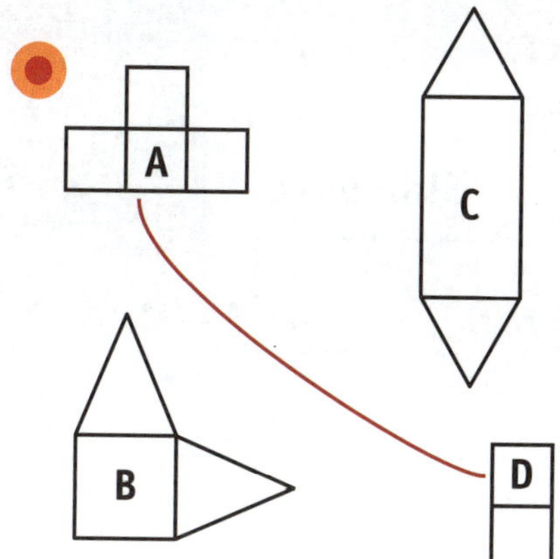

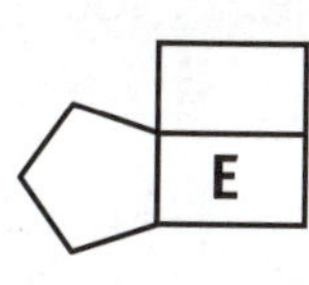
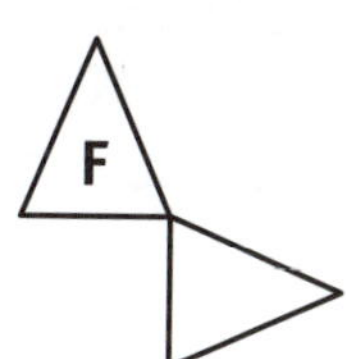

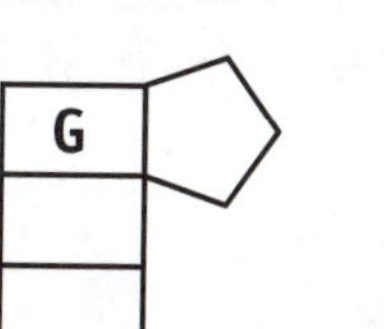

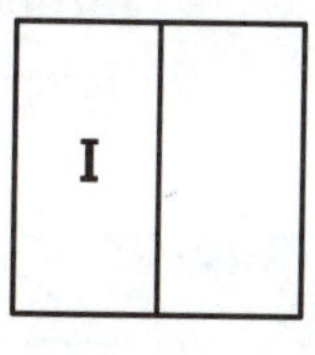

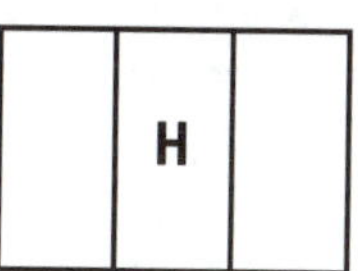

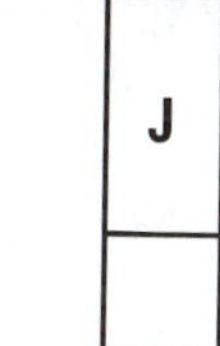

2 Use the six faces and the three views of the cube to colour the net correctly.

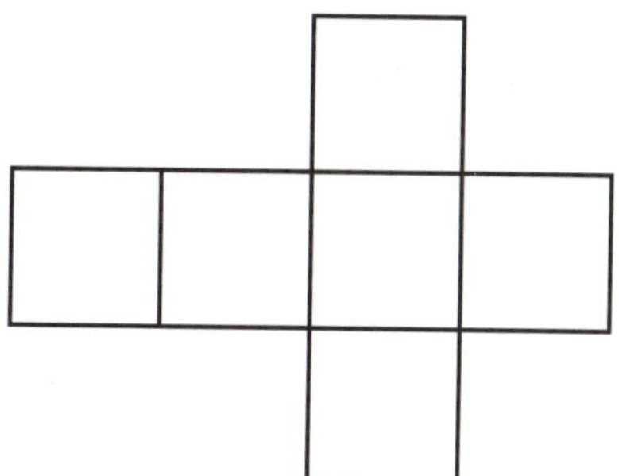

This net has been folded to make a triangular prism.

a What symbol is on the bottom? ___

b What symbol is opposite ★? ___

c What symbol is on the left side? ___

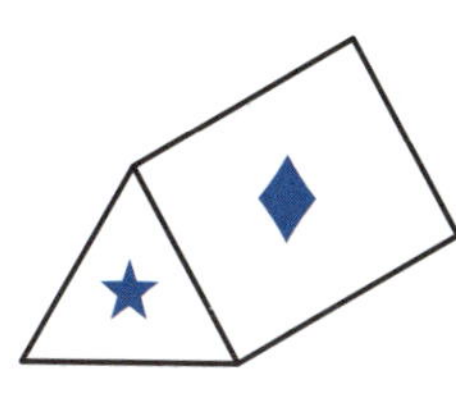

This net has been folded to make a pyramid.

a What letter is opposite D? ___

b What letter is opposite E? ___

c The apex is above letter ___.

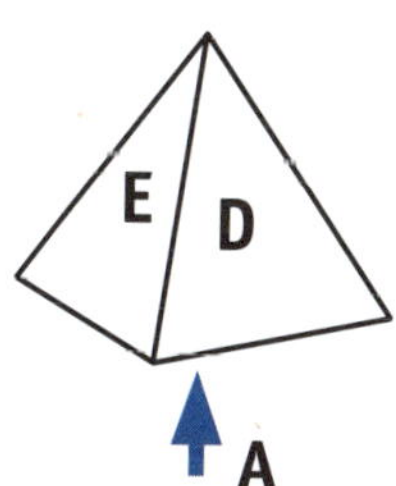

Use the net to label the cube.

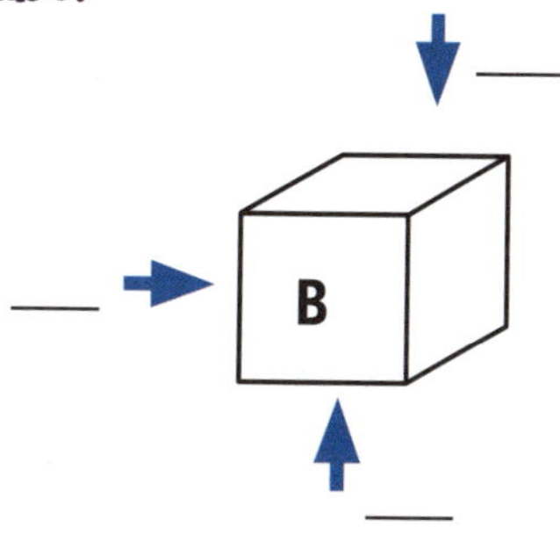

VIEWS

Three-dimensional (3D) objects look different when you look at them from different views.

		Top view	Front view	Side view
Rectangular prism	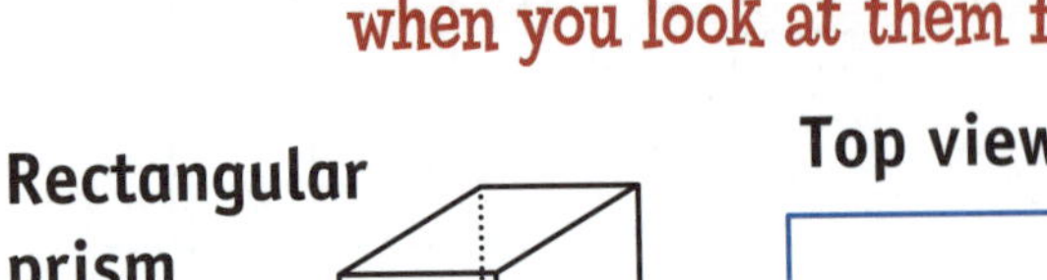			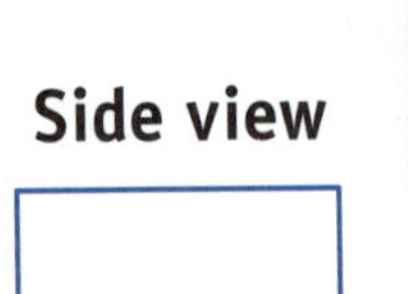
Rectangular pyramid	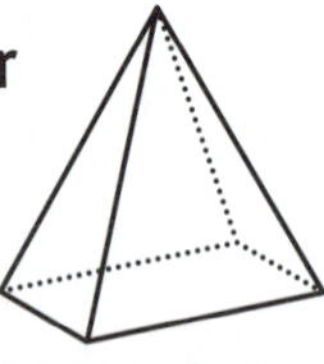	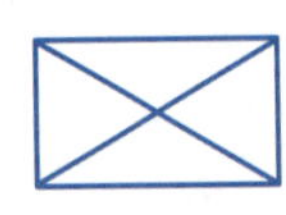		

Examples: Label the views as top, front or side.

a 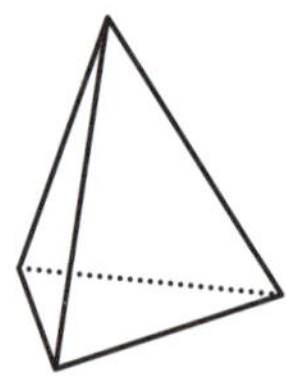 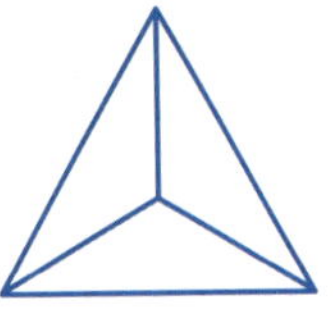 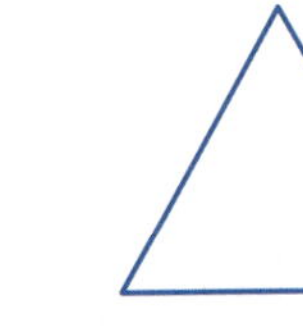

top ___ side ___ front ___

b 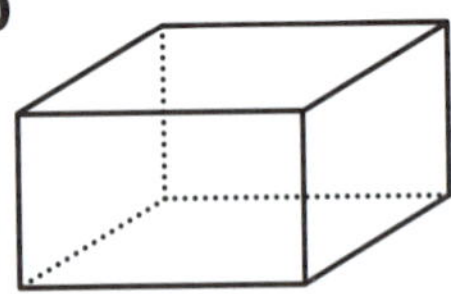   

______ ______ ______

Check the answers on the video!

Circle the shape that matches the view.

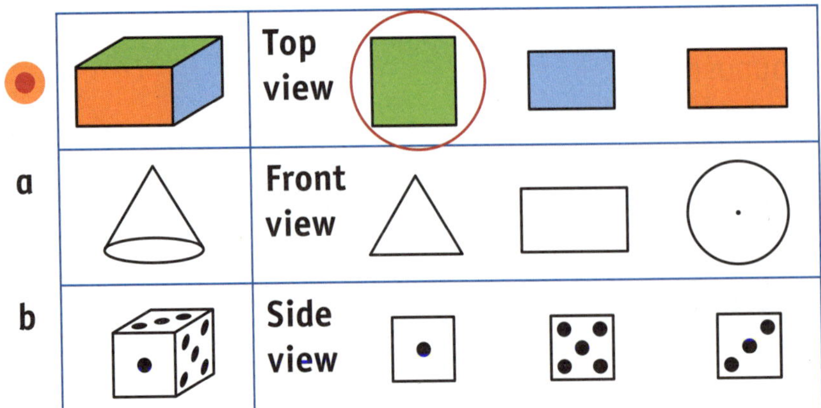

SELF CHECK Tick how you feel

Got it! ☐ Need help... ☐ I don't get it ☐

Check your answers
How many did you get correct? ☐

CATCH UP MATHS YEAR 6 BOOK B © PASCAL PRESS ISBN: 9781925726190

PRACTICE

Draw the view of each 3D object you would see from the arrow.

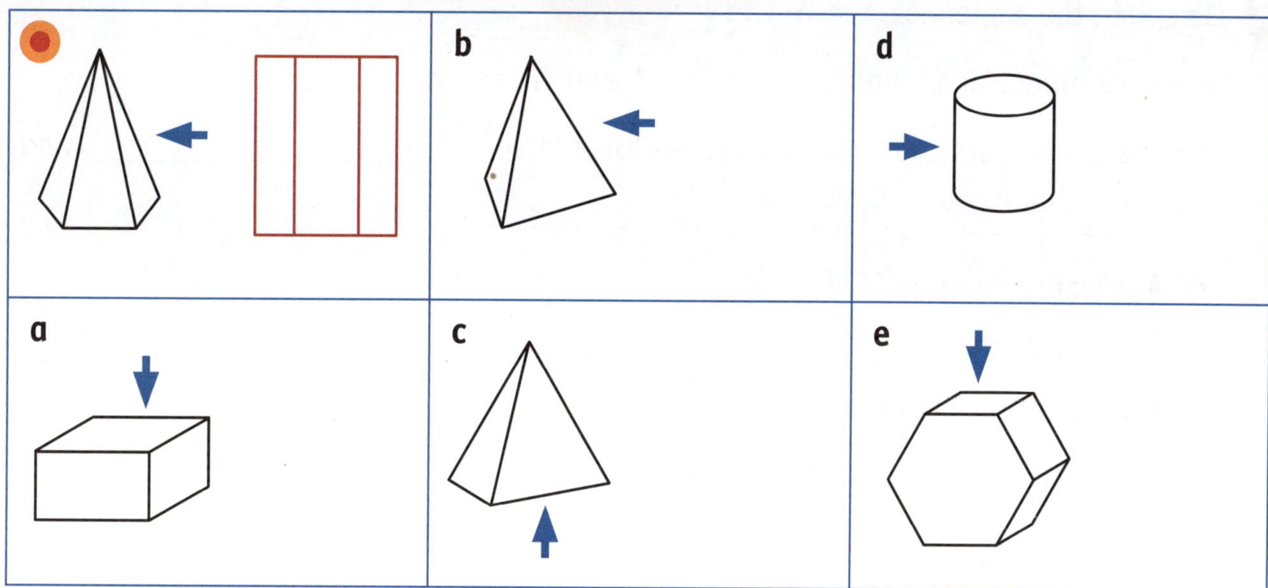

Draw the top, front and side view of each 3D object.

	Object	Top view	Front view	Side view
●				
a				
b				
c				
d				
e				

3D OBJECTS REVIEW

 **Complete the sentences.**

a A 3D object has height, ________ (length) and depth.

b The 3D objects that have curved surfaces are __________, __________ and __________.

c A prism has one pair of parallel _______.

The shape of the _____ names the prism.

All the other faces are ___________.

d A pyramid is named by the shape of the _____.

All the other faces are ___________ that meet at a point called the _____.

 **Label each 3D object.**

a 

b 

c 

d 

e 

f 

g 

h ________________

i 

j 

k 

l 

m 

n 

  ISBN: 9781925726190

3 Use the 3D objects in Question 2 to complete the following.

a Draw a red dot on each apex.

b Colour all the bases blue.

4 Draw the 3D objects.

a hexagonal prism	**d** cone	**g** hexagonal pyramid
b triangular prism	**e** rectangular prism	**h** cylinder
c octagonal pyramid	**f** pentagonal prism	**i** sphere

5 On your drawings in Question 4, trace the edges with orange, colour the faces yellow and draw a blue dot on the vertices.

6 Complete the sentences

A face is a ________ surface on a ___ object.

An ________ is the line made where two faces join.

A ________ is the point where faces meet.

  ISBN: 9781925726190

REVIEW

 **7** Complete the table.

	Name	Diagram	Number of vertices	Number of edges	Number of faces
a					
b	triangular pyramid				
c	octagonal prism				
d					
e	hexagonal prism				
f					
g	cone				
h					

CATCH UP MATHS YEAR 6 BOOK B © PASCAL PRESS ISBN: 9781925726190

8 Draw the cross-section you would see if you cut the 3D object along the orange line.

a	c	e
b	d	f

9 Draw a net for each 3D object.

a cube	d cone	g rectangular pyramid
b hexagonal pyramid	e cylinder	h square pyramid
c octagonal prism	f pentagonal prism	i hexagonal prism

This net has been folded to make a cube.

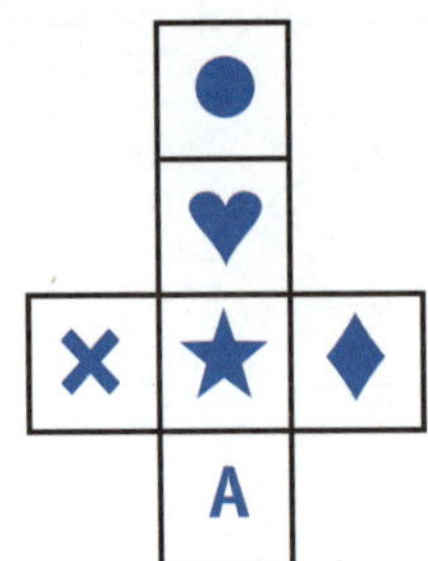

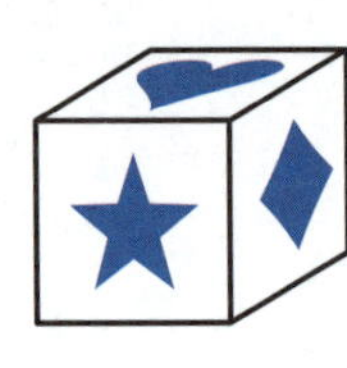

a What symbol is at the back? ____

b What symbol is on the bottom? ____

c What symbol is on the left side? ____

Name the 3D object that could have the top view shown.

a ____________	**c** ____________	**e** ____________
b ____________	**d** ____________	**f** ____________

Name a 3D object that could have the side view shown.

a ____________	**c** ____________	**e** ____________
b ____________	**d** ____________	**f** ____________

CATCH UP MATHS YEAR 6 BOOK B © PASCAL PRESS ISBN: 9781925726190

13 Draw the different views for each 3D object.

	Object	Top view	Front view	Side view
a	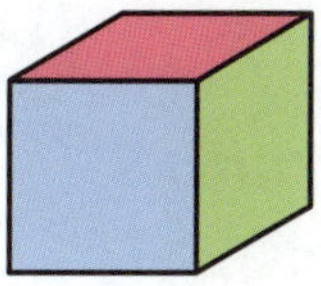			
b				
c				
d				
e				
f				

THE SQUARE CENTIMETRE

Small areas are measured using square centimetres (cm^2).

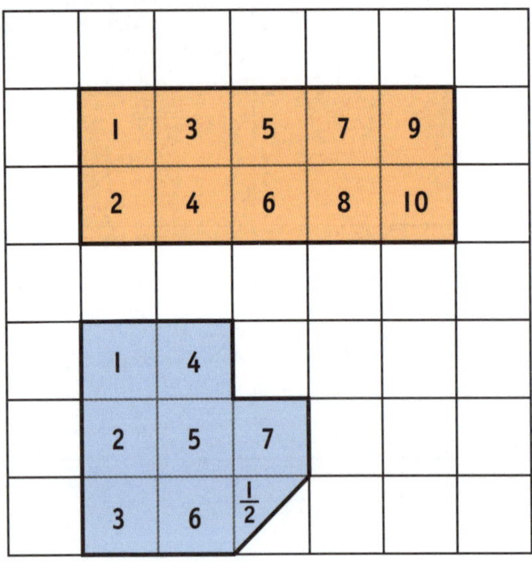

The orange rectangle has an area of 10 square centimetres (cm^2).

Area = 10 cm^2

The blue shape has an area of $7\frac{1}{2}$ square centimetres ($7\frac{1}{2}$ cm^2).

Area = $7\frac{1}{2}$ cm^2

Examples: Find the area of the shapes.

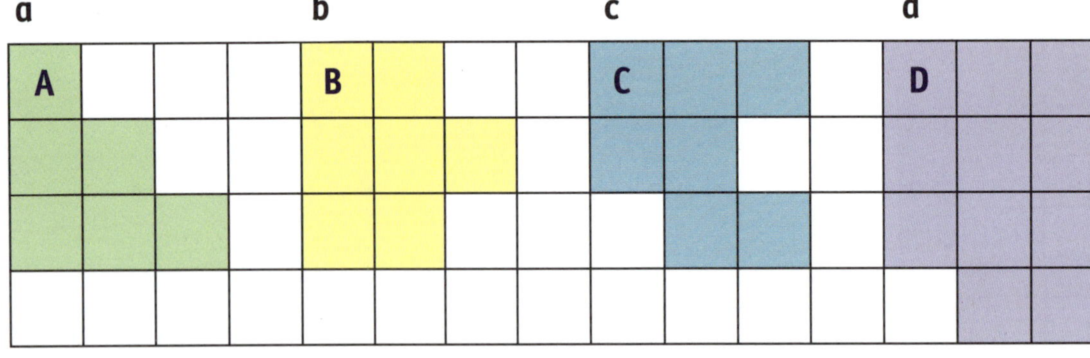

Area = 6 cm^2 Area = __ cm^2 Area = __ cm^2 Area = __ cm^2

Check the answers on the video!

Use the shapes in the Examples to answer the following questions.

- Which shape has the largest area? D

a Which shape has the smallest area? ___

b Which shapes have the same areas? ___ and ___

c What is the total area of the four shapes? ___________

Check your answers
How many did you get correct?

CATCH UP MATHS YEAR 6 BOOK B © PASCAL PRESS ISBN: 9781925726190

PRACTICE

1 Work out the area of these 2D shapes.

Area = 4 cm^2 b Area = ______ d Area = ______ f Area = ______

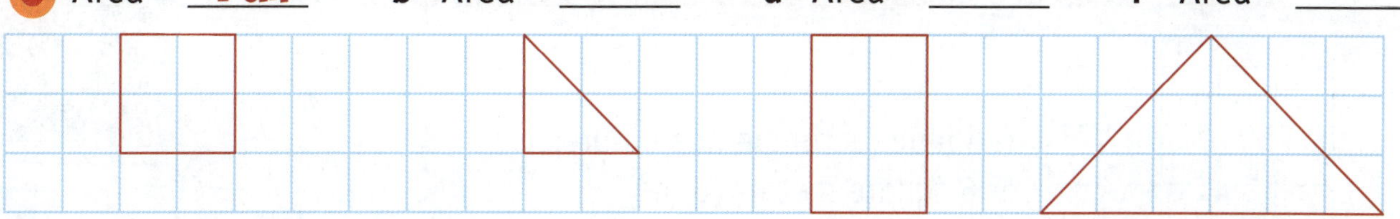

a Area = ______ c Area = ______ e Area = ______ g Area = ______

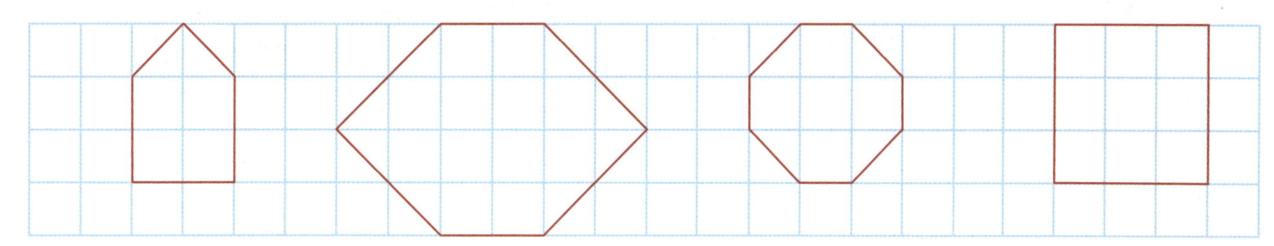

= 1 cm^2

2 Name each 2D shape in Question 1.

square b ______ d ______ f ______

a ______ c ______ e ______ g ______

3 Which shape in Question 1 has:

a the smallest area? ___

b the largest area? ___

c the same area as another shape? ___

4 Write the difference in area between each pair of shapes in Question 1.

large triangle, small square 5 cm^2 b pentagon, rectangle ______

a hexagon, octagon ______ c large square, small triangle ______

5 Draw five different shapes each with an area of 10 cm^2. = 1 cm^2

THE SQUARE METRE

Larger areas are measured using square metres (m^2).

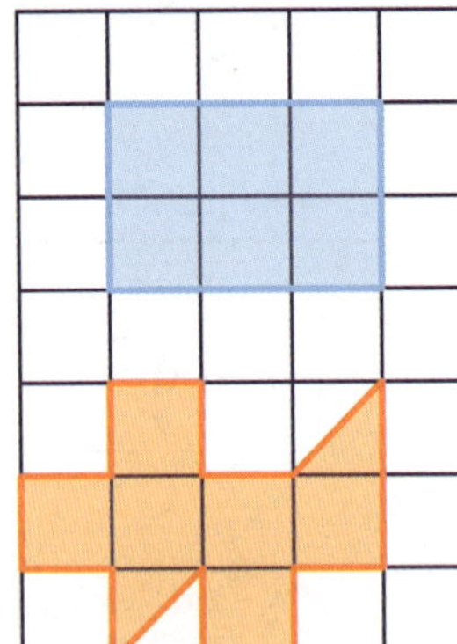

The blue rectangle has an area of 6 square metres (6 m^2).
Area = 6 m^2

The orange shape has an area of 7 square metres (7 m^2).
Area = 7 m^2

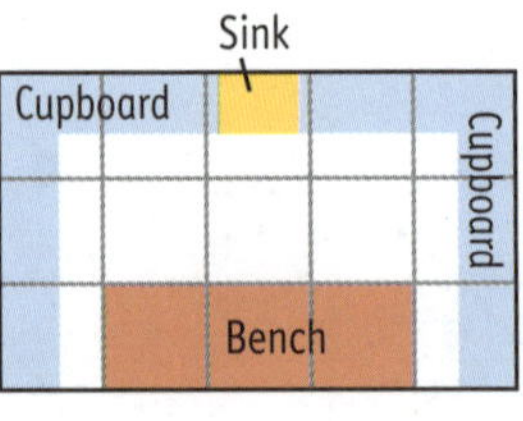

☐ = 1 m^2

Your kitchen would be measured using square metres.

Examples: Calculate the area. ☐ = 1 m^2

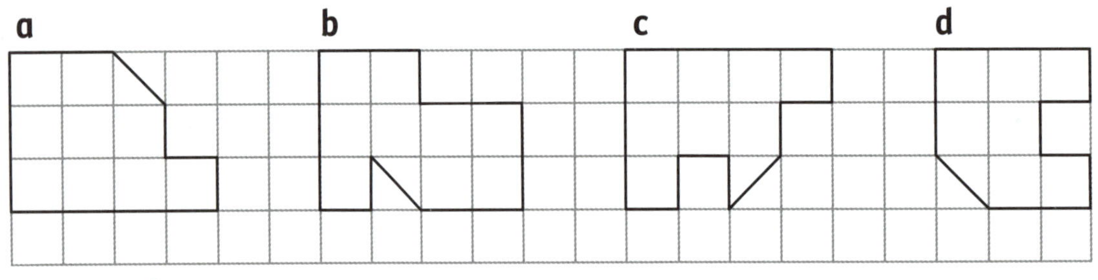

a Area = $9\frac{1}{2}$ m^2

b Area = ____ m^2

c Area = ____ m^2

d Area = ____ m^2

Check the answers on the video!

1 Colour the shapes in the Examples.

a Blue: The shape with an area of $8\frac{1}{2}$ m^2

b Red: The shape with the smallest area

c Green: The shapes with the same area

2 Write the total area.

a Blue shape + red shape ________

b Two green shapes ________

c One green shape + the blue shape ________

d One green shape + the red shape ________

SELF CHECK Tick how you feel

Got it!	Need help...	I don't get it
☐	☐	☐

Check your answers
How many did you get correct? ☐

CATCH UP MATHS YEAR 6 BOOK B © PASCAL PRESS ISBN: 9781925726190

PRACTICE

1 **Tick the items that would be measured using square metres (m^2).**

- ☑ a classroom floor
- **a** ☐ a small book cover
- **b** ☐ a garage floor
- **c** ☐ a notebook
- **d** ☐ a picture of your face
- **e** ☐ a sticky note
- **f** ☐ a packet of playing cards
- **g** ☐ a large window
- **h** ☐ a bathroom wall

2 **Rewrite using m^2.**

- seventeen square metres 17 m^2
- **a** nine square metres ______
- **b** fifty-eight square metres ______
- **c** forty square metres ______
- **d** one hundred and ten square metres ______
- **e** three hundred and sixty-three square metres ______

3 **Write the correct unit of measurement, cm^2 or m^2, for these items.**

- workbook cover: 600 cm^2
- **a** door mat: 1 _____
- **b** beach towel: 2 _____
- **c** classroom floor: 64 _____
- **d** sheet of paper: 420 _____
- **e** thumbnail: $1\frac{1}{2}$ _____

4 **In Super B-Mart, how many m^2 is each section?**

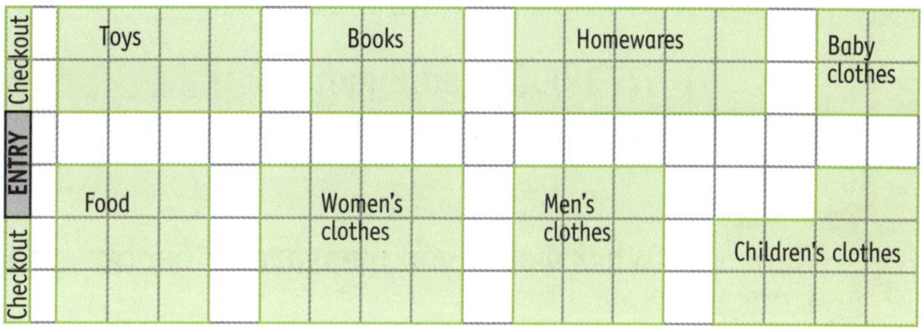

- toys 8 m^2
- **a** women's clothes ____
- **b** children's clothes ____
- **c** men's clothes ____
- **d** food ____
- **e** books ____
- **f** homewares ____
- **g** baby clothes ____
- **h** checkouts ____
- **i** entry ____

Super B-Mart

☐ = 1 m^2

Checkout
Toys
Books
Homewares
Baby clothes
ENTRY
Food
Women's clothes
Men's clothes
Children's clothes
Checkout

5 **What is the total area of Super B-Mart?** ______

6 **Which section(s) in Super B-Mart takes up:**

- the most space? women's clothes
- **a** the least space? ______________
- **b** 8 m^2? ______________
- **c** 6 m^2? ______________
- **d** 4 m^2? ______________
- **e** 9 m^2? ______________

  ISBN: 9781925726190

HECTARES

A hectare is a unit of measurement used to measure large areas.
A hectare (ha) is 10 000 m^2.

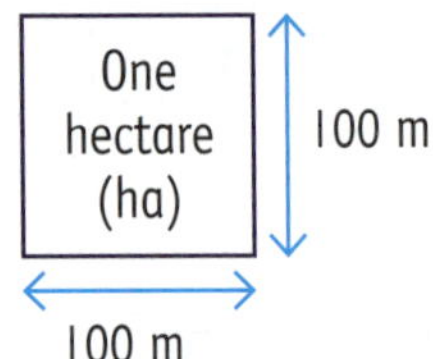

This small farm is 10 hectares.

SCAN to watch video

Examples: Write the short form for each measurement in m^2 and ha.

- **a** 40 000 square metres 40 000 m^2 = 4 ha
- **b** 25 000 square metres ________ = _____ ha
- **c** 92 000 square metres ________ = _____ ha
- **d** 68 000 square metres ________ = _____ ha
- **e** 147 000 square metres ________ = _____ ha
- **f** 803 000 square metres ________ = _____ ha
- **g** 920 000 square metres ________ = _____ ha
- **h** 469 000 square metres ________ = _____ ha
- **i** 773 000 square metres ________ = _____ ha

What would you measure in hectares (ha)? Colour the boxes.

- ● ■ a school ground
- **a** ☐ a kitchen
- **b** ☐ an airport
- **c** ☐ a cricket pitch
- **d** ☐ an apple farm
- **e** ☐ a showground
- **f** ☐ a baby pen
- **g** ☐ a bathroom
- **h** ☐ a basketball court
- **i** ☐ a suburb

SELF CHECK Tick how you feel

Got it!	Need help...	I don't get it
☐	☐	☐

Check your answers
How many did you get correct? ☐

CATCH UP MATHS YEAR 6 BOOK B © PASCAL PRESS ISBN: 9781925726190

PRACTICE

1 Write the long form for these measurements.

- ● 5 ha 5 hectares
- a 49 ha ______
- b 79 ha ______
- c 66 ha ______
- d 352 ha ______
- e 210 ha ______
- f 149 ha ______
- g 3764 ha ______
- h 4935 ha ______
- i 3980 ha ______
- j 6002 ha ______
- k 5963 ha ______

2 Complete the tables.

	m^2	hectares (ha)
●	30 000	3
a	24 000	
b	71 000	
c	149 000	
d	872 000	
e	209 000	
f	142 000	

	hectares (ha)	m^2
●	5	50 000
g	9	
h	46	
i	369	
j	481	
k	156	
l	639	

Use the map of Koby's orchard to complete the following.

3 How many hectares is each section?

- ● limes 6 ha
- a apples ______
- b lemons ______
- c mandarins ______
- d pears ______
- e oranges ______

Koby's Orchard

4 How many hectares altogether is Koby's Orchard? ______

THE SQUARE KILOMETRE

A square kilometre (km^2) is 1 000 000 m^2.
Very large areas are measured in square kilometres.

1 000 000 m^2

1 km (1000 m)

1 km (1000 m)

Countries are measured in km^2.

National Parks are measured using km^2.

Examples: Which places would be measured in km^2? Circle them.

a Sahara Desert	**e** Sydney	**i** a backyard
b a netball court	**f** a suburb	**j** a golf course
c Canada	**g** a soccer field	**k** a national park
d a grocery store	**h** a city	**l** an island

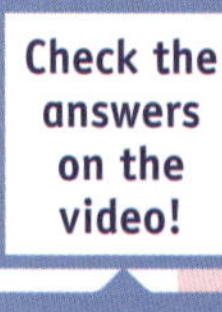

Write these areas in short form.

- twenty-three square kilometres 23 km^2

a forty square kilometres ________

b nine square kilometres ________

c seven hundred square kilometres ________

d one thousand, six hundred and thirty-four square kilometres ________

e four thousand, nine hundred and two square kilometres ________

f eight thousand and seventy-six square kilometres ________

Check your answers
How many did you get correct?

CATCH UP MATHS YEAR 6 BOOK B © PASCAL PRESS ISBN: 9781925726190

PRACTICE

Here are the areas in square kilometres of ten of the world's largest countries.

 China 9 596 960

 India 3 287 263

 Russia 17 098 242

 Brazil 8 515 770

 Australia 7 741 220

 Canada 9 984 670

 Argentina 2 780 400

 Kazakhstan 2 724 900

 USA 9 833 517

Algeria 2 381 741

1 List the countries in order from largest area (1) to smallest (10).

1 Russia

2 ____________

3 ____________

4 ____________

5 ____________

6 ____________

7 ____________

8 ____________

9 ____________

10 ____________

2 What is the difference in area?

- Canada and USA 151 153 km²

a Russia and Australia ____________

b India and Kazakhstan ____________

c Argentina and China ____________

Use the table of desert areas to complete the following.

3 What is the area?

- Antarctica Desert 14 000 000 km²

a Arabian Desert ____________

b Chihuahan Desert ____________

c Patagonian Desert ____________

d Great Basin Desert ____________

Desert	Area (km²)
1 Antarctica (Antarctica)	14 000 000
2 Sahara (Africa)	9 000 000
3 Australian (Australia)	2 700 000
4 Arabian (Western Asia)	2 330 000
5 Gobi (Eastern Asia)	1 295 000
6 Kalahari (Southern Africa)	900 000
7 Patagonian (South America)	620 000
8 Syrian (Western Asia)	520 000
9 Great Basin (USA)	492 000
10 Chihuahuan (Mexico, USA)	450 000

4 How much larger?

- Australian Desert than the Syrian Desert

  2 180 000 km²

a Sahara Desert than the Kalahari Desert ____________

b Gobi Desert than the Patagonian Desert ____________

USING MULTIPLICATION TO CALCULATE AREA – SQUARES AND RECTANGLES

You can use multiplication to calculate the area of squares and rectangles.

Area of a square

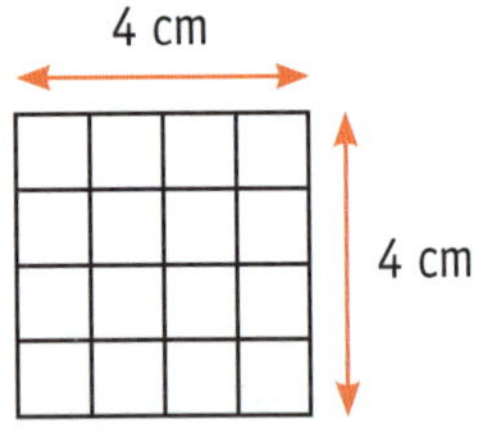

Area = length × width

= 4 cm × 4 cm

= 16 cm^2

Area of a rectangle

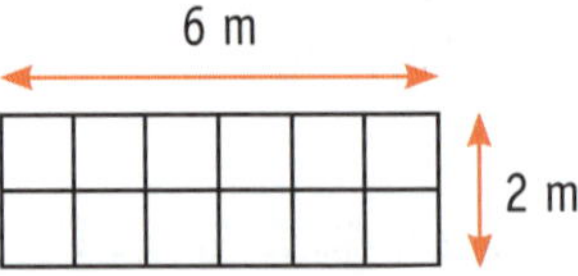

Area = length × width

= 6 m × 2 m

= 12 m^2

Area = length × width is the formula to calculate the area of a square or rectangle.

Examples: Calculate the area.

a

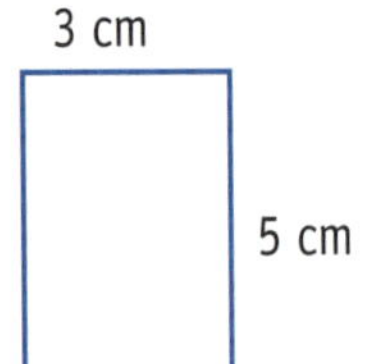

Area = length × width

= 3 cm × 5 cm

= 15 cm^2

b

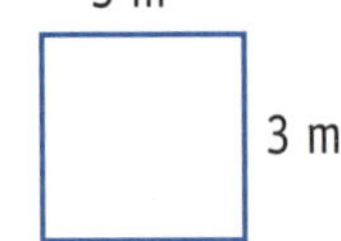

Area = ________ × width

= 3 m × 3 m

= ____ m^2

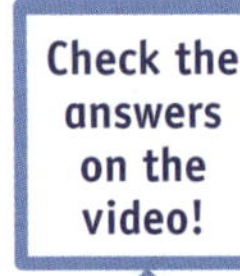

Your turn Calculate the area.

●

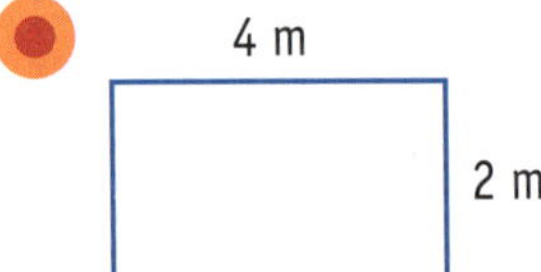

Area = length × width

= 2 m × 4 m

= 8 m^2

a

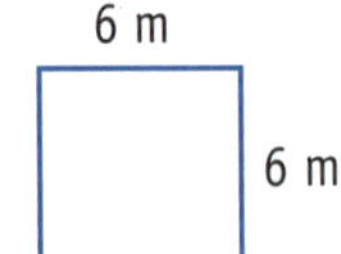

Area = ________ × ________

= ____ × ____

= ________

SELF CHECK Tick how you feel

Got it!	Need help...	I don't get it
 ☐	 ☐	 ☐

Check your answers

How many did you get correct? ☐

CATCH UP MATHS YEAR 6 BOOK B © PASCAL PRESS ISBN: 9781925726190

PRACTICE

1 Calculate the area of these rectangles and squares.

1 cm

Area = length × width
= 1 cm × 1 cm
= 1 cm²

a

7 m

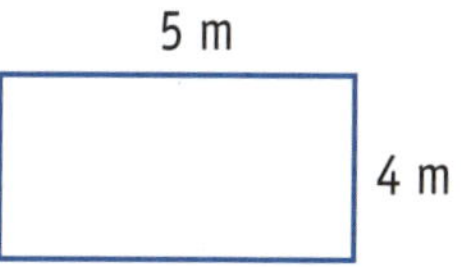

Area = ________ × ________
= _____ × _____
= _______

b

5 m

4 m

Area = ________ × ________
= _____ × _____
= _______

c

1 cm

8 cm

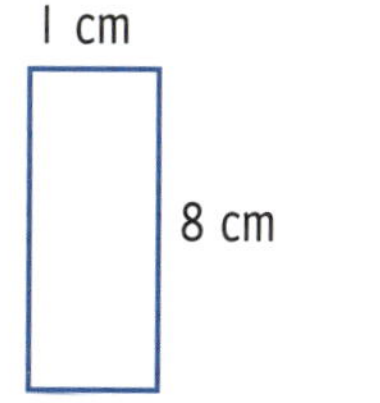

Area = ________ × ________
= _____ × _____
= _______

d

10 cm

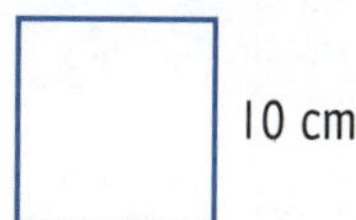

Area = ________ × ________
= _____ × _____
= _______

e

7 cm

6 cm

Area = ________ × ________
= _____ × _____
= _______

f

2 m

Area = ________ × ________
= _____ × _____
= _______

g

12 m

8 m

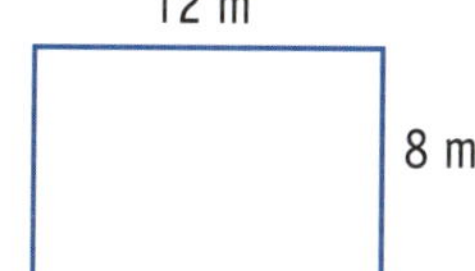

Area = ________ × ________
= _____ × _____
= _______

  ISBN: 9781925726190

Find the area of these irregular shapes by subtracting the shaded areas.

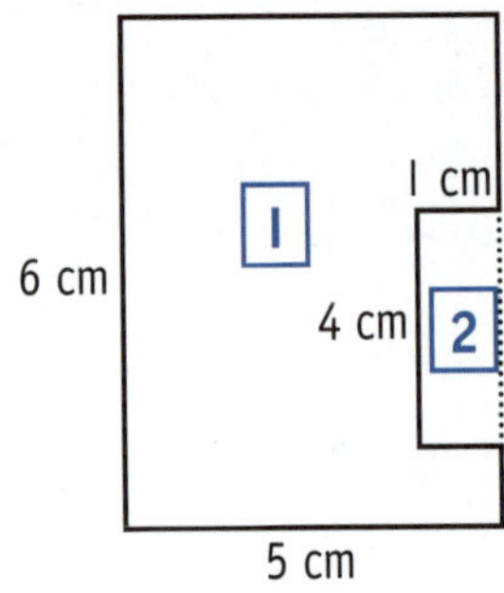

Area 1 = length × width

= 6 cm × 5 cm

= 30 cm^2

Area 2 = length × width

= 1 cm × 4 cm

= 4 cm^2

Area = Area 1 − Area 2

= 30 cm^2 − 4 cm^2

= 26 cm^2

a

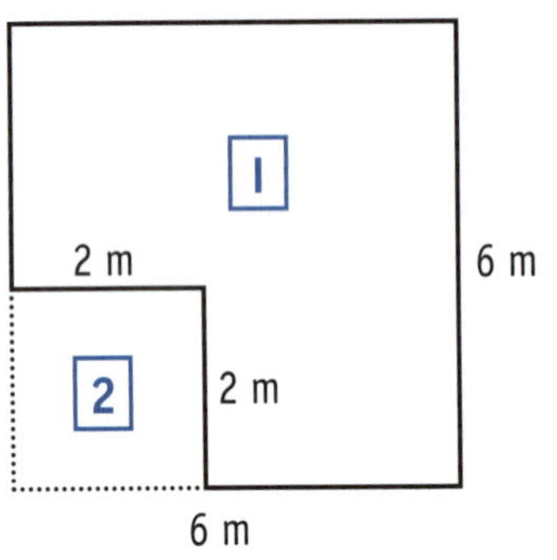

Area 1 = ________ × ________

= ______ × ______

= _______

Area 2 = ________ × ________

= ______ × ______

= _______

Area = Area 1 − Area 2

= ________________

= _______

b

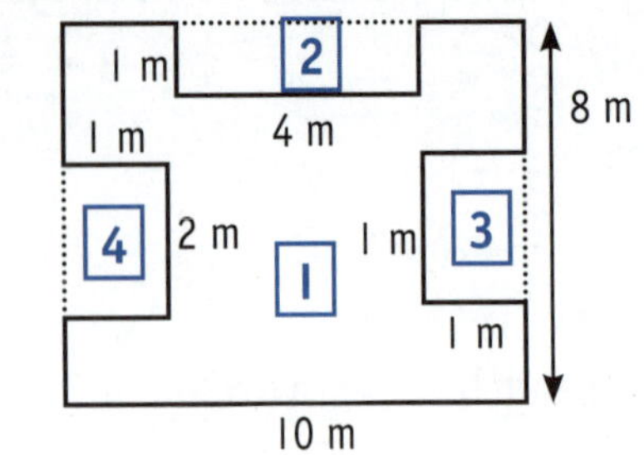

Area 1 = ________ × ________

= ______ × ______

= _______

Area 2 = ________ × ________

= ______ × ______

= _______

Area 3 = ________ × ________

= ______ × ______

= _______

Area 4 = ________ × ________

= ______ × ______

= _______

Area = Area 1 − (Area 2 + Area 3 + Area 4)

= ____________________

= ____________________

= _______

  ISBN: 9781925726190

USING MULTIPLICATION TO CALCULATE AREA – TRIANGLES

SCAN to watch video

You can use multiplication to calculate the area of a triangle. A triangle is half a rectangle.

Area of a rectangle

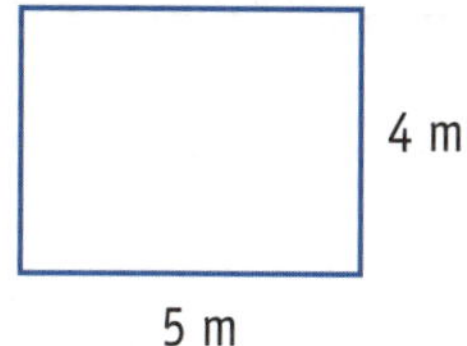

Area = length × width

= 5 m × 4 m

= 20 m^2

Area of a triangle

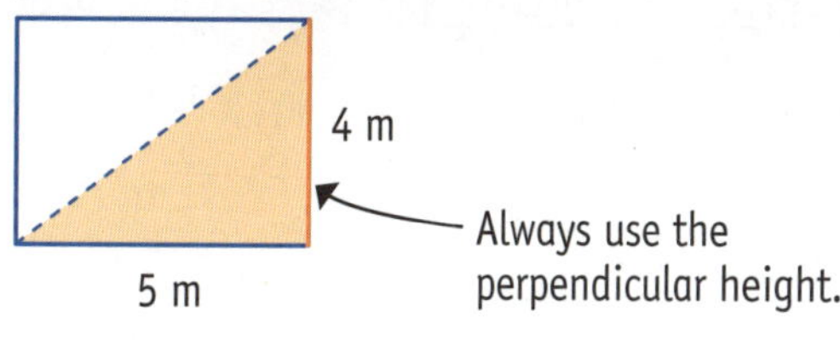

Area = $\frac{1}{2}$ × length × width

= $\frac{1}{2}$ × 5 m × 4 m

= 10 m^2

Area = $\frac{1}{2}$ × length × width is the formula to calculate the area of a triangle.

Examples: Use the formula A = $\frac{1}{2} \times l \times w$ to calculate the area of these triangles.

a

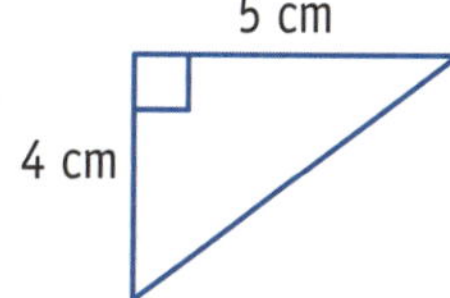

Area = $\frac{1}{2}$ × length × width

= $\frac{1}{2}$ × 5 cm × 4 cm

= 10 cm^2

b

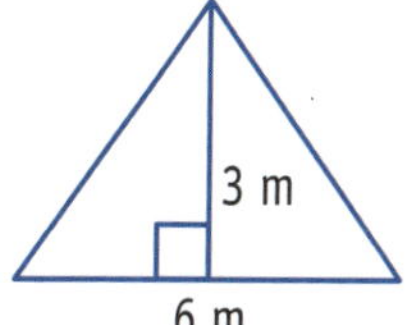

Area = ___ × _______ × _______

= ___ × _______ × _______

= _______

What is the area of the triangles?

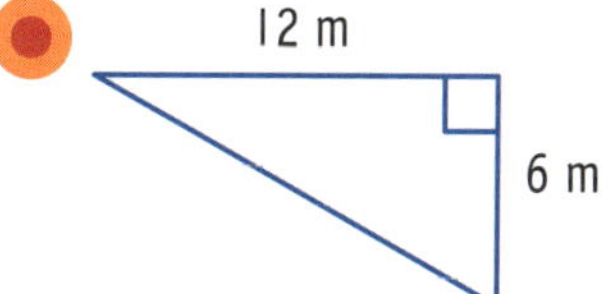

Area = $\frac{1}{2}$ × length × width

= $\frac{1}{2}$ × 12 m × 6 m

= 36 m^2

a

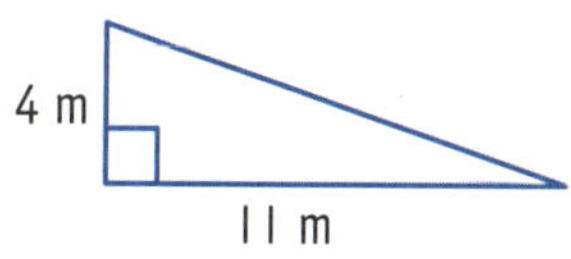

Area = ___ × _______ × _______

= ___ × _______ × _______

= _______

Check your answers

How many did you get correct?

  ISBN: 9781925726190

PRACTICE

1 Find the area of each triangle.

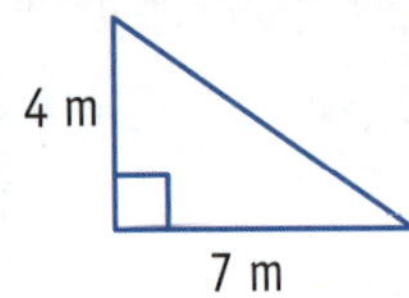

Area = $\frac{1}{2}$ × length × width

= $\frac{1}{2}$ × 4 m × 7 m

= 14 m^2

a

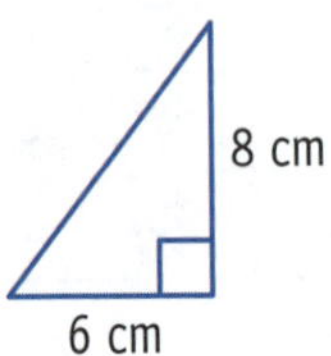

Area = ___ × ______ × ______

= ___ × ______ × ______

= ______

b

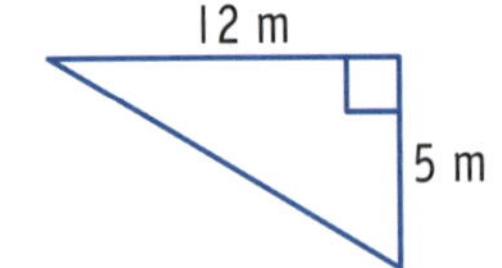

Area = ___ × ______ × ______

= ___ × ______ × ______

= ______

c

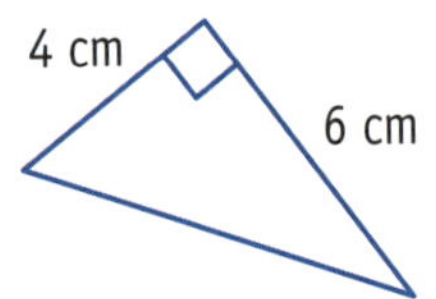

Area = ___ × ______ × ______

= ___ × ______ × ______

= ______

d

Area = ___ × ______ × ______

= ___ × ______ × ______

= ______

e

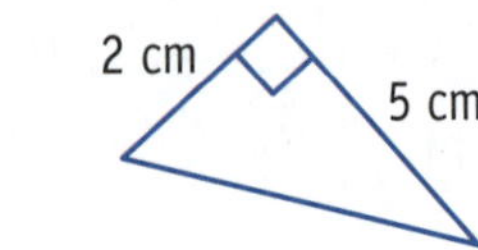

Area = ___ × ______ × ______

= ___ × ______ × ______

= ______

f

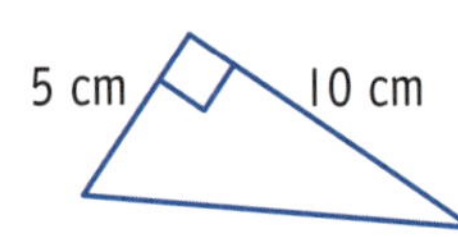

Area = ___ × ______ × ______

= ___ × ______ × ______

= ______

g

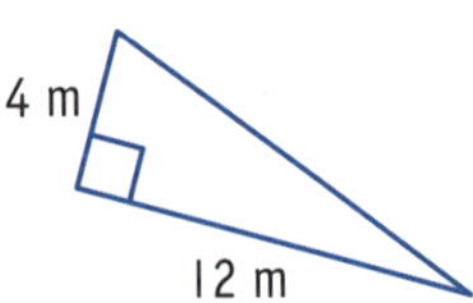

Area = ___ × ______ × ______

= ___ × ______ × ______

= ______

  ISBN: 9781925726190

PERIMETER

The distance around the outside of a shape is the perimeter (P). To calculate the perimeter, add the lengths of all the sides.

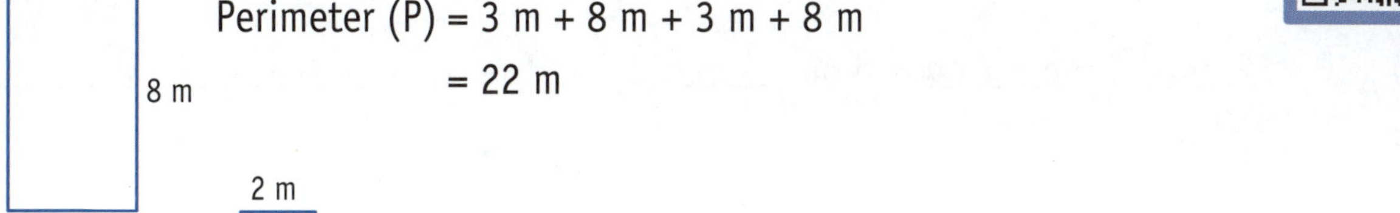

Perimeter (P) = 3 m + 8 m + 3 m + 8 m
= 22 m

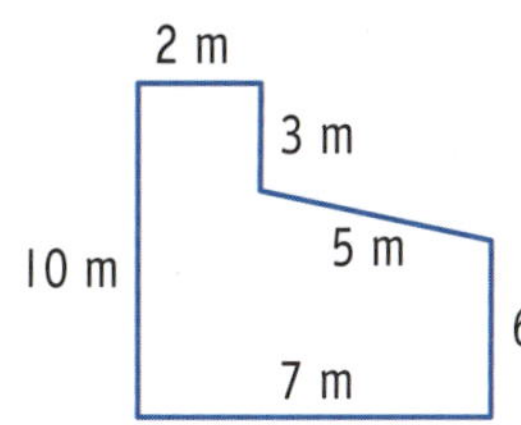

Perimeter (P) = 10 m + 2 m + 3 m + 5 m + 6 m + 7 m
= 33 m

Examples: Work out the perimeter.

a
6 cm
2 cm

P = 2 cm + 6 cm + 2 cm + 6 cm

= 16 cm

b

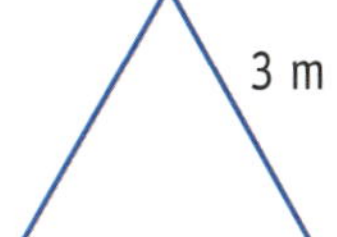

P = ____ + ____ + ____
= ______

Check the answers on the video!

Calculate the perimeter of these shapes.

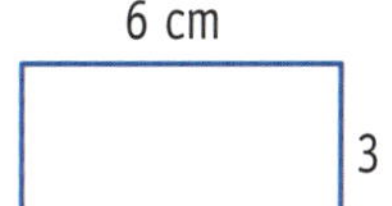

P = 3 cm + 6 cm + 3 cm + 6 cm
= 18 cm

a

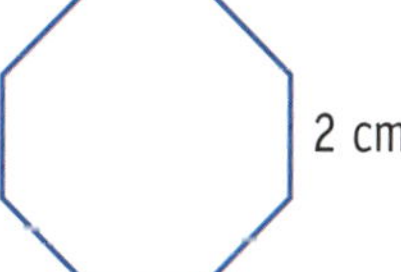

P = ____ + ____ + ____ + ____ + ____ + ____ + ____ + ____
= ______

b

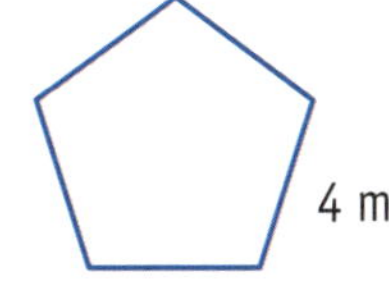

P = ____ + ____ + ____ + ____ + ____
= ______

SELF CHECK Tick how you feel

Got it!	Need help...	I don't get it
 ☐	 ☐	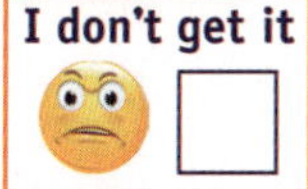 ☐

Check your answers
How many did you get correct? ☐

PRACTICE

1 Calculate the perimeter of these 2D shapes.

4 cm
2 cm   3 cm
5 cm

P = 2 cm + 4 cm + 3 cm + 5 cm

= 14 cm

c

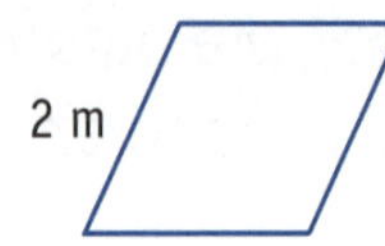

P = ______________________

= ________

a

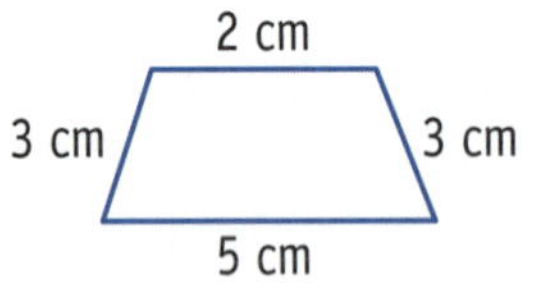

P = ______________________

= ________

d

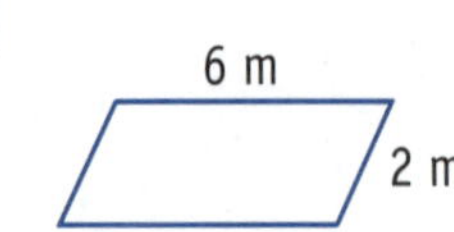

P = ______________________

= ________

b

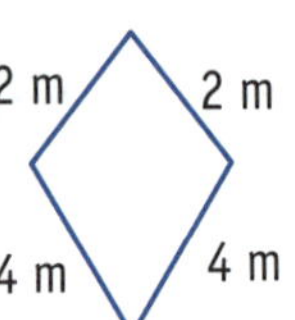

P = ______________________

= ________

e

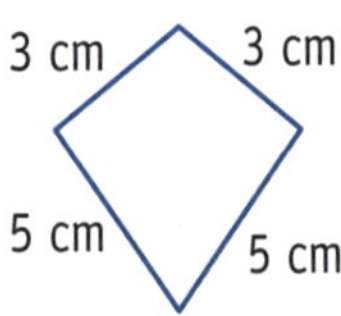

P = ______________________

= ________

2 Work out the perimeter of these shapes.

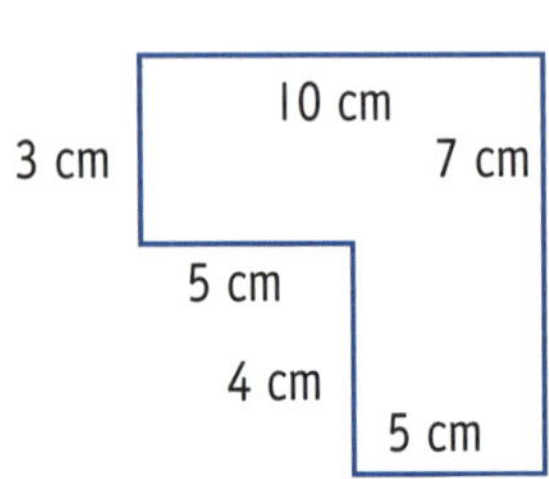

P = 10 cm + 7 cm + 5 cm + 4 cm + 5 cm + 3 cm

= 34 cm

a

4 m   2 m
8 m   6 m
6 m
10 m

P = ______________________

= ________

b

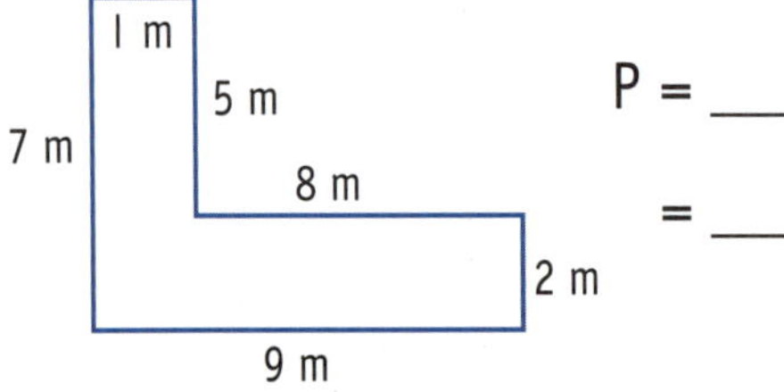

P = ______________________

= ________

CATCH UP MATHS YEAR 6 BOOK B © PASCAL PRESS ISBN: 9781925726190

Work out the lengths of the sides without measurements and then find the perimeter.

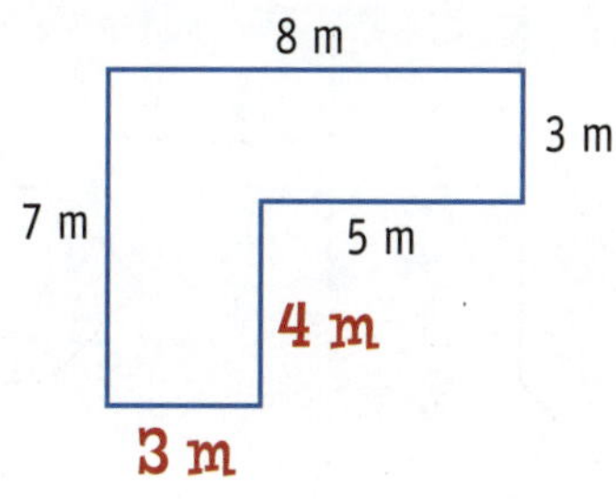

P = 7 m + 8 m + 3 m + 5 m + 4 m + 3 m

= 30 m

a

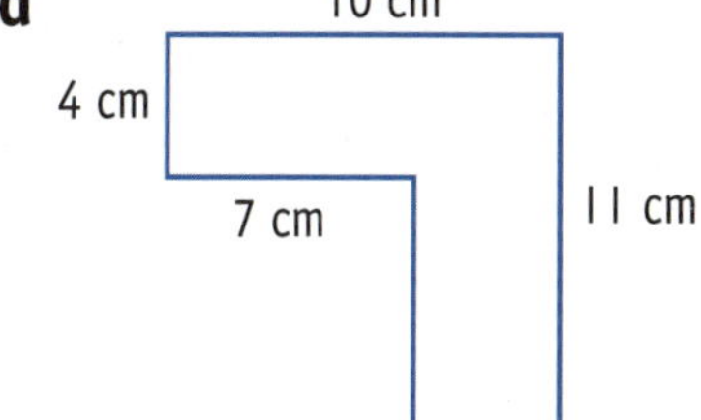

P = ____________________

= ________

b

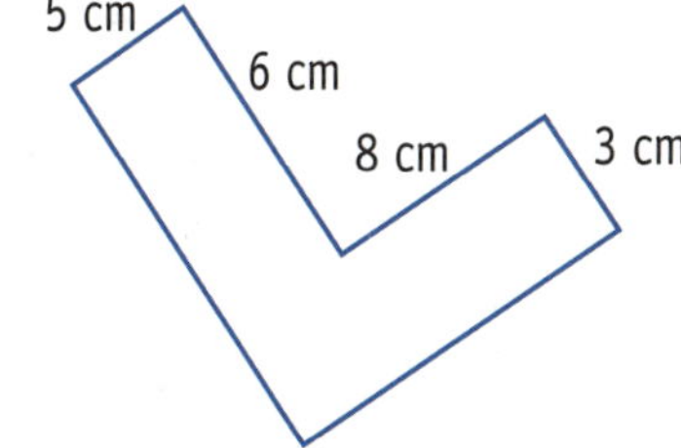

P = ____________________

= ________

c

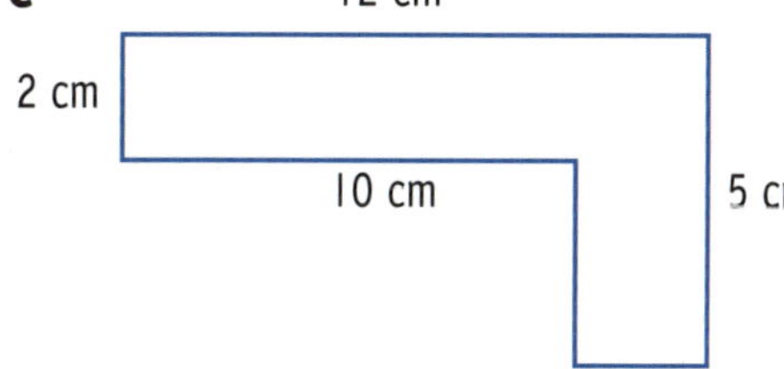

P = ____________________

= ________

d

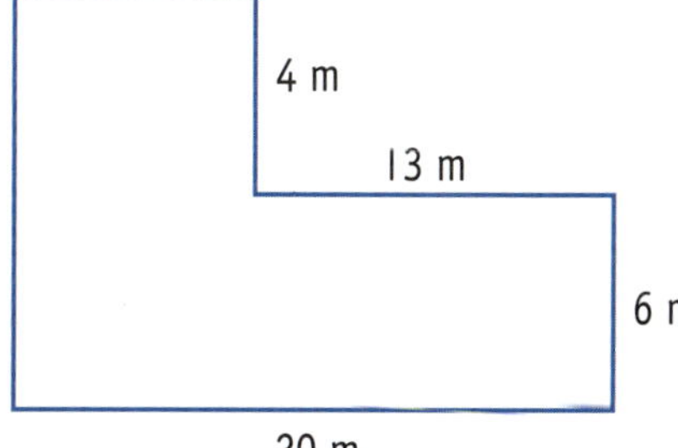

P = ____________________

= ________

e

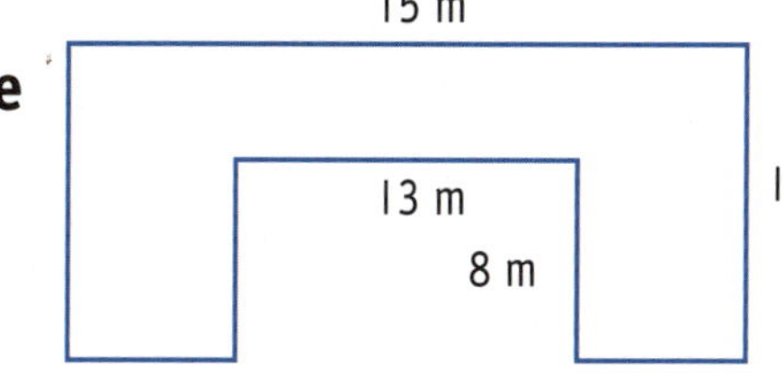

P = ____________________

= ________

AREA REVIEW

What is the area of each 2D shape? □ = 1 cm^2

a ________ d ________ g ________ j ________

b ________ e ________ h ________ k ________

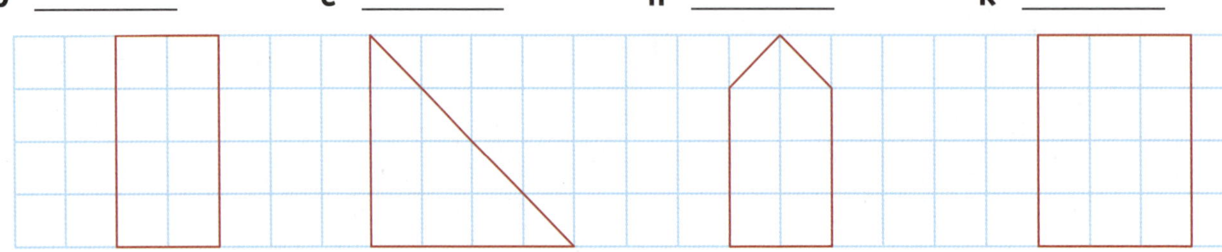

c ________ f ________ i ________ l ________

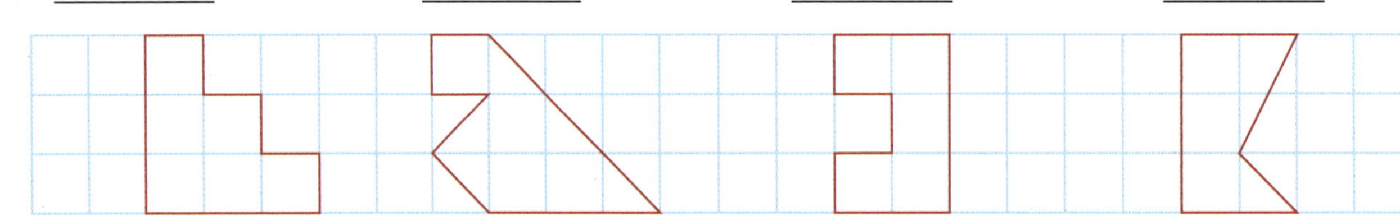

Name each 2D shape in Question 1.

a ________________ e ________________ i ________________

b ________________ f ________________ j ________________

c ________________ g ________________ k ________________

d ________________ h ________________ l ________________

Name five areas that would be measured in square metres.

____

Use the map for Shen's Garden Centre to complete the following.

How many m^2 is each area?

a roses ____ d indoor plants ____

b trees ____ e flowering plants ____

c succulents ____ f border plants ____

Shen's Garden Centre

Roses
Indoor plants
Succulents
Flowering plants
Border plants
Trees

□ = 1 m^2

CATCH UP MATHS YEAR 6 BOOK B © PASCAL PRESS ISBN: 9781925726190

5 Which section takes up:

a the most space? ______________ b the least space? ______________

6 How many m² altogether is Shen's Garden Centre? ____

7 How much bigger?

a border plants than trees _____

b indoor plants than succulents _____

c roses than trees _____

8 Fill in the gaps.

__ hectare = ________ m²

100 ___

100 ___

9 Write the short form for these measurements.

a 70 000 square metres ___________

b 50 000 square metres ___________

c 10 000 square metres ___________

d 100 000 square metres ___________

10 Name five areas that hectares could be used to measure.

____

11 How many hectares (ha)?

a 29 000 m² ________

b 46 000 m² ________

c 82 000 m² ________

d 10 000 m² ________

e 77 000 m² ________

f 114 000 m² ________

12 Name five things you would use square kilometres to measure.

____

REVIEW

13 Write the short form for these measurements.

a twenty-seven square kilometres __________

b sixty-four square kilometres __________

c one hundred square kilometres __________

d six hundred and two square kilometres __________

e two thousand, eight hundred and ninety-four square kilometres __________

14 Use multiplication to work out the area of these squares.

a

3 m

Area = ________ × ________

= ______ × ______

= _______

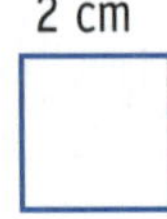

c

2 cm

Area = ________ × ________

= ______ × ______

= _______

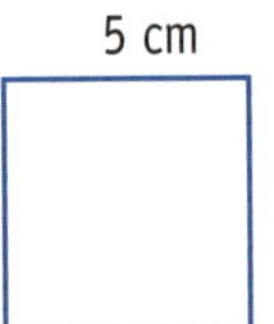

b

5 cm

Area = ________ × ________

= ______ × ______

= _______

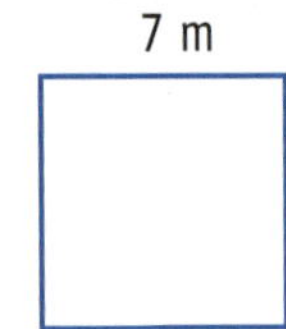

d

7 m

Area = ________ × ________

= ______ × ______

= _______

15 Use multiplication to work out the area of these rectangles.

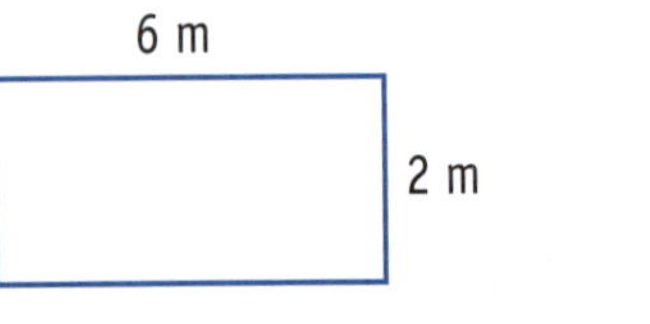

a

6 m

2 m

Area = ________ × ________

= ______ × ______

= _______

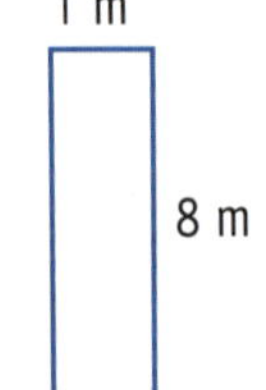

b

1 m

8 m

Area = ________ × ________

= ______ × ______

= _______

  ISBN: 9781925726190

c

14 cm

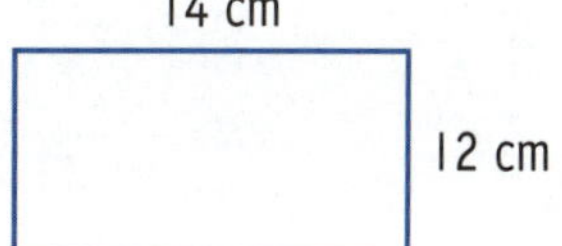

12 cm

Area = ________ × ________

= ______ × ______

= _______

e

5 cm

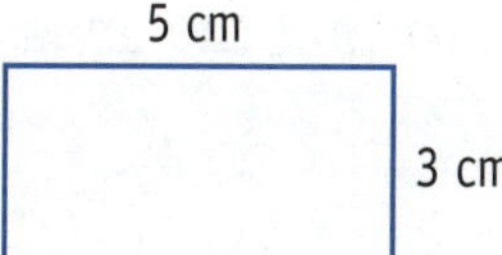

3 cm

Area = ________ × ________

= ______ × ______

= _______

d

15 m

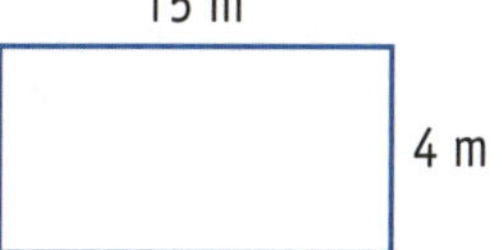

4 m

Area = ________ × ________

= ______ × ______

= _______

f

20 m

15 m

Area = ________ × ________

= ______ × ______

= _______

Find the area of these irregular shapes by subtracting the shaded area.

a

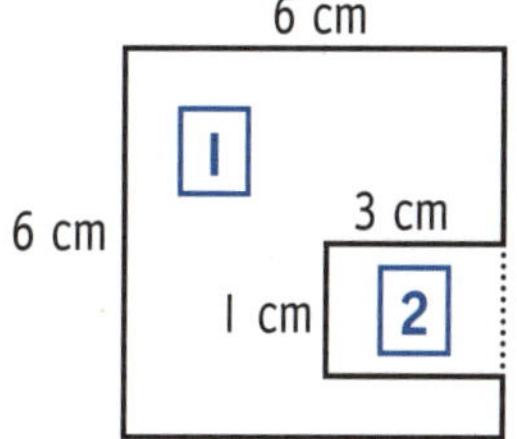

Area 1 = ________ × ________

= ______ × ______

= _______

Area 2 = ________ × ________

= ______ × ______

= _______

Area = Area 1 – Area 2

= ________________

= _______

b

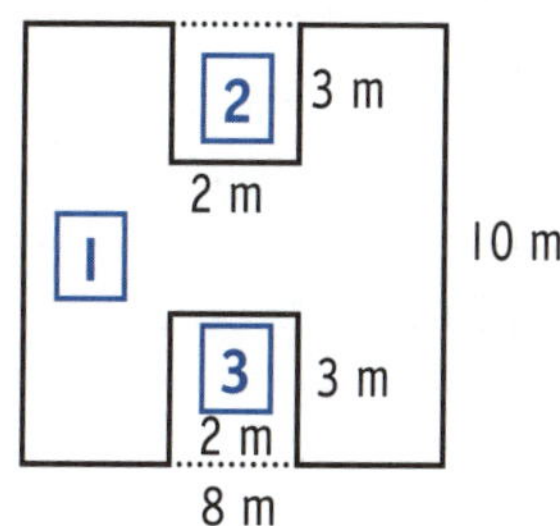

Area 1 = ________ × ________

= ______ × ______

= _______

Area 2 = ________ × ________

= ______ × ______

= _______

Area 3 = ________ × ________

= ______ × ______

= _______

Area = Area 1 – (Area 2 + Area 3)

= ________________

= _______

REVIEW

17 Use multiplication to find the area of each triangle.

a

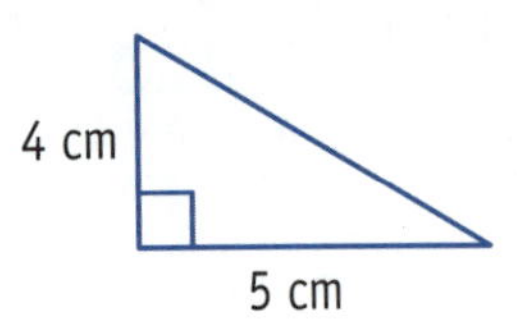

Area = ___ × ________ × ________

= ___ × _______ × _______

= ________

b

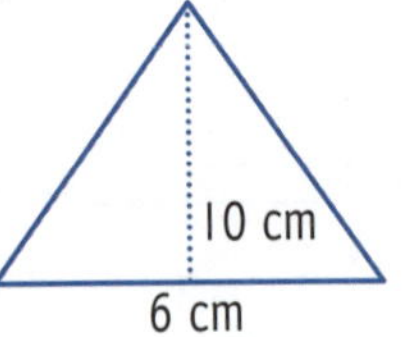

Area = ___ × ________ × ________

= ___ × _______ × _______

= ________

c

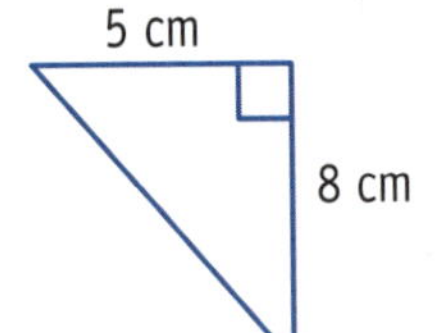

Area = ___ × ________ × ________

= ___ × _______ × _______

= ________

d

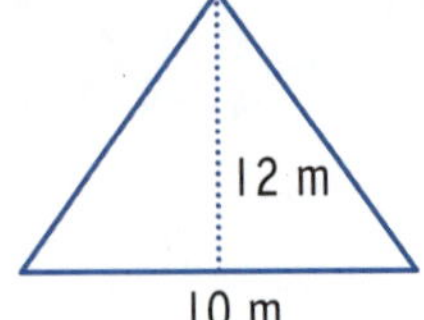

Area = ___ × ________ × ________

= ___ × _______ × _______

= ________

e

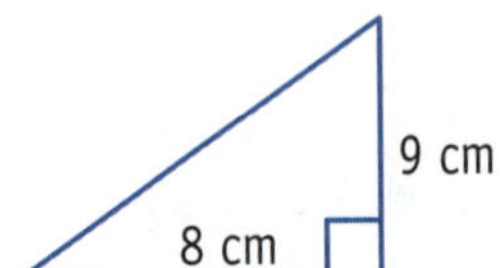

Area = ___ × ________ × ________

= ___ × _______ × _______

= ________

f

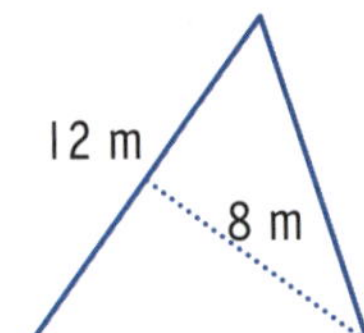

Area = ___ × ________ × ________

= ___ × _______ × _______

= ________

18 What is the perimeter of each 2D shape?

a

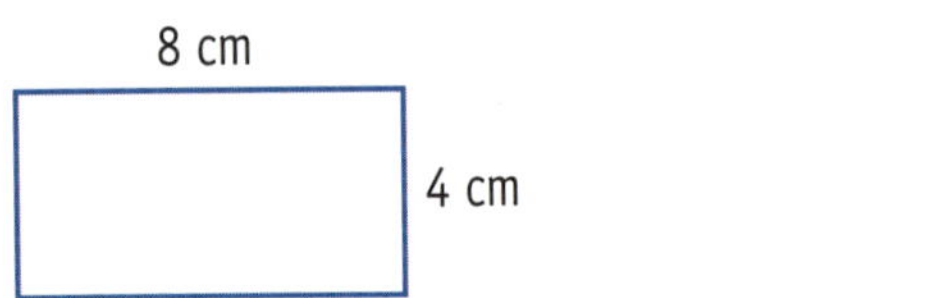

P = ____________________________________

= ________

b

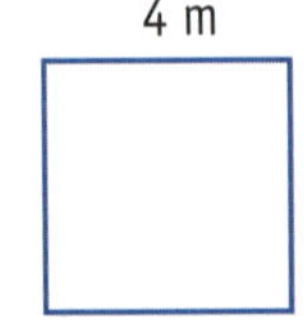

P = ____________________________________

= ________

CATCH UP MATHS YEAR 6 BOOK B © PASCAL PRESS ISBN: 9781925726190

c

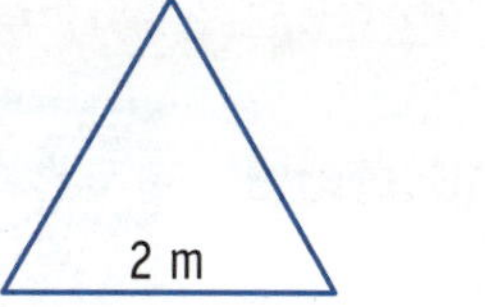

P = ____________________

= ______

d

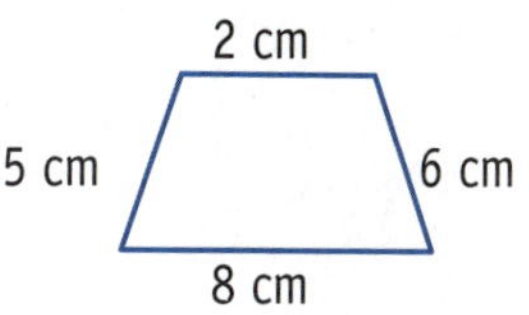

P = ____________________

= ______

e

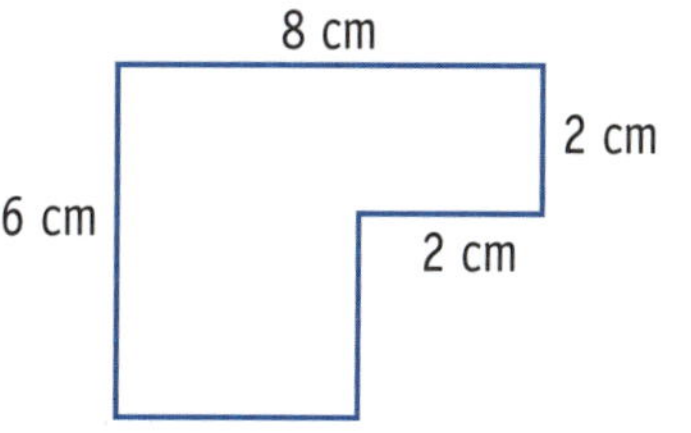

P = ____________________

= ______

f

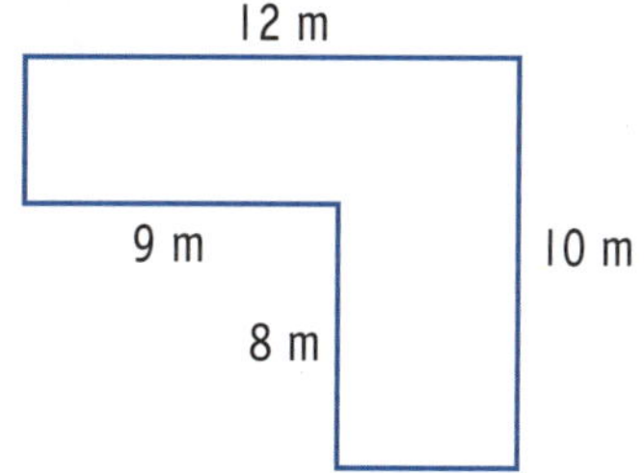

P = ____________________

= ______

g

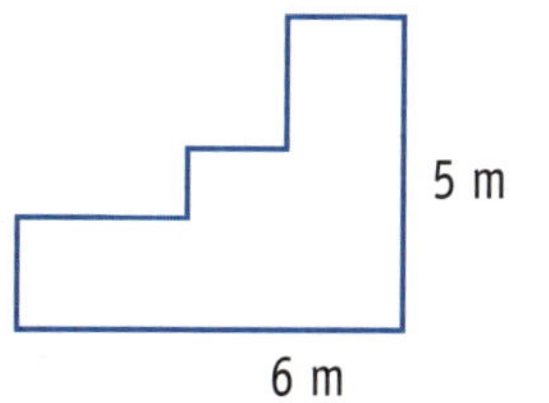

P = ____________________

= ______

h

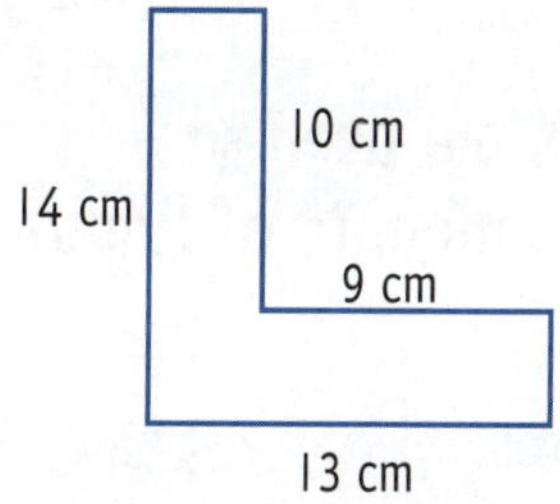

P = ____________________

= ______

i

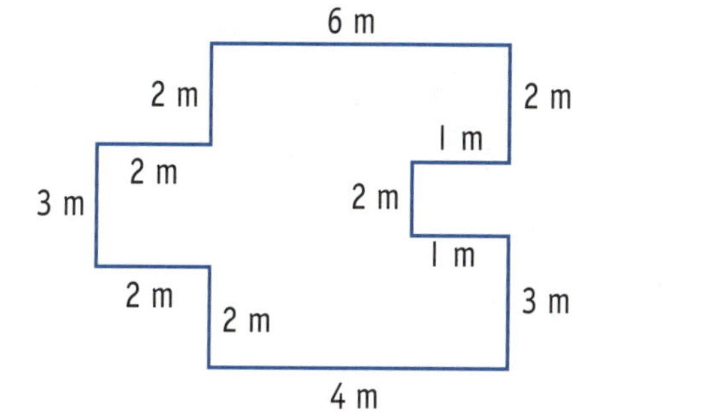

P = ____________________

= ______

j

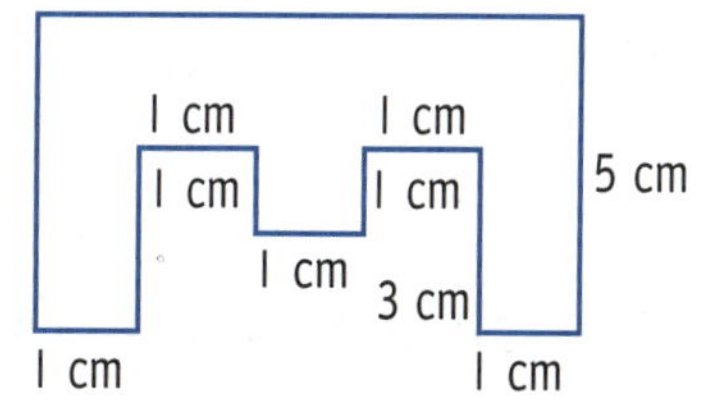

P = ____________________

= ______

k

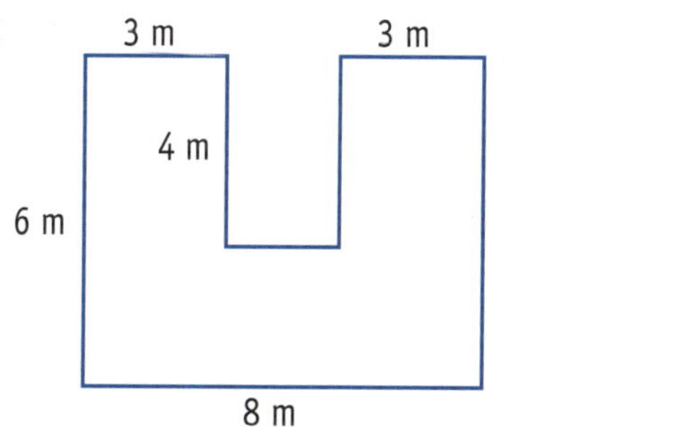

P = ____________________

= ______

MILLILITRES AND LITRES

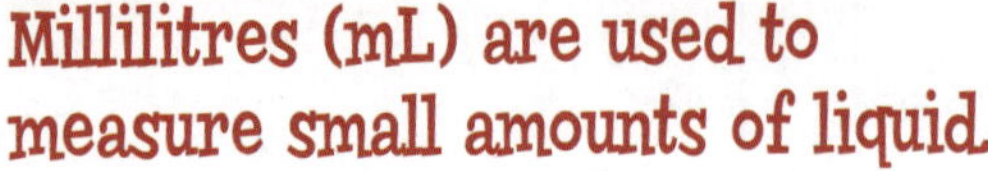

Millilitres (mL) are used to measure small amounts of liquid.

These containers measure millilitres:

Litres (L) are used to measure large amounts of liquid.

These containers measure litres:

Examples:

Would the containers hold mL or L? Write the units.

a 

250 mL

b 

10 ___

c 

2 ___

d 

3 ___

Check the answers on the video!

Your turn

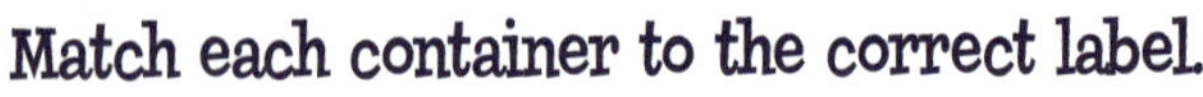

Match each container to the correct label.

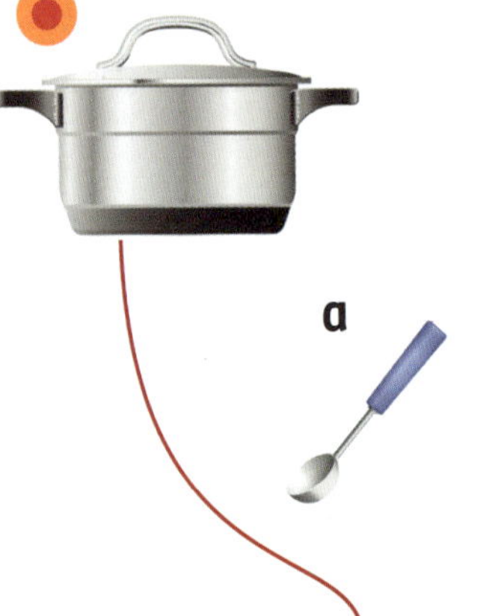

a

b 

c

d 

e

250 mL	4 L	15 mL	500 mL	2 L	10 L

SELF CHECK Tick how you feel

Got it!	Need help...	I don't get it

Check your answers

How many did you get correct?

CATCH UP MATHS YEAR 6 BOOK B © PASCAL PRESS ISBN: 9781925726190

PRACTICE

1 Write in short form.

- 7 millilitres 7 mL
- a 6 litres ______
- b 39 litres ______
- c 88 millilitres ______
- d 100 millilitres ______
- e 380 millilitres ______
- f 8 litres ______
- g 52 litres ______
- h 2000 millilitres ______
- i 5000 millilitres ______

2 How many millilitres?

- 1 litre 1000 mL
- a $\frac{1}{4}$ litre ______
- b $\frac{1}{2}$ litre ______
- c $\frac{3}{4}$ litre ______

3 Write as litres.

- 3500 mL $3\frac{1}{2}$ L
- a 2000 mL ______
- b 2250 mL ______
- c 1750 mL ______
- d 8250 mL ______
- e 4500 mL ______
- f 10 500 mL ______
- g 6750 mL ______
- h 5000 mL ______
- i 7750 mL ______

4 Write as millilitres.

- $8\frac{1}{2}$ L 8500 mL
- a $3\frac{1}{4}$ L ______
- b 4 L ______
- c $6\frac{1}{4}$ L ______
- d $9\frac{1}{2}$ L ______
- e $7\frac{3}{4}$ L ______
- f $5\frac{1}{2}$ L ______
- g $8\frac{3}{4}$ L ______
- h $12\frac{1}{4}$ L ______
- i $10\frac{1}{2}$ L ______

5 Write as litres and millilitres.

- 3200 mL 3 L 200 mL
- a 490 mL __ L ______ mL
- b 1810 mL __ L ______ mL
- c 3590 mL __ L ______ mL
- d 2660 mL __ L ______ mL
- e 8750 mL __ L ______ mL
- f 6480 mL __ L ______ mL
- g 5370 mL __ L ______ mL
- h 7230 mL __ L ______ mL
- i 4930 mL __ L ______ mL

6 Add these amounts.

- 4 L 250 mL + 6 L 500 mL = 10 L 750 mL

a 5 L 750 mL + 3 L 500 mL = ___ L _____ mL

b 7 L 200 mL + 5 L 300 mL = ___ L _____ mL

c 8 L 150 mL + 7 L 290 mL = ___ L _____ mL

d 1 L 450 mL + 3 L 800 mL = ___ L _____ mL

e 2 L 650 mL + 9 L 540 mL = ___ L _____ mL

7 Complete the table.

	Litres (decimal)	Millilitres
•	0.25 L	250 mL
a	1.5 L	
b		3575 mL
c	7.425 L	
d		2650 mL
e	3.20 L	

	Litres (decimal)	Millilitres
f		4900 mL
g	8.75 L	
h	9.325 L	
i	6.660 L	
j		5850 mL
k		2700 mL

8 Convert to mL.

- 6.35 L 6350 mL

a 4.375 L ________

b 2.75 L ________

c 7.125 L ________

d 5.45 L ________

e 6.9 L ________

f 9.575 L ________

g 15.6 L ________

h 11.820 L ________

i 12.655 L ________

9 Convert to L.

- 1085 mL 1.085 L

a 3125 mL ________

b 7520 mL ________

c 5570 mL ________

d 2320 mL ________

e 4645 mL ________

f 6450 mL ________

g 8280 mL ________

h 9795 mL ________

i 2590 mL ________

  ISBN: 9781925726190

10 Number the boxes to put these amounts in descending order.

●	5 L 350 mL	4 L 430 mL	1500 mL	700 mL	9565 mL
	2	3	4	5	1
a	6250 mL	9820 mL	5 L 50 mL	7 L 660 mL	1495 mL
	☐	☐	☐	☐	☐
b	4 L 40 mL	6250 mL	4450 mL	6.5 L	9.3 L
	☐	☐	☐	☐	☐

11 What amount of liquid is in each container?

● 450 mL

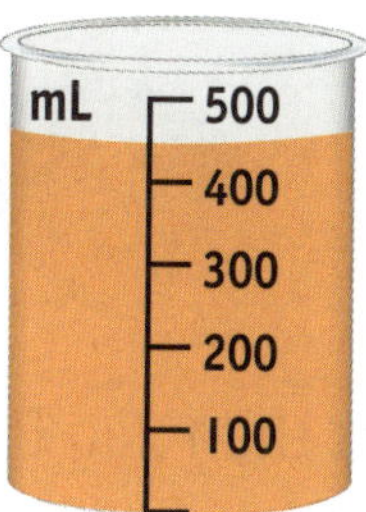

b ________

d ________

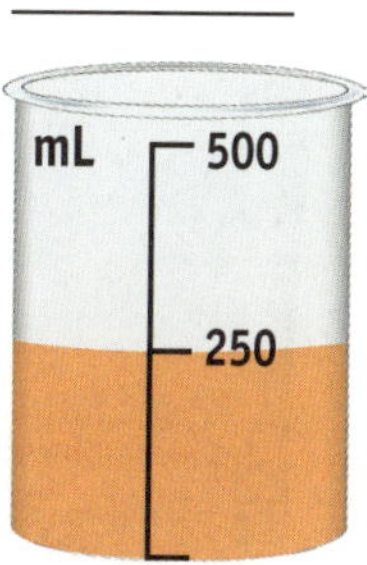

f ________

a ________

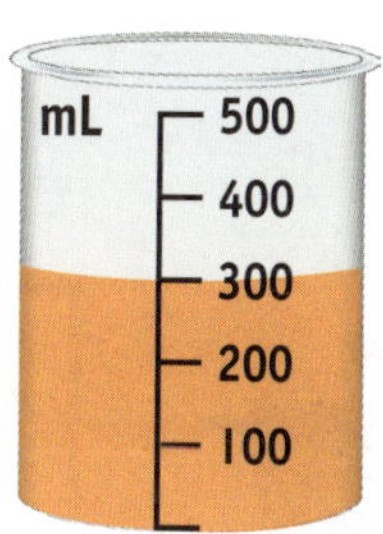

c ________

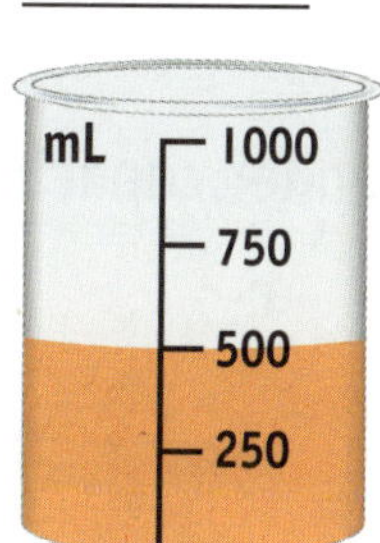

e ________

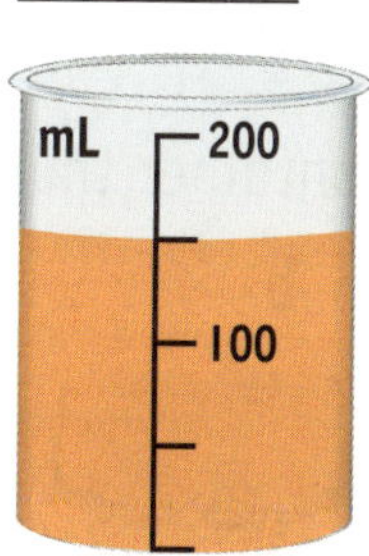

g ________

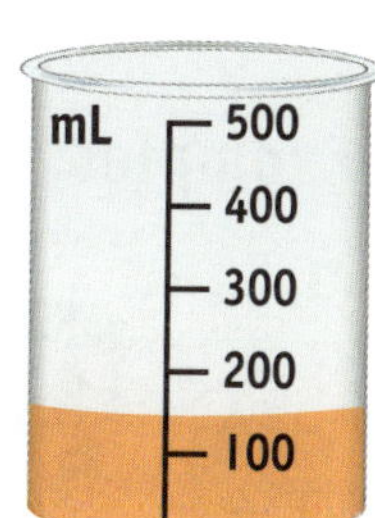

12 Colour the containers to show the amount.

● 3 L

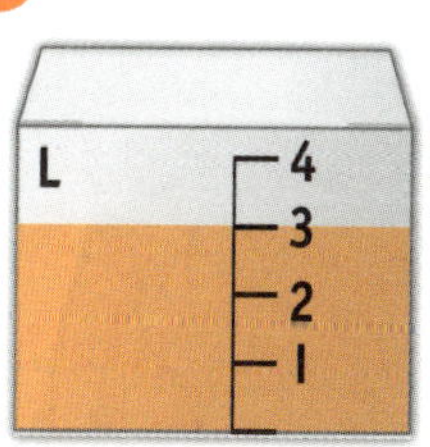

b 350 mL

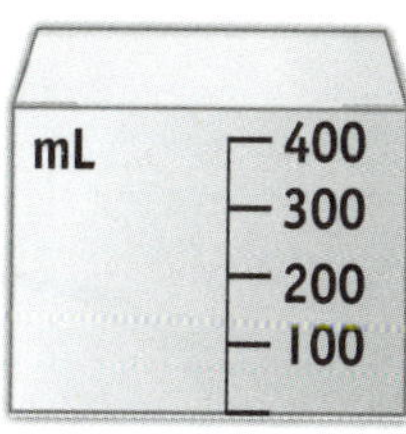

d 250 mL

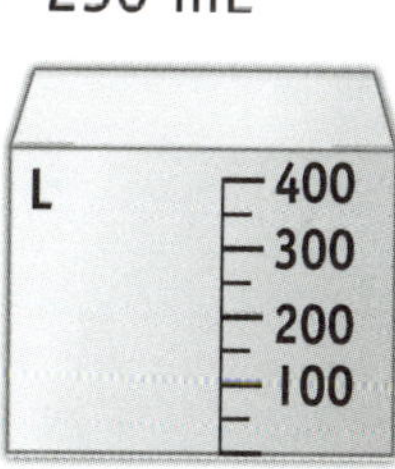

f 150 mL

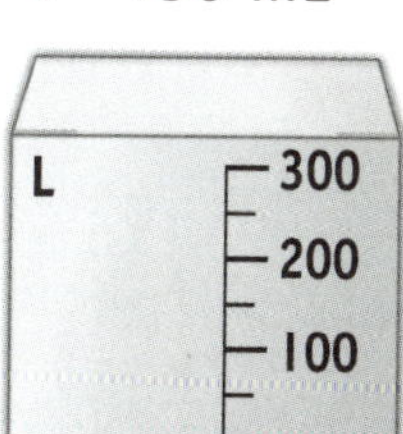

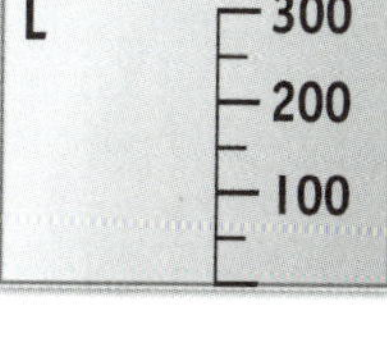

a $2\frac{1}{2}$ L

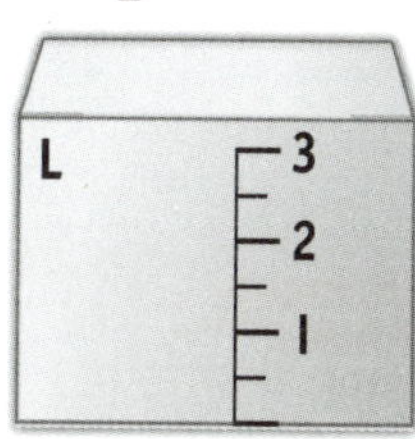

c 5 L

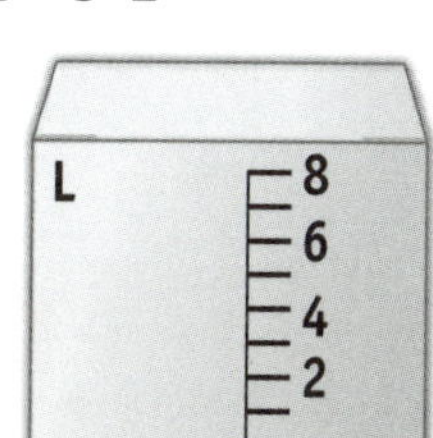

e 500 mL

g 3500 mL

CAPACITY

Capacity is the amount of liquid a container can hold.

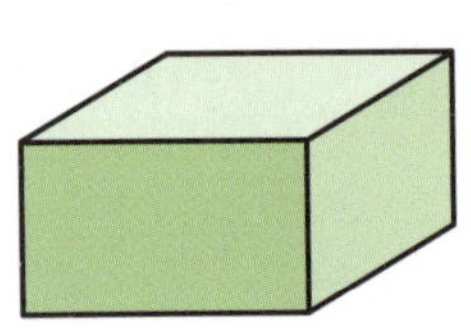
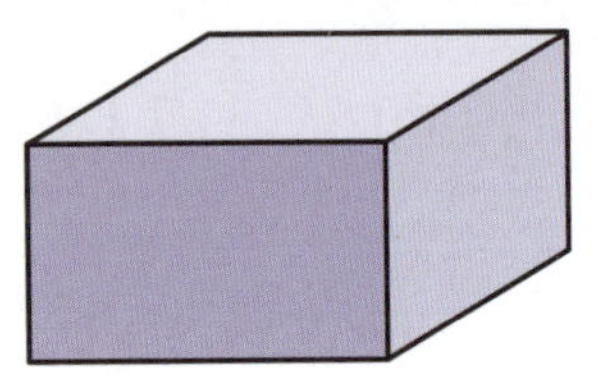
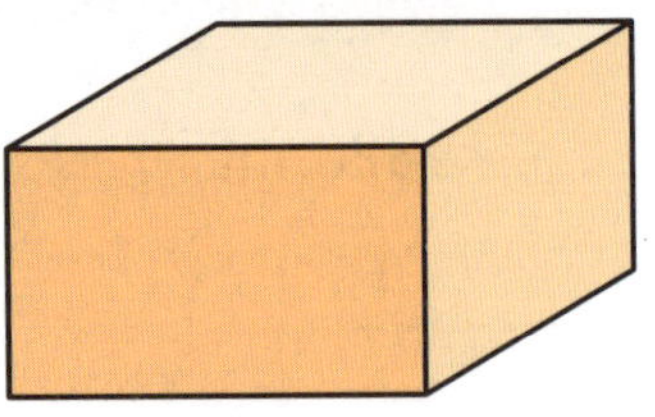

The green container has the smallest capacity because it can hold the least liquid.

The orange container has the largest capacity because it can hold the most liquid.

smaller container = smaller capacity
bigger container = bigger capacity

Jug A has more juice in it than Jug B. But Jug B can hold more juice than Jug A, so it has a larger capacity.

Examples: Circle the container that has the smaller capacity in each pair.

a

b

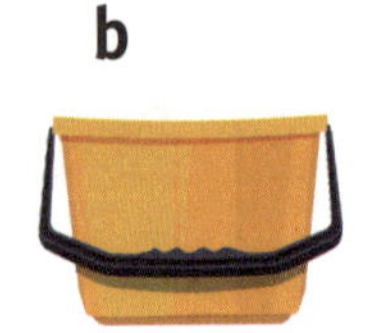

c

Check the answers on the video!

Number the boxes to order the containers from smallest capacity (1) to largest capacity (6).

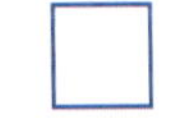

1

SELF CHECK Tick how you feel

Got it!	Need help...	I don't get it
 ☐	 ☐	 ☐

Check your answers
How many did you get correct? ☐

CATCH UP MATHS YEAR 6 BOOK B © PASCAL PRESS ISBN: 9781925726190

PRACTICE

Draw a similar container that has a larger capacity.

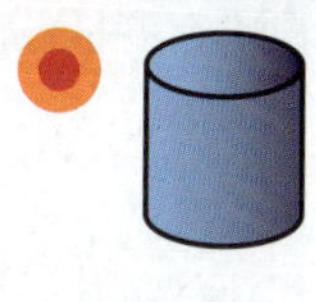
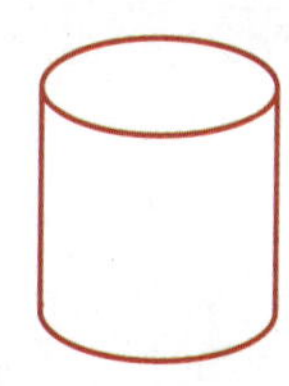

a
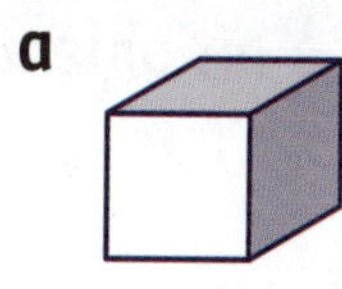

b

Draw a similar container that has a smaller capacity.

a
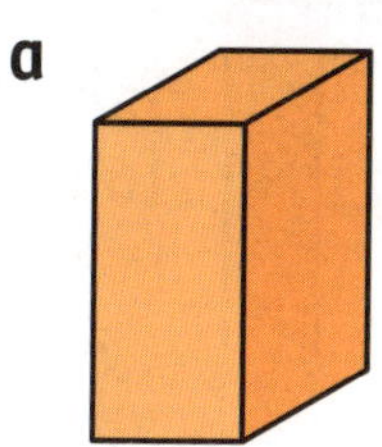

b
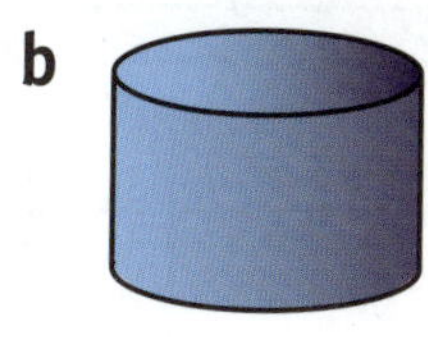

Would you use millilitres (mL) or litres (L) to measure these?

30 mL

b 5 ___

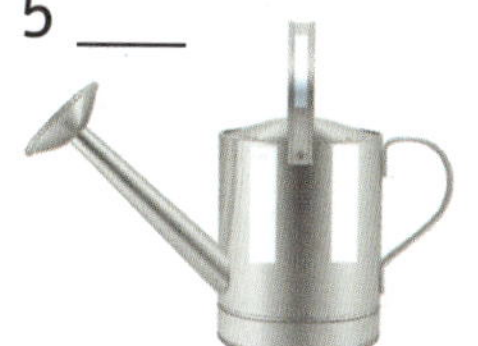

d 300 ___

f 5 ___

a 4 ___

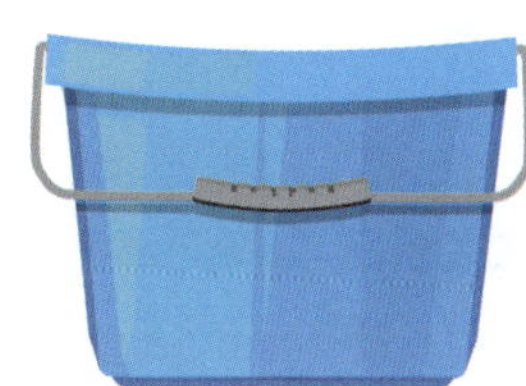

c 1 ___

e 3 ___

g 4 ___

4 What is the total capacity in litres of each group of containers?

12 L

b ______

d ______

a ______

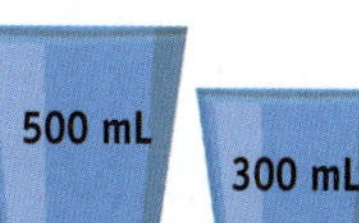

c ______

e ______

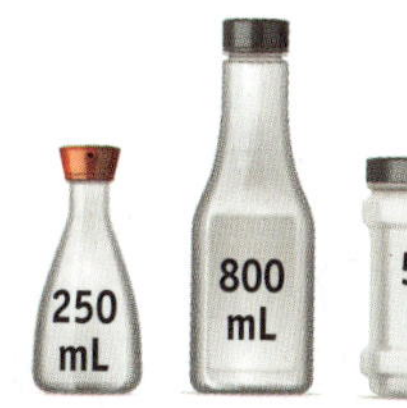

VOLUME

Volume (V) is the amount of space an object takes up. It is measured in cubic units.

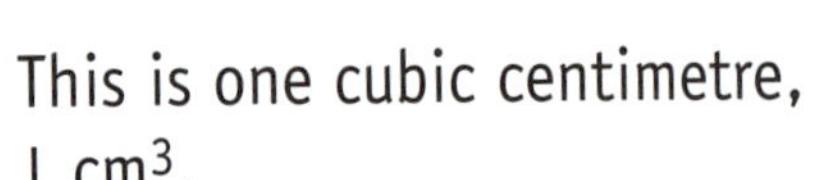
This is one cubic centimetre, 1 cm^3.

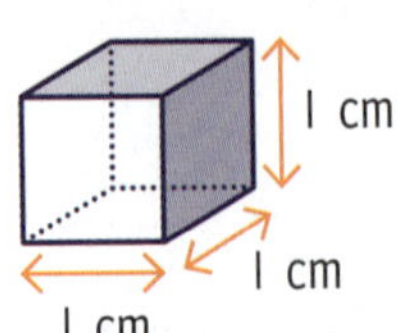

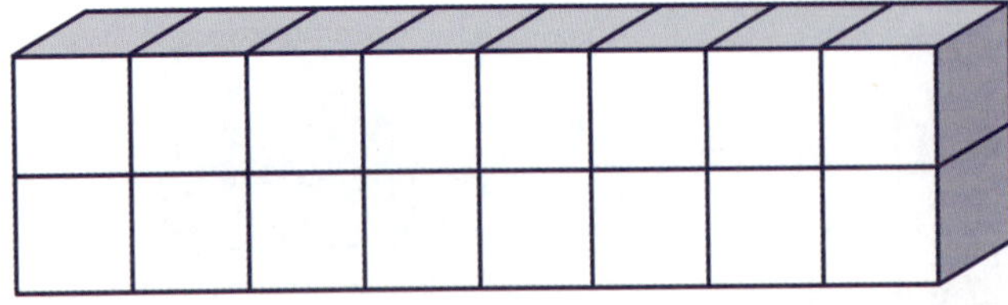

This object was made with 16 one-centimetre cubes. It has a volume of 16 cm^3.

Examples: Count the cubic centimetres to find the volume.

a

V = 12 cubic centimetres

= 12 cm^3

b

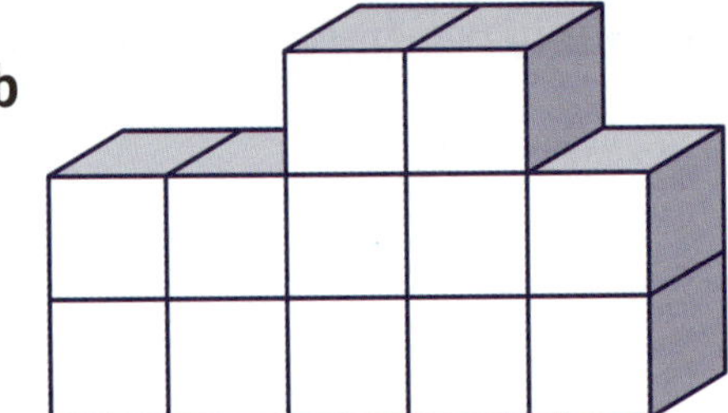

V = ___ cubic centimetres

= ___ cm^3

Check the answers on the video!

Your turn

Colour the objects that have a volume of 8 cubic centimetres.

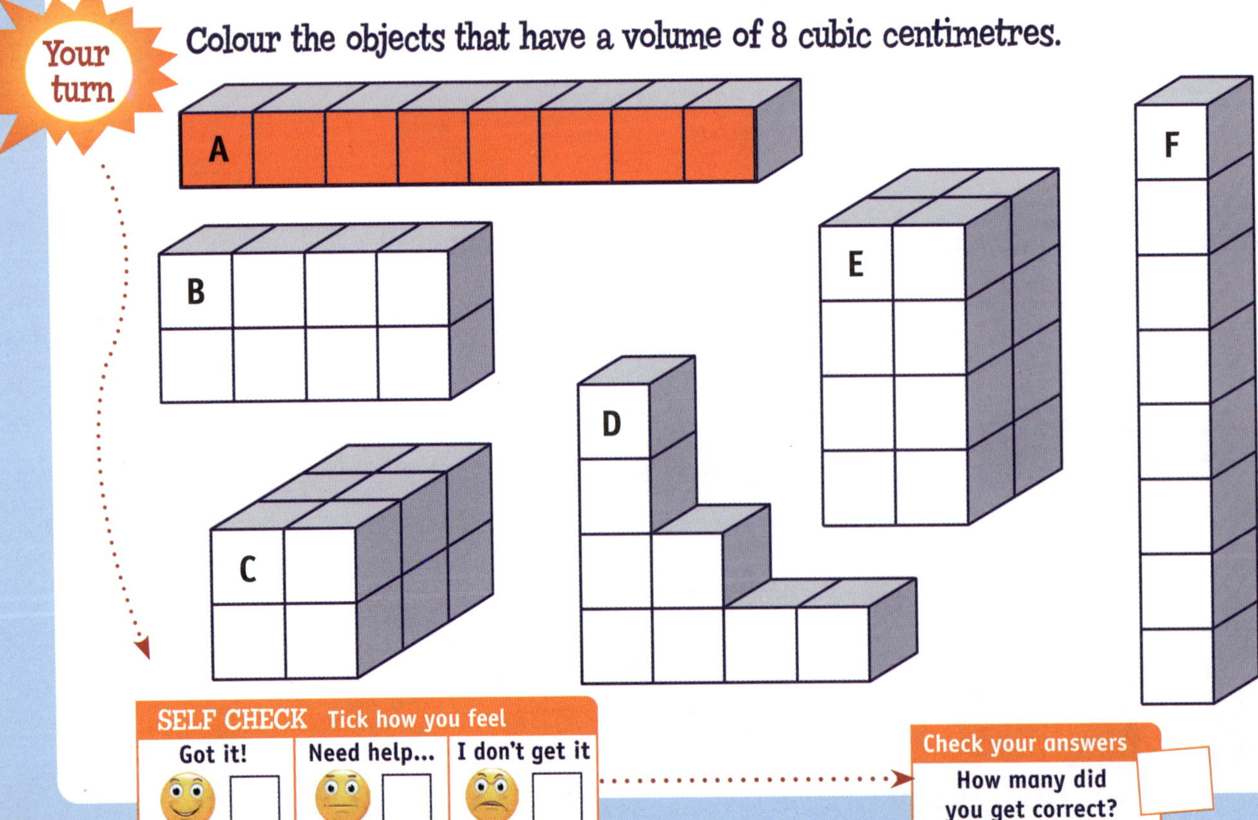

SELF CHECK Tick how you feel		
Got it! ☐	Need help... ☐	I don't get it ☐

Check your answers
How many did you get correct? ☐

CATCH UP MATHS YEAR 6 BOOK B © PASCAL PRESS ISBN: 9781925726190

PRACTICE

1 The symbol for cubic centimetres is _____.

2 How many cubic centimetres were used to make the objects?

● 10 cubic centimetres   a ___ cubic centimetres   b ___ cubic centimetres

1   2   3

3 Use the objects in Question 2 to answer the following questions.

a Which object has the least volume? ___

b Which object has the greatest volume? ___

c List the objects in descending order: ☐ ☐ ☐

d What is the difference in volume between objects 2 and 3? ______

e What is the difference in volume between objects 1 and 2? ______

f What is the difference in volume between objects 1 and 3? ______

g How many cubic centimetres altogether were used to make the objects? ______

4 Draw four different objects that have a volume of 12 cm^3.

CUBIC METRES

Cubic metres (m^3) are used to measure the volume of large objects.

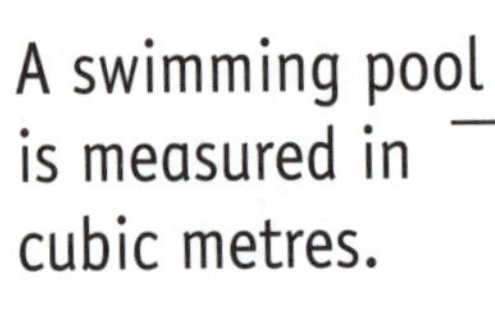
A swimming pool is measured in cubic metres.

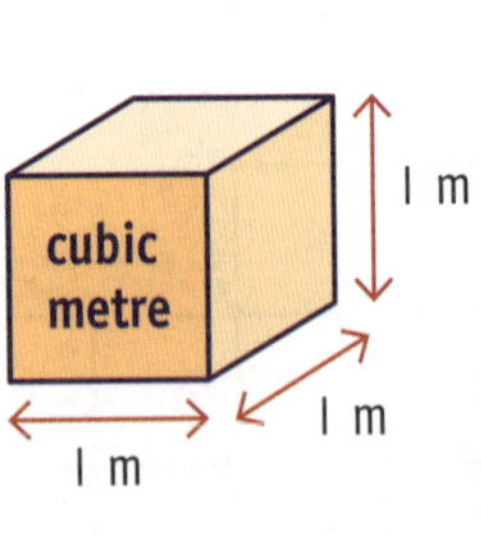

This spa is measured in cubic metres.

Examples: Use red to colour the labels of the items that would be measured in cubic metres (m^3) and use yellow for cubic centimetres (cm^3).

a	a lunchbox	**d**	a truck	**g**	a classroom
b	a suitcase	**e**	a fridge	**h**	a dictionary
c	a car	**f**	a computer tower	**i**	a clothes dryer

Check the answers on the video!

Your turn

Complete the table by listing items measured in cm^3 and m^3.

	cubic centimetres (cm^3)	cubic metres (m^3)
●	a large pencil case	a bedroom
a		
b		
c		
d		

SELF CHECK Tick how you feel

Got it!	Need help...	I don't get it
☐	☐	☐

Check your answers
How many did you get correct? ☐

CATCH UP MATHS YEAR 6 BOOK B © PASCAL PRESS ISBN: 9781925726190

PRACTICE

1 Label the cubic centimetre and the cubic metre.

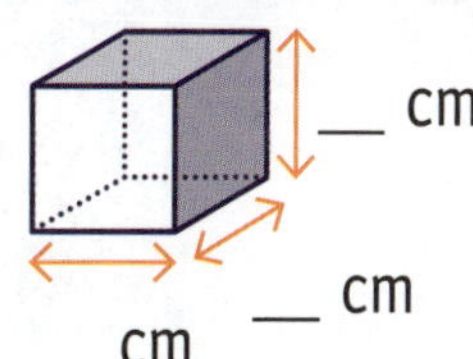

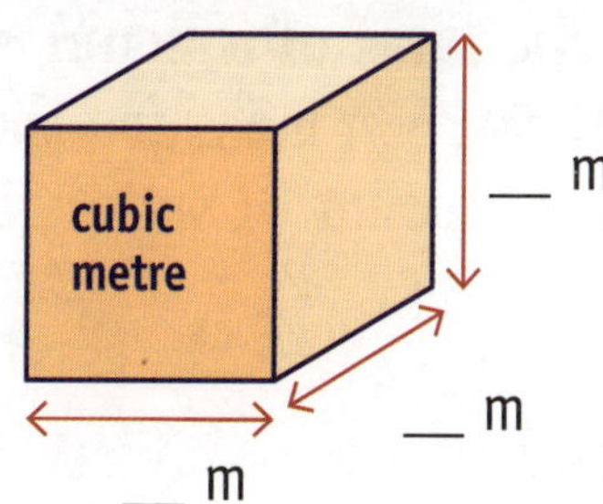

2 Write in short form.

- 6 cubic metres 6 m^3
- **a** 15 cubic metres ______
- **b** 24 cubic metres ______
- **c** 38 cubic metres ______
- **d** 157 cubic metres ______
- **e** 295 cubic metres ______
- **f** 1495 cubic metres ______
- **g** 6258 cubic metres ______

3 Write as cubic metres.

- 8 m^3 8 cubic metres
- **a** 10 m^3 ______
- **b** 19 m^3 ______
- **c** 26 m^3 ______
- **d** 392 m^3 ______
- **e** 748 m^3 ______
- **f** 1264 m^3 ______
- **g** 8976 m^3 ______

4 Match the labels to the boxes by colouring them the same colour.

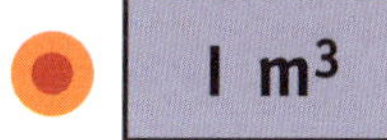

- **a** 12 m^3
- **b** 5 m^3
- **c** 7 m^3
- **d** 10 m^3
- **e** 14 m^3

DISPLACEMENT

Displacement is how much the level of a liquid rises when an object is placed into it. You can measure the volume of an object by measuring the amount of water it displaces.

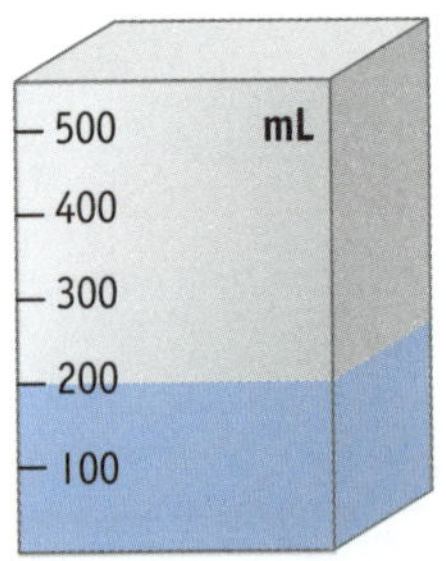

The water level is 200 mL.

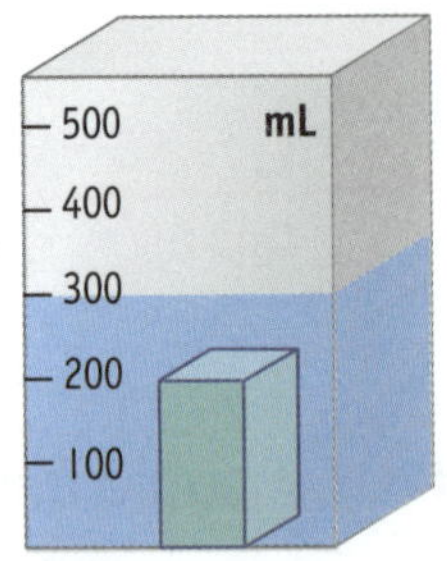

Now the water level is 300 mL.

The water level rose 100 mL when the block was placed into it, so the displacement is 100 mL.

Check the answers on the video!

Examples: Work out the displacement.

a

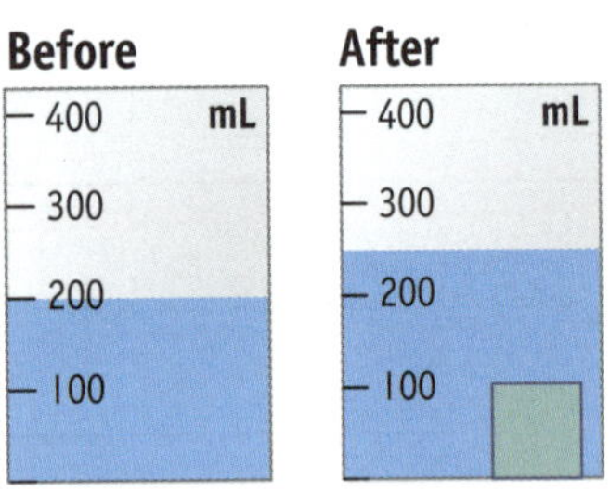

200 mL 250 mL

Displacement = 50 mL

b

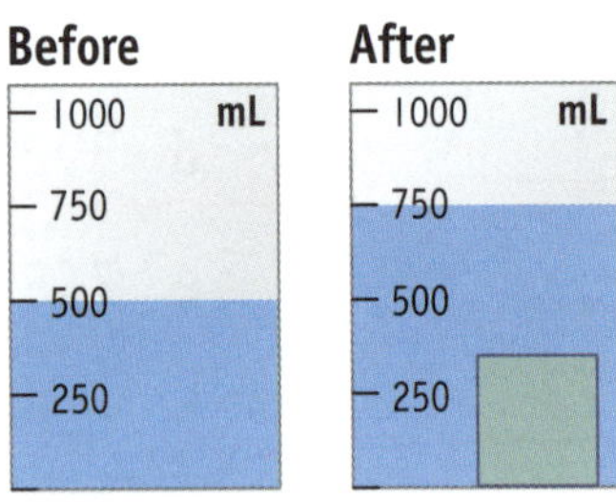

____ mL ____ mL

Displacement = ____ mL

c

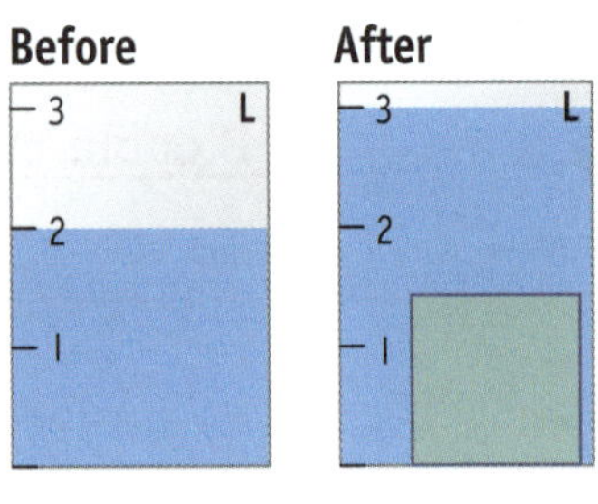

____ L ____ L

Displacement = ____ L

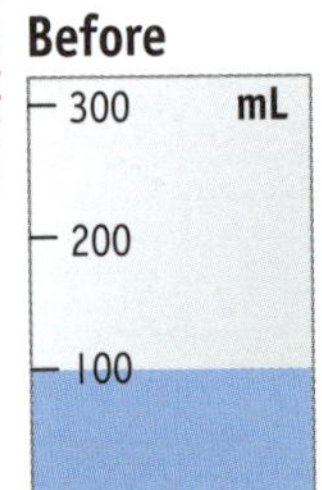

Work out the displacement.

Before After

300 200 100 mL

100 mL 300 mL

Displacement = 200 mL

a

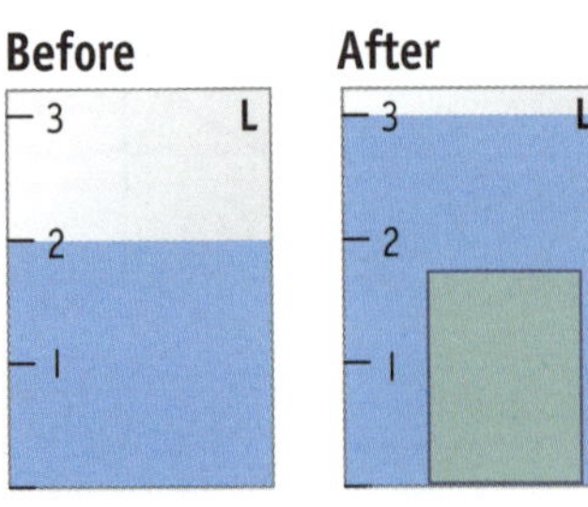

____ L ____ L

Displacement = ____ L

b

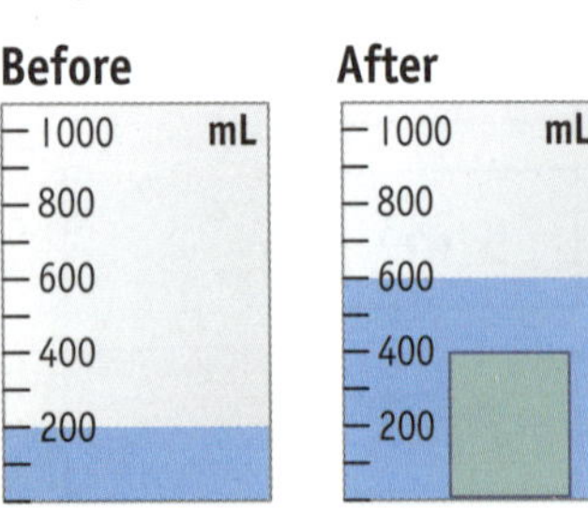

____ mL ____ mL

Displacement = ____ mL

SELF CHECK Tick how you feel

Got it!	Need help...	I don't get it
☐	☐	☐

Check your answers
How many did you get correct? ☐

CATCH UP MATHS YEAR 6 BOOK B © PASCAL PRESS ISBN: 9781925726190

PRACTICE

How much water was displaced? What is the volume of the object?
Remember: 1 mL water = 1 cm^3.

● (example)

Before: 500 mL scale | After: 500 mL scale

Displacement 300 mL

Volume of object $300\ cm^3$

a

Before: 500 mL scale | After: 500 mL scale

Displacement ________

Volume of object ________

b

Before: 300 mL scale | After: 300 mL scale

Displacement ________

Volume of object ________

c

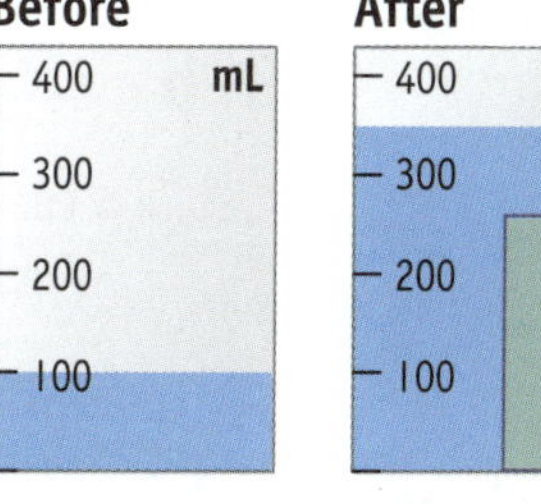

Displacement ________

Volume of object ________

d

Before: 3 L scale | After: 3 L scale

Displacement ________

Volume of object ________

e

Before: 1000 mL scale | After: 1000 mL scale

Displacement ________

Volume of object ________

Complete the table.

	Water Level		Displacement	Volume of object
	Before	After		
●	400 mL	650 mL	250 mL	$250\ cm^3$
a	2500 mL	4000 mL		
b	3200 mL		1600 mL	
c		230 mL	80 mL	
d		1000 mL	500 mL	
e		5 L	750 mL	
f	6 L	8.250 L		
g	450 mL		160 mL	

MEASURING VOLUME

To work out the volume of a 3D object, find the area of the base (length × width) and then multiply by the height.

SCAN to watch video

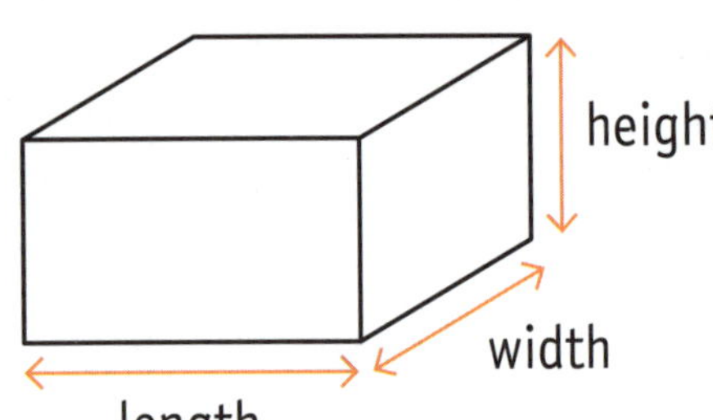

Volume (V) = length × width × height
$= l \times w \times h$

Remember, volume is the amount of space an object takes up.

Examples: Find the volume.

a

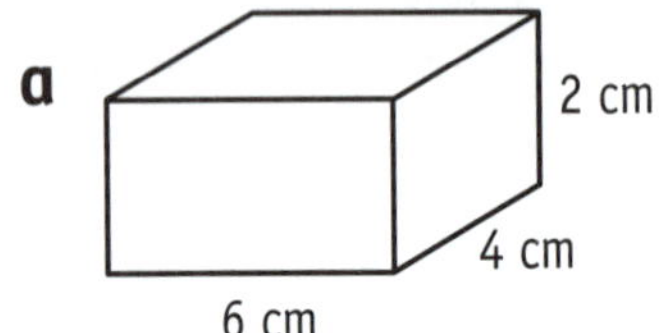

Volume = length × width × height

$= l \times w \times$ h

= 6 cm × 4 cm × 2 cm

= 48 cm^3

b

5 cm
2 cm
3 cm

Volume = length × width × ______

$= l \times$ __ $\times h$

= 2 cm × 3 cm × 5 cm

= ___ cm^3

Check the answers on the video!

Work out the volume.

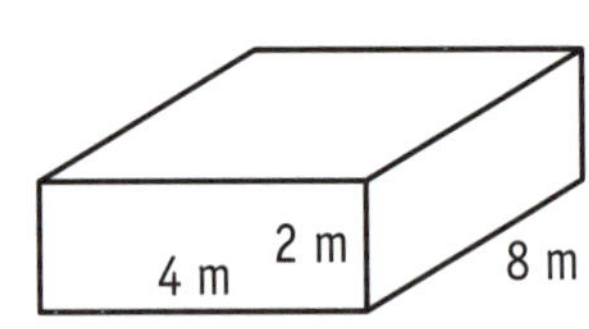

Volume = length × width × height

= l × w × h

= 8 m × 4 m × 2 m

= 64 m^3

a

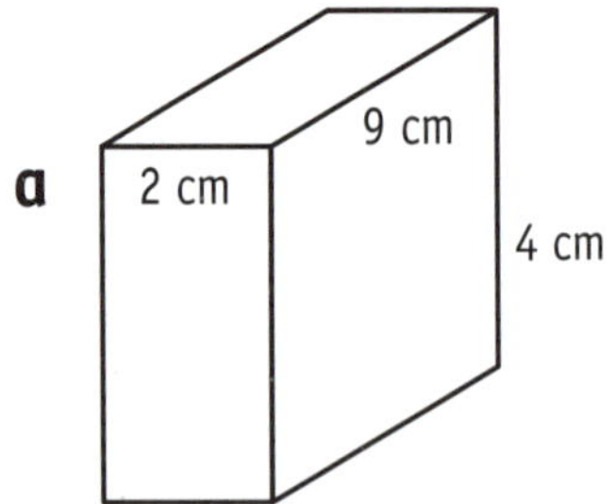

Volume = ______ × ______ × ______

= __ × __ × __

= _____ × _____ × _____

= _________

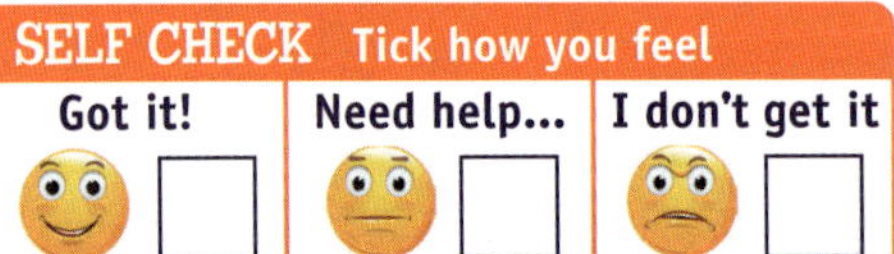

Check your answers
How many did you get correct?

CATCH UP MATHS YEAR 6 BOOK B © PASCAL PRESS ISBN: 9781925726190

1 Work out the volume.

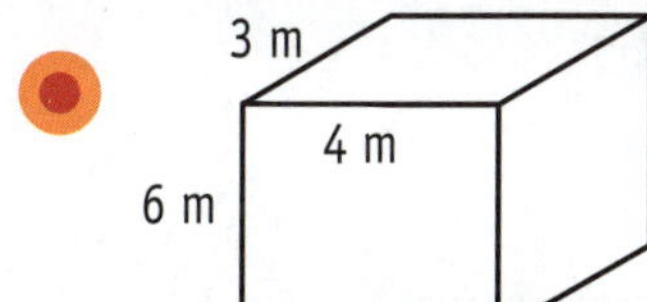

Volume = length × width × height

= l × w × h

= 4 m × 3 m × 6 m

= 72 m^3

b

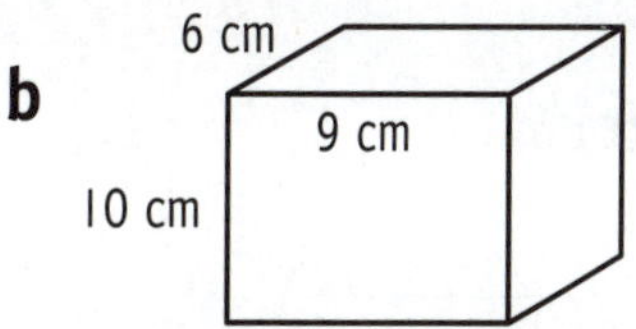

Volume = ______ × ______ × ______

= __ × __ × __

= _____ × _____ × _____

= ______

a

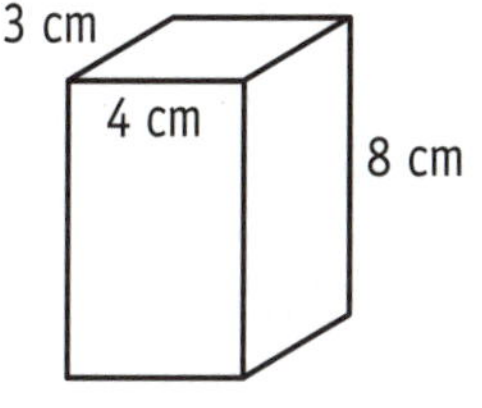

Volume = ______ × ______ × ______

= __ × __ × __

= _____ × _____ × _____

= ______

c

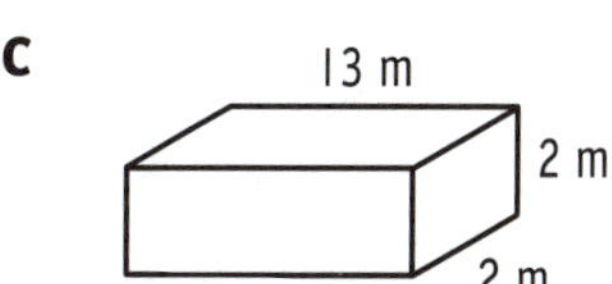

Volume = ______ × ______ × ______

= __ × __ × __

= _____ × _____ × _____

= ______

2 Complete the table for rectangular prisms with these dimensions.

	Length (*l*)	Width (*w*)	Height (*h*)	Volume (V)
●	6 cm	8 cm	0.5 cm	24 cm^3
a	2 cm	4 cm	3 cm	
b		8 cm	2 cm	64 cm^3
c	3 m	12 m	7 m	
d	1.5 cm	30 cm		540 cm^3
e	10 m	3 m	2.5 m	
f	5 cm		7.5 cm	262.5 cm^3
g	8 m	6.5 m		182 m^3
h		4	2.5 m	95 m^3
i	4.5 m	1 m	8 m	
j	2 cm	0.5 cm		8 cm^3

VOLUME AND CAPACITY REVIEW

1 Write in short form.

a 8 millilitres ________

b 6 litres ________

c 12 litres ____

d 24 millilitres ________

e 97 millilitres ________

f 105 litres ________

g 254 litres ________

h 19 litres ________

i 430 millilitres ________

j 380 millilitres ________

2 Write as litres.

a 2500 mL ______

b 3000 mL ______

c 1250 mL ______

d 6750 mL ______

e 2250 mL ______

f 6000 mL ______

3 How many millilitres?

a 1 litre ________

b $\frac{3}{4}$ litre ________

c $\frac{1}{2}$ litre ________

d $\frac{1}{4}$ litre ________

4 Write as millilitres.

a $8\frac{1}{2}$ L ________

b $5\frac{3}{4}$ L ________

c $2\frac{1}{4}$ L ________

d 4 L ________

e $7\frac{1}{2}$ L ________

f $1\frac{3}{4}$ L ________

5 Write as litres and millilitres.

a 6200 mL ___ L ______ mL

b 3490 mL ___ L ______ mL

c 1810 mL ___ L ______ mL

d 2730 mL ___ L ______ mL

e 10 365 mL ___ L ______ mL

f 23 428 mL ___ L ______ mL

6 Convert to L.

a 2805 mL = _______ L

b 6230 mL = _______ L

c 8460 mL = _______ L

d 4375 mL = _______ L

e 3710 mL = _______ L

f 12 436 mL = _______ L

  ISBN: 9781925726190

7 Convert to mL.

a 10.325 L = ________ mL

b 4.984 L = ________ mL

c 6.485 L = ________ mL

d 9.61 L = ________ mL

e 23.12 L = ________ mL

f 8.6 L = ________ mL

8 Number the boxes to put these amounts in descending order.

a	430 mL	5349 mL	90 mL	6.7 L	1.8 L
	☐				
b	3.75 L	400 mL	3250 mL	4.3 L	4 L
		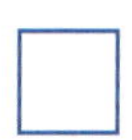		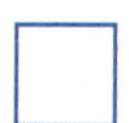	
c	8.35 L	8.6 L	8.10 L	8420 mL	8010 mL
	 		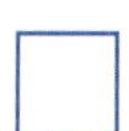		

9 What amount of liquid is in each container?

a ____________ b ____________ c ____________ d ____________

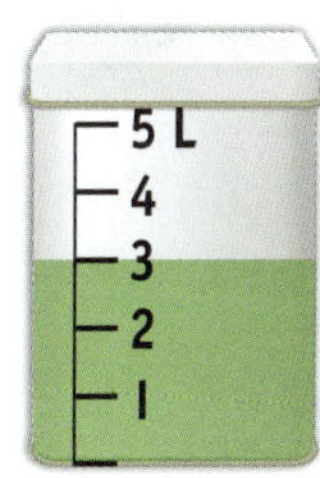

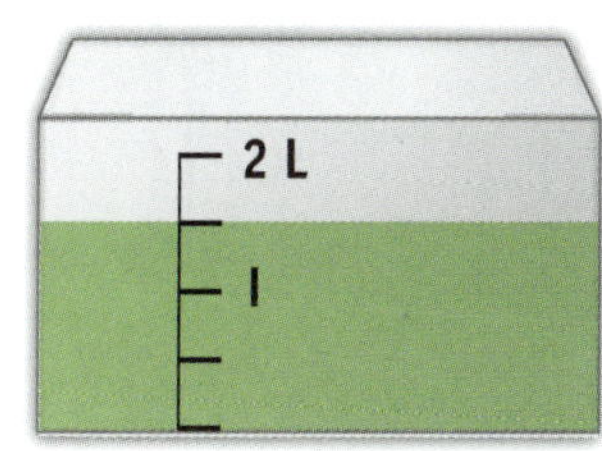

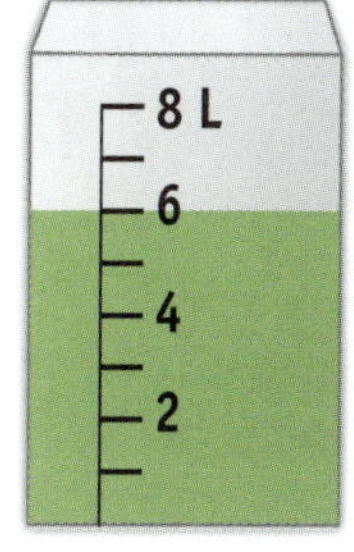

10 Colour the containers to show the amount.

a 4 litres b 500 mL c 300 mL d 3 litres

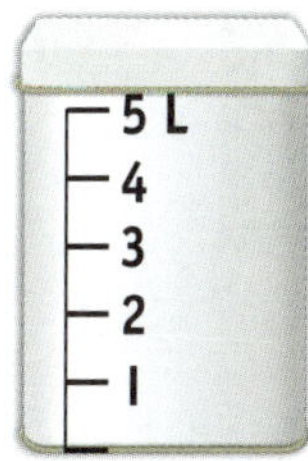

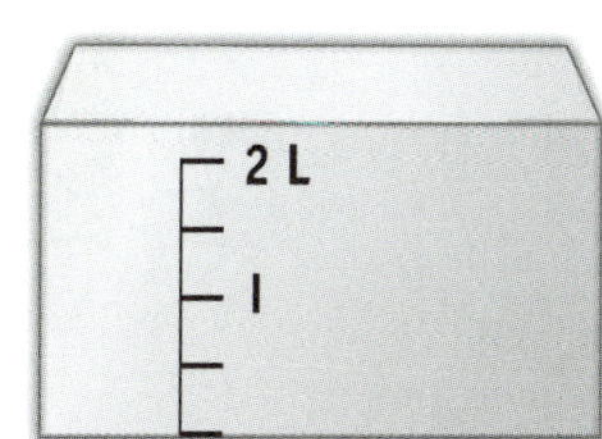

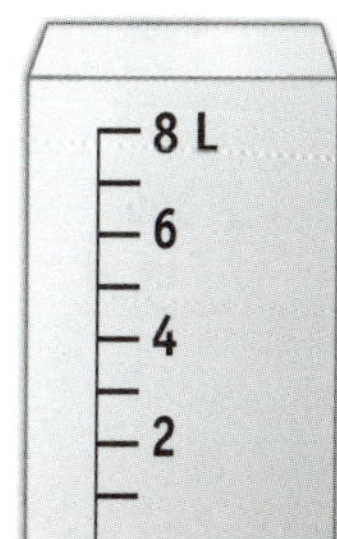

11 Complete the sentence:

Capacity is the amount of ____________ a container can ____________.

REVIEW

12 Circle the container with the larger capacity in each pair.

a

b

c

13 Draw a similar container with a smaller capacity.

a

b

c

14 Are these measurements in millilitres (mL) or litres (L)?

a 15 ____

b 1 ____

c 4 ____

d 1 ____

15 What is the total capacity of each group of containers?

a ________

b ________

500 mL

c ________

d ________

4500 mL

CATCH UP MATHS YEAR 6 BOOK B © PASCAL PRESS ISBN: 9781925726190

16 **Complete the sentence:**

The symbol for cubic centimetres is ____.

17 **How many cubic centimetres were used to make these objects?**

a __________

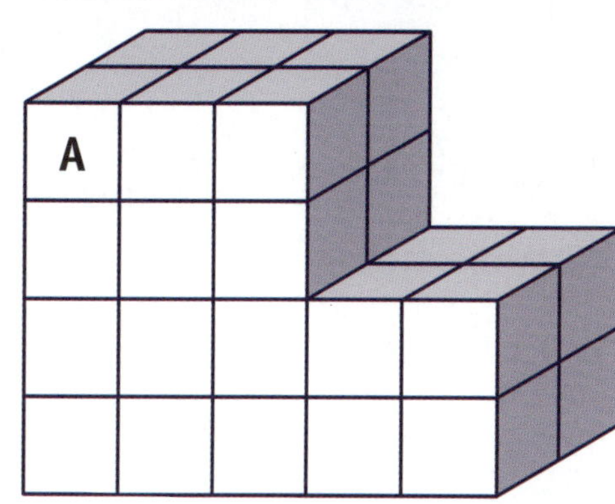

c __________

b __________

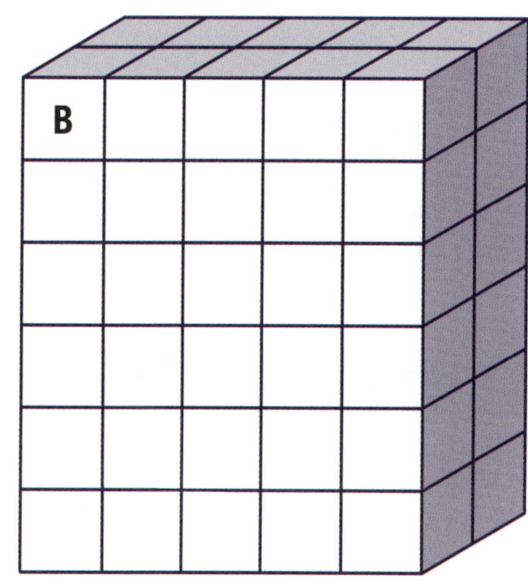

d __________

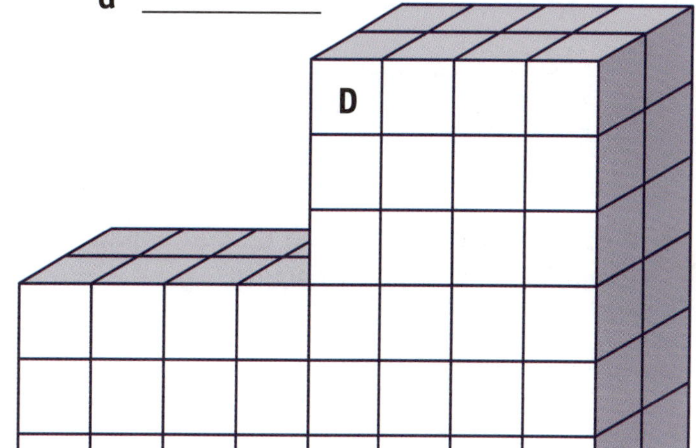

Use the objects in Question 17 to complete the following.

18 **What is the difference in volume?**

a Objects D and A ________

b Objects C and A ________

c Objects D and B ________

d Objects C and B ________

e Objects C and D ________

f Objects A and B ________

19 **How many cubic centimetres were used to make all the objects?** ________

20 Complete the sentence

One cubic metre is written ______.

21 What is measured in cubic metres? Name three things.

____

22 Write in short form.

a 9 cubic metres ________
b 374 cubic metres ________
c 73 cubic metres ________
d 1428 cubic metres ________
e 25 cubic metres ________
f 47 cubic metres ________
g 169 cubic metres ________
h 862 cubic metres ________

23 Write as cubic metres in long form.

a 2 m^3 ______________
b 63 m^3 ______________
c 12 m^3 ______________
d 128 m^3 ______________
e 8 m^3 ______________
f 25 m^3 ______________
g 57 m^3 ______________
h 198 m^3 ______________

24 Christian has been using displacement to measure the weight of some objects. Complete his table of results.

	Water Level		Displacement	Volume of object
	Before	After		
a	300 mL	650 mL		
b	500 mL	1000 mL		
c	2000 mL	2130 mL		
d	1000 mL	3200 mL		
e	400 mL	1850 mL		
f	3250 mL	4930 mL		
g	4650 mL		395 mL	
h		1895 mL	620 mL	

  ISBN: 9781925726190

What is the volume?

a

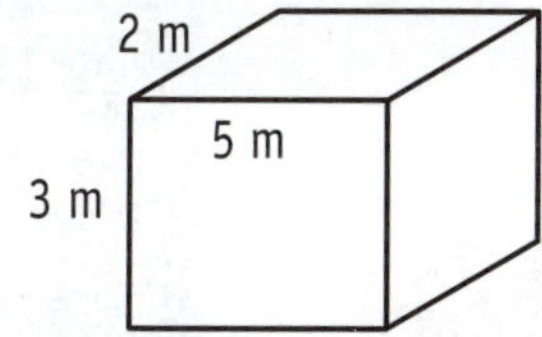

Volume = ______ × ______ × ______

= __ × __ × __

= ______ × ______ × ______

= ________

c

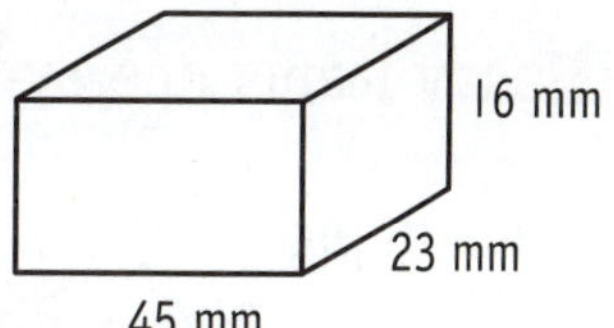

Volume = ______ × ______ × ______

= __ × __ × __

= ______ × ______ × ______

= ________

b

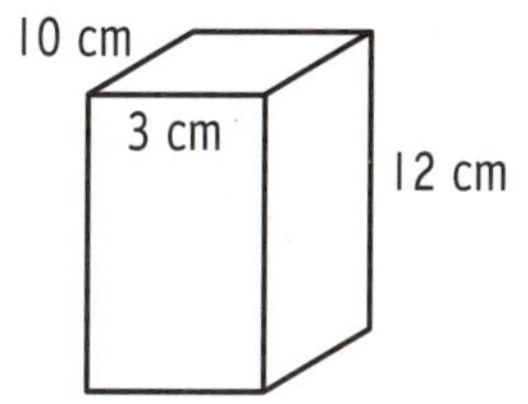

Volume = ______ × ______ × ______

= __ × __ × __

= ______ × ______ × ______

= ________

d

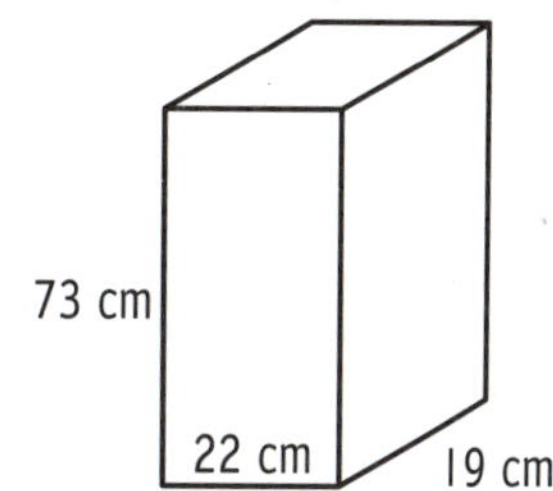

Volume = ______ × ______ × ______

= __ × __ × __

= ______ × ______ × ______

= ________

Complete the table for rectangular prisms with these dimensions.

	Length (l)	Width (w)	Height (h)	Volume (V)
a	8 cm	0.5 cm		24 cm^3
b		6 m	7 m	126 m^3
c	4 mm	8 mm	1.5 mm	
d	7 cm	1.5 cm		84 cm^3
e	13 m		12 m	624 m^3
f	4 m	16 m	15 m	
g	5 cm	8 cm	12 cm	
h	4 mm		8 mm	288 mm^3
i		6.5 m	7.2 m	159.12 m^3
j	9 m		8 m	446.4 m^3

KILOGRAMS

Heavy items are measured in kilograms (kg).

SCAN to watch video

This pineapple weighs 1 kilogram (1 kg).

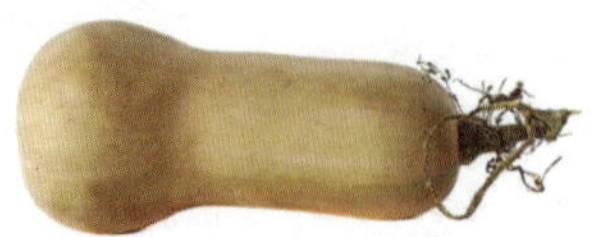

This pumpkin weighs 2 kilograms (2 kg).

This net of oranges weighs 3 kilograms (3 kg).

Kilo means 1000.

There are 1000 grams in one kilogram.

Examples: Circle the objects that we buy in kilograms.

a apples	**e** sugar	**i** cement
b milk	**f** shampoo	**j** orange juice
c sauce	**g** potatoes	**k** sand
d sausages	**h** lemons	**l** shoes

Check the answers on the video!

Your turn

1 Circle the masses that have been written correctly.

● 9 kg	**c** 37KG	**f** 172 kG	**i** 158 kg
a 12 Kg	**d** 44 Kg	**g** 88KG	**j** 22kG
b 24 kg	**e** 125 kg	**h** 100 kg	**k** 758 kg

2 Write the masses in short form.

● 2 kilograms	2 kg	**d** 73 kilograms	______
a 9 kilograms	______	**e** 105 kilograms	______
b 16 kilograms	______	**f** 250 kilograms	______
c 20 kilograms	______	**g** 748 kilograms	______

SELF CHECK Tick how you feel

Got it! ☐ Need help... ☐ I don't get it ☐

Check your answers

How many did you get correct? ☐

CATCH UP MATHS YEAR 6 BOOK B © PASCAL PRESS ISBN: 9781925726190

PRACTICE

1 Circle the most likely mass for each item.

- a man
  - 15 kg
  - 45 kg
  - (90 kg)
- a a packet of flour
  - 2 kg
  - 45 kg
  - 240 kg
- b a bag of potatoes
  - 60 kg
  - 5 kg
  - 125 kg
- c a net of apples
  - 32 kg
  - 15 kg
  - 1 kg

2 What could be getting weighed? Colour the correct label.

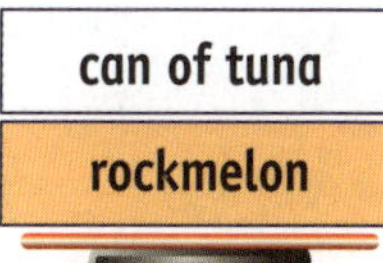

a

b

c

3 How many grams?

- $2\frac{1}{2}$ kg 2500 g
- b $6\frac{3}{4}$ kg ______
- d $5\frac{1}{4}$ kg ______
- a $3\frac{1}{4}$ kg ______
- c 4 kg ______
- e $7\frac{1}{2}$ kg ______

4 How many of each weight make up 1 kilogram?

- 100 g × 10 = 1 kg
- b 200 g × ___ = 1 kg
- d 500 g × ___ = 1 kg
- a 50 g × ___ = 1 kg
- c 250 g × ___ = 1 kg
- e 1000 g × ___ = 1 kg

5 Number the masses to order them from lightest (1) to heaviest (5).

●	1 kg 200 g	3 kg 400 g	2600 g	6250 g	$3\frac{1}{2}$ kg
	1	3	2	5	4
a	$4\frac{1}{4}$ kg	2 kg 300 g	5300 g	1470 g	$1\frac{1}{2}$ kg
b	6500 g	$2\frac{3}{4}$ kg	$6\frac{1}{4}$ kg	2700 g	5000 g
c	$3\frac{1}{2}$ kg	1250 g	$4\frac{1}{4}$ kg	1000 g	2750 g

6 Join the matching masses.

a

kg	grams
$\frac{1}{2}$ kg	2000 g
5 kg	7500 g
2 kg	5000 g
$3\frac{3}{4}$ kg	500 g
$7\frac{1}{2}$ kg	3750 g

b

kg	grams
0.25 kg	250 g
1.68 kg	300 g
2.75 kg	1680 g
0.75 kg	2750 g
0.3 kg	750 g

c

kg, g	grams
$2\frac{3}{4}$ kg	4250 g
2 kg 300 g	2750 g
5 kg 625 g	4750 g
4 kg 750 g	5625 g
$4\frac{1}{4}$ kg	2300 g

d

kg	grams
12 kg	14 250 g
$14\frac{1}{4}$ kg	11 750 g
$13\frac{1}{2}$ kg	12 750 g
$12\frac{3}{4}$ kg	13 500 g
$11\frac{3}{4}$ kg	12 000 g

e

kg	grams
8.125 kg	8300 g
8.25 kg	8125 g
8.3 kg	8980 g
8.75 kg	8250 g
8.98 kg	8750 g

f

kg, g	grams
4 kg 4 g	4425 g
4 kg 8 g	4004 g
4 kg 400 g	4400 g
4 kg 425 g	4008 g
4 kg 800 g	4800 g

7 Calculate the total mass. Answer in both grams and kilograms.

- 40 g + 60 g + 1.2 kg = 1300 g = 1.3 kg

a 8.6 kg − 425 g − 125 g = ______ g = ______ kg

b 427 g + 4 kg = ______ g = ______ kg

c 510 g × 6 = ______ g = ______ kg

d 8.2 kg ÷ 4 = ______ g = ______ kg

8 What is the average mass? Answer in both grams and kilograms.

- 42 kg, 38 kg, 46 kg, 34 kg — Average mass = 40 000 g = 40 kg

a 37 g + 42 g + 1.45 kg + 1.375 kg — Average mass = ______ g = ______ kg

b 85.6 g, 92.2 g, 74.5 g, 88.5 g — Average mass = ______ g = ______ kg

c 1.2 kg, 1.3 kg, 1.25 kg, 2.1 kg, 1.7 kg — Average mass = ______ g = ______ kg

d 3.8 kg, 4.1 kg, 4.9 kg, 3.7 kg — Average mass = ______ g = ______ kg

e 76.3 g, 84.9 g — Average mass = ______ g = ______ kg

  ISBN: 9781925726190

9 Complete the table.

	Item	Mass of each item	How many items?	Total Mass	
				grams	kilograms
●	apple	250 g	5	1250 g	1.25 kg
a	a loaf of bread	780 g	4		
b	can of salmon	315 g	5		
c	chocolate block	225 g	3		
d	orange	200 g	6		
e	bag of nuts	1.2 kg	4		
f	box of pasta	150 g	15		
g	muesli bar	35.5 g	6		
h	noodle pack	52.5 g	8		

Use the construction items below to complete the following.

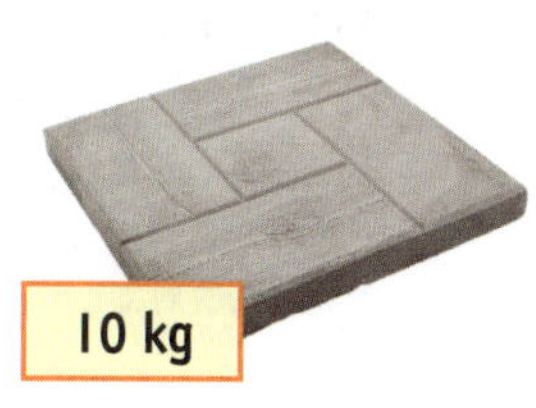

10 Write the mass in grams for each construction item.

● paver 10 000 g b toolbox ________ d boots ________

a brick ________ c lunchbox ________ e ladder ________

11 What is the mass in kilograms?

● lightest item 2.46 kg

a heaviest item ________

b the total of all the items ________

c paver and toolbox ________

d lunchbox and toolbox ________

e paver and ladder ________

f brick, boots and ladder ________

g paver, lunchbox and brick ________

h boots and lunchbox ________

12 Which items weigh less than 5000 g?

____

GRAMS

Light objects are measured using grams (g).
There are 1000 grams in 1 kilogram.

An egg weighs about 60 grams (60 g).

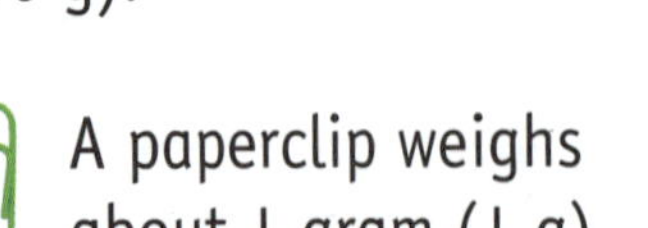

A paperclip weighs about 1 gram (1 g).

A metal fork weighs about 100 grams (100 g).

Examples: List ten things that would be measured in grams.

a pencil ________ ________

________ ________

________ ________

________ ________

________ ________

Check the answers on the video!

1 Write in short form.

● 6 grams	6 g	c 99 grams	______	f 375 grams	______
a 24 grams	______	d 170 grams	______	g 420 grams	______
b 37 grams	______	e 250 grams	______	h 750 grams	______

2 Estimate how many grams each item weighs.

● Estimate: 5 g

a Estimate: ______

b Estimate: ______

c Estimate: ______

SELF CHECK Tick how you feel

Got it!	Need help...	I don't get it
 ☐	 ☐	 ☐

Check your answers

How many did you get correct?

CATCH UP MATHS YEAR 6 BOOK B © PASCAL PRESS ISBN: 9781925726190

PRACTICE

1 Tick or cross the masses to show if they are written correctly.

- ☑ 75 g
- a ☐ 12 G
- b ☐ 5gms
- c ☐ 76 g
- d ☐ 881G
- e ☐ 30 GMS
- f ☐ 649 Gr
- g ☐ 95 gs
- h ☐ 470 g
- i ☐ 10 g

2 Number the items to order them from lightest (1) to heaviest (7).

☐ ☐ ☐ ☐ ☐ ☐ ☐

3 How many more grams to make 1 kilogram?

	Item	Weight	Grams to make 1 kg
●	chocolate	250 g	750 g
a	tuna	125 g	
b	sardines	100 g	
c	jam	500 g	

	Item	Weight	Grams to make 1 kg
d	cookies	755 g	
e	spaghetti	245 g	
f	muesli bar	25 g	
g	chips	75 g	

4 What is the mass? Use the tables above. Answer in both grams and kilograms.

● 4 jars of jam
= 2000 g = 2 kg

a 7 muesli bars
= ________ g = ______ kg

b 5 cans of sardines
= ________ g = ______ kg

c 2 packets of spaghetti
= ________ g = ______ kg

d 5 bags of cookies
= ________ g = ______ kg

e 6 cans of tuna
= ________ g = ______ kg

5 How many grams?

● 4.238 kg 4238 g

a 0.156 kg ________

b 2.735 kg ________

c 6.7 kg ________

d 3.07 kg ________

e 5.803 kg ________

f 10.42 kg ________

g 0.2 kg ________

h 0.53 kg ________

READING SCALES

We use different scales to measure different masses.

This is a balance scale.

We use it to measure small masses.

When it is level, the mass on one side is equal to the mass on the other side.

The mass of the bag of marbles is 600 g.

This is a scale you step on.

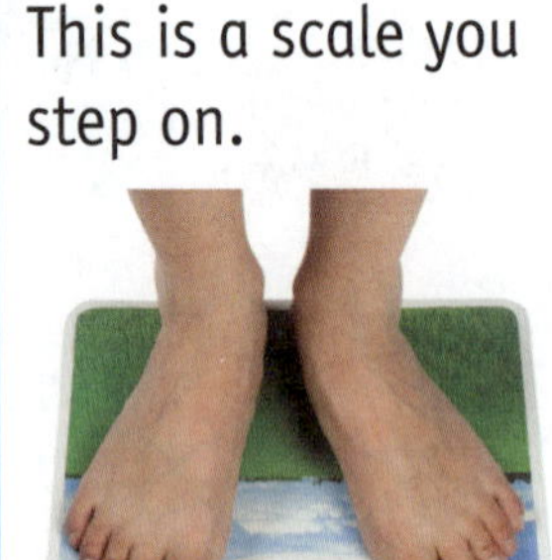

Ivan weighs 50 kg.

These scales can weigh heavier items. The mass of the watermelon is 6 kg.

Examples: Read the scales and write the mass.

a 400 g

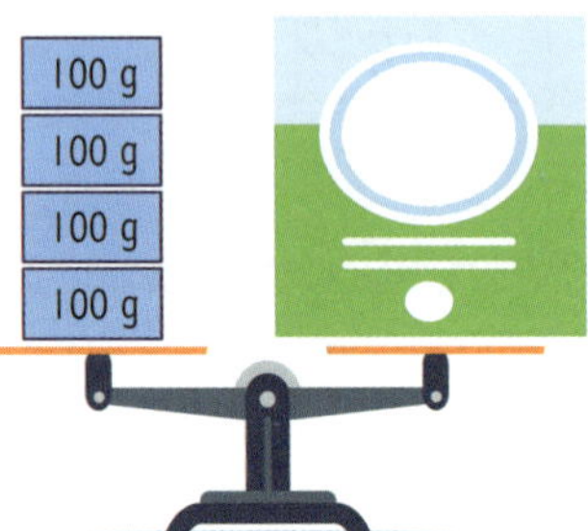

b _____ g

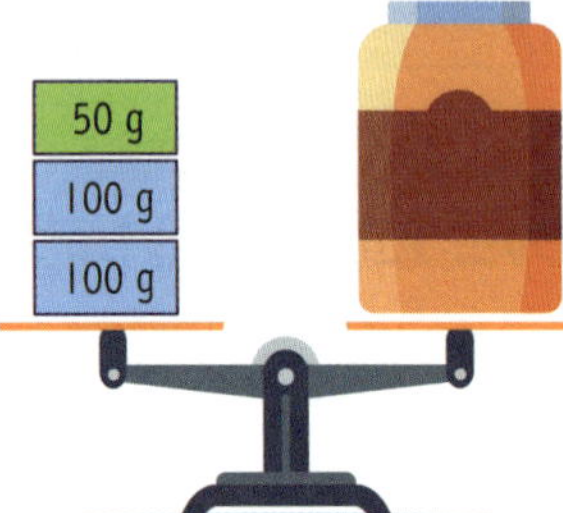

c _____ g

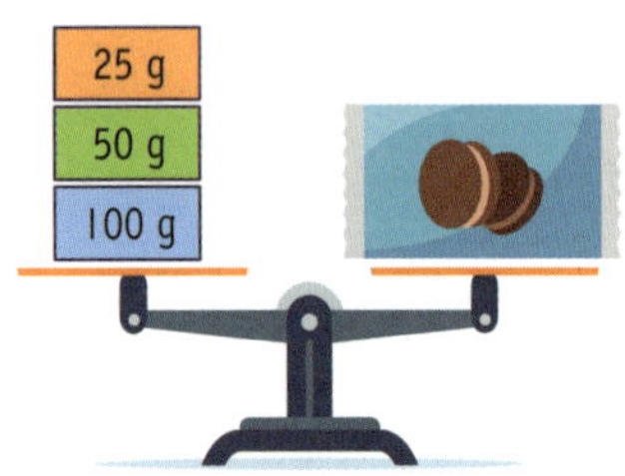

Check the answers on the video!

Your turn

Draw 100 g, 50 g and 25 g weights to show the total mass of each item.

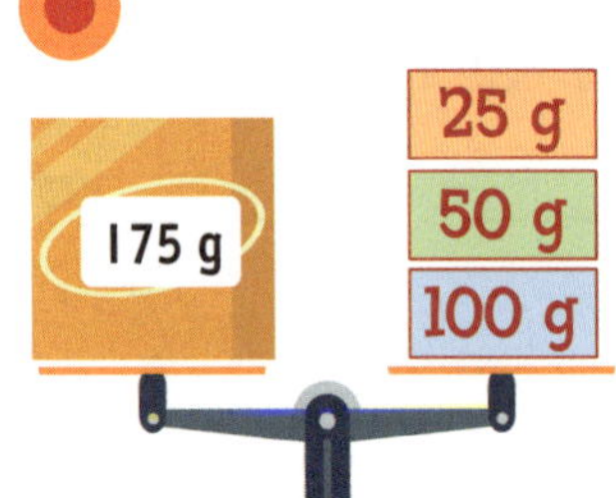

a

b

SELF CHECK Tick how you feel

Got it!	Need help...	I don't get it
 ☐	 ☐	☐

Check your answers

How many did you get correct?

CATCH UP MATHS YEAR 6 BOOK B © PASCAL PRESS ISBN: 9781925726190

PRACTICE

Read the scales and write the mass of each bag.

● 4 kg a ________ b ________ c ________

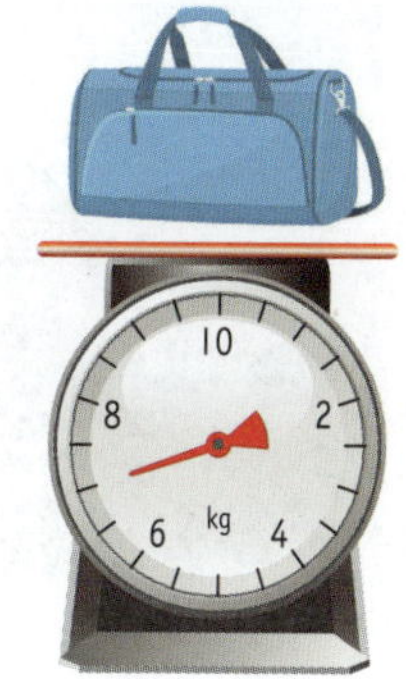

2 Write the mass of each bag in grams.

● green bag 4000 g

a brown bag ________

b orange bag ________

c blue suitcase ________

Order the bags from lightest (1) to heaviest (4).

1 ________ 2 ________ 3 ________ 4 ________

Draw the needles on each scale to match the item(s) being weighed.

●

b

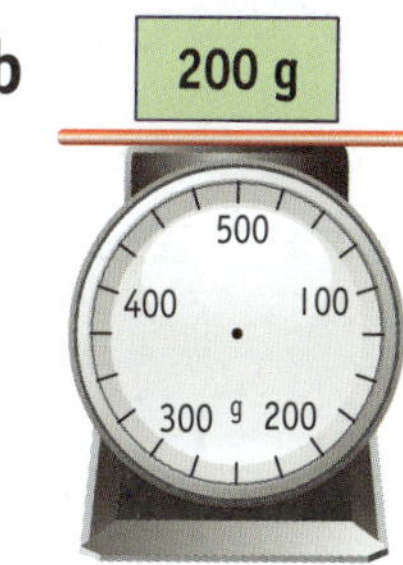

d

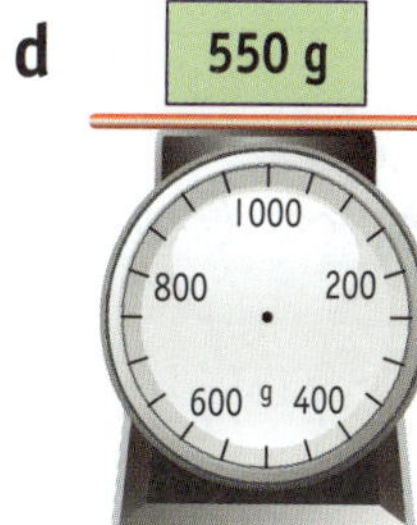

f

a

c

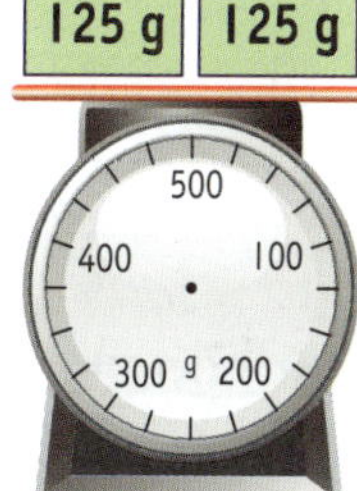

e

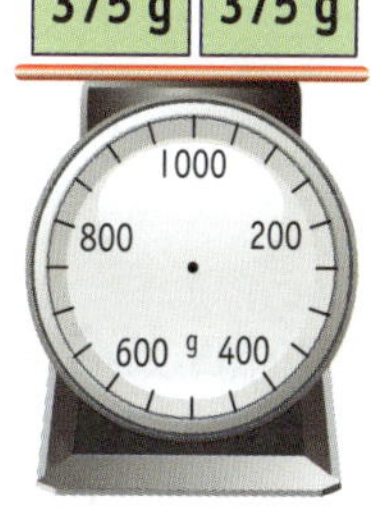

g

5 Write the total mass in grams and in kilograms for the items weighed in Question 4.

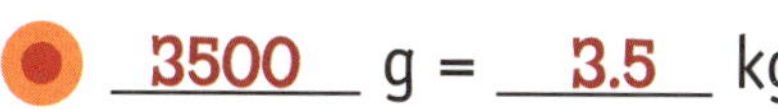

● 3500 g = 3.5 kg

a ________ g = ________ kg

b ________ g = ________ kg

c ________ g = ________ kg

d ________ g = ________ kg

e ________ g = ________ kg

f ________ g = ________ kg

g ________ g = ________ kg

THE TONNE

Tonnes (t) are used for measuring very heavy items.
One tonne is 1000 kilograms.

SCAN to watch video

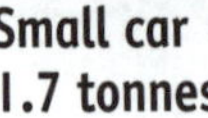
Small car
1.7 tonnes

Hippopotamus
1.5 tonnes

Remember,
1 tonne = 1000 kg

Examples: Write in short form.

a 1 tonne	1 t	d 10 tonnes	______
b 6 tonnes	______	e 8 tonnes	______
c 5 tonnes	______	f 100 tonnes	______

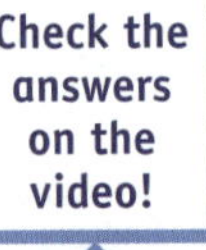

1 Colour the objects that would weigh one tonne or more.

● elephant	c tiger	f salmon
a truck	d blue whale	g bus
b sumo wrestler	e horse	h bicycle

2 If 1 tonne weighs 1000 kg, how many kilograms in these masses?

● $\frac{1}{4}$ tonne = 250 kg

a $\frac{1}{2}$ tonne = ________

b $\frac{3}{4}$ tonne = ________

c $4\frac{1}{2}$ tonnes = ________

d $3\frac{3}{4}$ tonnes = ________

e $5\frac{1}{4}$ tonnes = ________

SELF CHECK Tick how you feel

Got it!	Need help...	I don't get it
☐	☐	☐

Check your answers
How many did you get correct? ☐

CATCH UP MATHS YEAR 6 BOOK B © PASCAL PRESS ISBN: 9781925726190

PRACTICE

1 Complete the tables.

	Grams	Kilograms	Tonnes
●	10 000	10	0.01
a		50	
b			0.7
c		90	
d	400 000		
e		100	
f		1000	
g			6
h	7 000 000		

	Grams	Kilograms	Tonnes
i			$4\frac{1}{2}$
j	5 750 000		
k		13 250	
l			$9\frac{1}{2}$
m		3000	
n	4 000 000		
o	20 000		
p			$8\frac{1}{4}$
q		9750	

2 How many of each weight would equal one tonne?

● 250 kg × __4__ = 1 tonne

a 200 kg × _____ = 1 tonne

b 100 kg × _____ = 1 tonne

c 50 kg × _____ = 1 tonne

d 1000 kg × _____ = 1 tonne

e 25 kg × _____ = 1 tonne

3 Write T for true or F for false in the box.

● [T] 1 tonne = 1000 kg

a [ ] 0.10 t = 10 kg

b [ ] 1 000 000 g = 1 t

c [ ] 90 000 g = 90 kg

d [ ] 400 kg = 0.4 t

e [ ] 0.07 t = 70 kg

f [ ] 6 t = 6000 kg

g [ ] 0.80 t = 800 kg

h [ ] 525 000 g = $5\frac{1}{4}$ t

i [ ] 200 kg = 0.02 t

j [ ] 50 kg = 0.5 t

k [ ] 4 250 000 g = $4\frac{1}{2}$ t

l [ ] 9 tonnes = 9000 kg

m [ ] 675 000 kg = $6\frac{3}{4}$ t

n [ ] 250 kg = 0.25 t

o [ ] 0.2 t = 200 kg

GROSS MASS AND NET MASS

Gross mass is the mass of the container plus the mass of the contents. Net mass is the mass of the contents only.

The gross mass (container + contents) is 500 g.
The net mass of the coffee beans is 450 g.
The jar the coffee beans are in weighs 50 g.

Examples: Write the missing numbers.

a 

This tin of beans has a mass of 560 g.
The beans weigh 540 g.
The tin weighs 20 g.

b 

This tin of corn has a mass of 420 g.
The corn weighs 395 g.
The tin weighs ___ g.

c 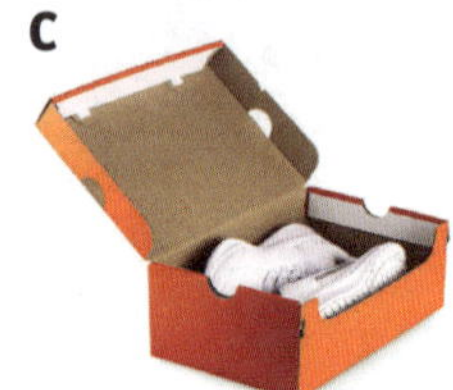

This box of shoes has a mass of 800 g.
The shoes weigh 725 g.
The box weighs ____ g.

d 

This parcel has a mass of 2 kg. The contents weigh 1810 g.
The packaging weighs _____ g.

Your turn

Write the missing numbers.

● The tin has a mass of 15 g.
The peas have a mass of 115 g.

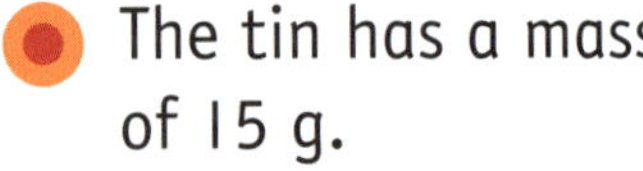

Gross mass = 15 g + 115 g
= 130 g
Net mass = 115 g

a The jar has a mass of 95 g.
The cookies have a mass of 435 g.

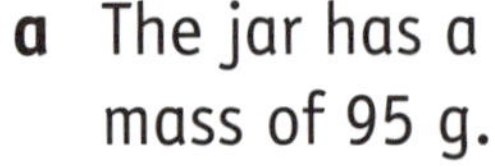

Gross mass = ___ g + ____ g
= _____ g

Net mass = ____ g

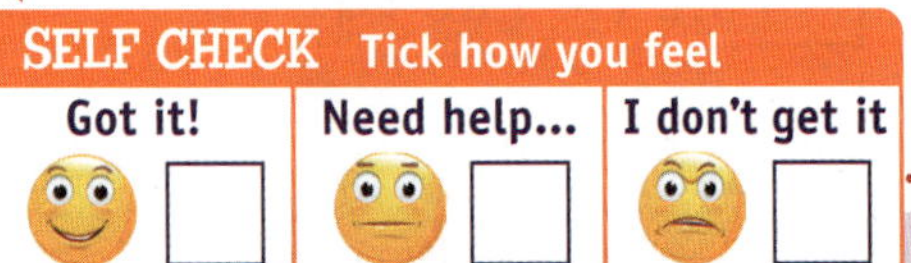

Check your answers
How many did you get correct?

CATCH UP MATHS YEAR 6 BOOK B © PASCAL PRESS ISBN: 9781925726190

PRACTICE

Complete the table.

	Item	Net Mass	Packaging	Gross Mass
●	lollies	200 g	20 g	220 g
a	chocolate		10 g	110 g
b	potato chips	155 g		160 g
c	nuts	275 g	35 g	
d	can of beans		25 g	600 g
e	box of pasta	155 g		170 g
f	box of cereal	880 g		900 g
g	bag of cookies		5 g	200 g
h	can of tuna	400 g		425 g

2 Write the gross mass on the items.

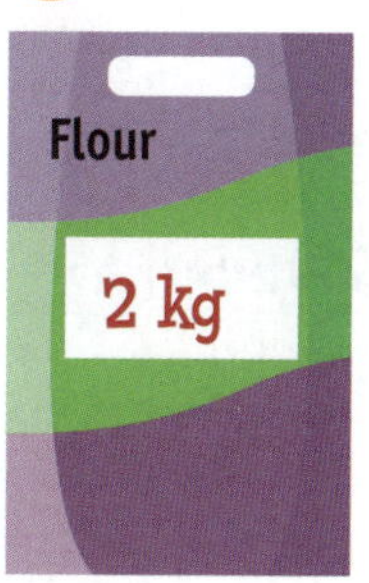

Net weight: 1.995 kg
Packaging: 5 g

a

Net weight: 260 g
Packaging: 20 g

b

Net weight: 625 g
Packaging: 25 g

c 

Net weight: 725 g
Packaging: 50 g

d

Net weight: 115 g
Packaging: 5 g

e

Net weight: 700 g
Packaging: 50 g

f 

Net weight: 250 g
Packaging: 5 g

g

Net weight: 520 g
Packaging: 50 g

h

Net weight: 205 g
Packaging: 10 g

BEST BUYS

To work out the best buy, work out how much each item would cost for the same mass, such as one kilogram.

1 kg $9.00

Best buy!

500 g $4.60

This would be $9.20 per kilogram, which is 20 cents more per kilogram.

Examples: Colour the cheaper label.

a | 1 L $2.40 |  | 1 L $2.60 |

b | 3 kg $14.60 |  | 3 kg $14.65 |

c | 1 kg $10.50 |  | 1 kg $9.95 |

d | $39.00 per kg | 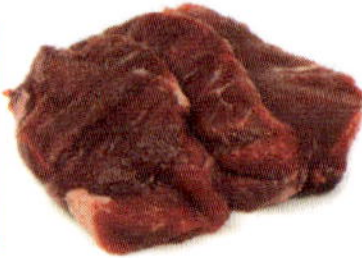 | $39.95 per kg |

Check the answers on the video!

Your turn

Look at Kira's working and then colour the best buy.

1 kg for $2.40	500 g for $1.25

1 kg for $2.50

b

1 kg for $5.80	2 kg for $11.50

1 kg for $5.75

a

5 kg for $24.00	1 kg for $5.00

1 kg for $4.80

c

2 kg for $8.50	1 kg for $4.15

1 kg for $4.25

SELF CHECK Tick how you feel

Got it!	Need help...	I don't get it

Check your answers
How many did you get correct?

CATCH UP MATHS YEAR 6 BOOK B © PASCAL PRESS ISBN: 9781925726190

PRACTICE

Compare the prices and weights or quantities. Then tick the best buy.

e.g.	✔ 100 g for $1.50	500 g for $8.00
a	250 g for 70c	1 kg for $3.00
b	250 g for $5.20	500 g for $10.30
c	4 kg for $5.00	2 kg for $2.75
d	900 g for $14.00	100 g for $1.55
e	200 g for $2.70	600 g for $8.00
f	250 g for $1.50	1 kg for $5.90
g	200 g for $2.00	1 kg for $9.00
h	500g for $1.50	1 kg for $2.95
i	1 kg for $2.50	5 kg for $12.00
j	100 g for $2.50	1 kg for $25.50
k	500 g for $9.50	250 g for $5.00
l	1 kg for $5.80	750 g for $4.20
m	200 g for $2.50	500 g for $5.50

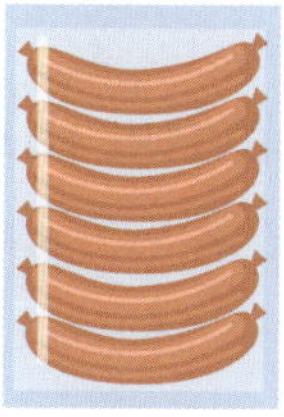

MASS REVIEW

1 Tick or cross the masses to show if they are written correctly.

a ☐ 9KG
b ☐ 16 kg
c ☐ 22 kgs
d ☐ 38 kg
e ☐ 49kg
f ☐ 83 kg
g ☐ 373 kg
h ☐ 211 ks
i ☐ 293 kg
j ☐ 839 kg

2 Write the masses in short form.

a three kilograms _______

b seventy-four kilograms _______

c fifty-six kilograms _______

d one hundred and eighty-nine kilograms _______

e seven hundred and twenty-five kilograms _______

3 Circle the most likely mass for each of the following.

a a woman
- 13 kg
- 80 kg
- 240 kg

b packet of sugar
- 2 kg
- 20 kg
- 130 kg

c net of oranges
- 5 kg
- 50 kg
- 500 kg

d a watermelon
- 8 kilograms
- 18 kilograms
- 28 kilograms

4 How many grams?

a $3\frac{1}{2}$ kg _______

b $6\frac{1}{4}$ kg _______

c $2\frac{3}{4}$ kg _______

d 5 kg _______

e $15\frac{1}{4}$ kg _______

f 12 kg _______

5 Complete the table.

Weight	25 g	50 g	100 g	125 g	200 g	250 g	500 g	1 kg
How much more to make 1 kg?								

6 Number the masses to order them from lightest (1) to heaviest (5).

a 2500 g ☐ | 2 kg ☐ | 2850 g ☐ | $2\frac{1}{4}$ kg ☐ | 2 kg 700 g ☐

b 3150 g ☐ | 3 kg 280 g ☐ | $3\frac{1}{2}$ kg ☐ | 3.2 kg ☐ | $3\frac{3}{4}$ kg ☐

c 4150 g ☐ | 4.1 kg ☐ | $4\frac{3}{4}$ kg ☐ | 4.38 kg ☐ | 4 kg 900 g ☐

  ISBN: 9781925726190

7 Join the matching masses.

a

kg	grams
$\frac{1}{4}$ kg	7250 g
$3\frac{3}{4}$ kg	5000 g
$\frac{1}{2}$ kg	250 g
5 kg	3750 g
$7\frac{1}{4}$ kg	8500 g
$8\frac{1}{2}$ kg	500 g

b

kg	grams
0.75 kg	1340 g
2.68 kg	2680 g
0.4 kg	750 g
1.34 kg	400 g
4 kg	3620 g
3.62 kg	4000 g

c

kg	grams
$3\frac{1}{4}$ kg	3750 g
3.5 kg	300 g
3 kg 200 g	3200 g
3 kg 750 g	3000 g
3 kg	3500 g
0.3 kg	3250 g

8 Calculate the total mass. Answer in both grams and kilograms.

a 30 g + 3.3 kg + 40 g = ________ g = ______ kg

b 7.6 kg – 400 g – 0.5 kg = ________ g = ______ kg

c 5 kg + 827 g = ________ g = ______ kg

d 210 g × 8 = ________ g = ______ kg

e 10.4 g ÷ 4 = ________ g = ______ kg

f 1.43 kg × 9 = ________ g = ______ kg

g Average mass of 32 kg, 41 kg, 38 kg and 43 kg = ________ g = ______ kg

h 36.6 kg ÷ 6 = ________ g = ______ kg

9 Complete the table.

	Item	Mass of each item	How many items?	Total Mass	
				grams	kilograms
a	pear	155 g	8		
b	mango	240 g	6		
c	can of tuna	425 g	5		
d	jar of jam	350 g	3		
e	chocolate bar	255 g	2		
f	pack of mints	64 g	4		
g	jar of olives	848 g	3		
h	muesli bar	126 g	7		

REVIEW

10 Write in short form.

a 8 grams ______ c 26 grams ______ e 173 grams ______

b 12 grams ______ d 59 grams ______ f 245 grams ______

11 Name five things measured in grams.

____

12 How many more grams to make one kilogram?

a 240 g ______ e 525 g ______ i 455 g ______

b 725 g ______ f 610 g ______ j 343 g ______

c 330 g ______ g 898 g ______ k 908 g ______

d 995 g ______ h 100 g ______ l 788 g ______

Use the items below to complete the following question.

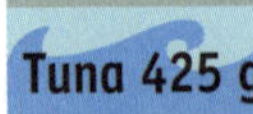

Protein bar 25 g

13 Complete the table.

	Item	Weight	Grams to make 1 kg
a	packet of chips		
b	tin of tuna		
c	protein bar		
d	bag of lollies		
e	jar of jam		

14 What is the mass? Use the table above. Answer in both grams and kilograms.

a 5 tins of tuna = ________ g = ______ kg

b 8 packets of chips = ________ g = ______ kg

c 6 bags of lollies = ________ g = ______ kg

d 9 protein bars = ________ g = ______ kg

e 4 jars of jam = ________ g = ______ kg

CATCH UP MATHS YEAR 6 BOOK B © PASCAL PRESS ISBN: 9781925726190

15 How many grams?

a	4.3 kg ______	e	6.805 kg ______	i	2.87 kg ______
b	$7\frac{1}{4}$ kg ______	f	$9\frac{3}{4}$ kg ______	j	1.675 kg ______
c	3.25 kg ______	g	10.030 kg ______	k	5.5 kg ______
d	5.02 kg ______	h	8.468 kg ______	l	6.0 kg ______

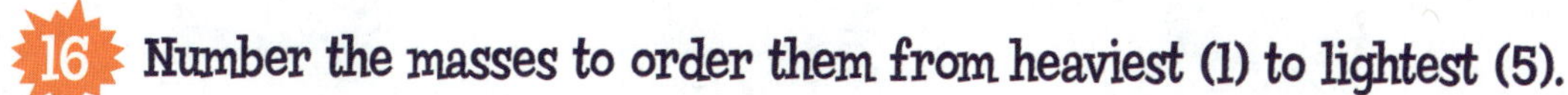

16 Number the masses to order them from heaviest (1) to lightest (5).

a	2 kg 500 g ☐	5200 g ☐	2505 g ☐	3595 g ☐	3 kg 955 g ☐
b	6495 g ☐	4 kg 695 g ☐	6 kg 549 g ☐	4 kg 950 g ☐	9564 g ☐
c	3 kg 420 g ☐	3402 g ☐	4 kg 340 g ☐	4300 g ☐	4430 g ☐

17 How much do these items weigh?

a apples ____ b bananas ____ c oranges ____ d cabbage ____

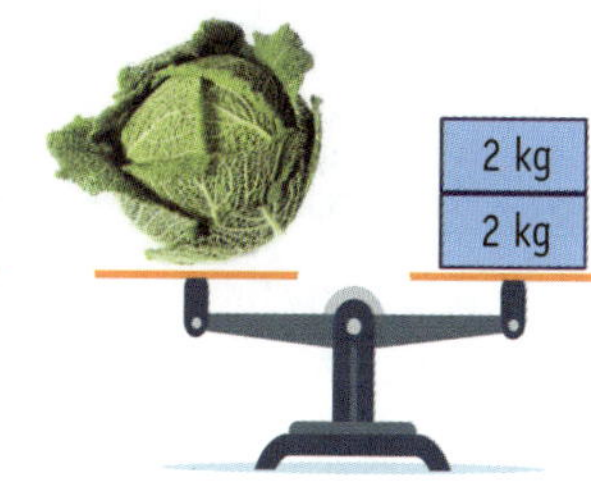

18 If the bananas above are put on the same side of the scale as the apples, how many cabbages will balance the scale? ________

19 Read the scales and write the mass.

a ________ b ________ c ________

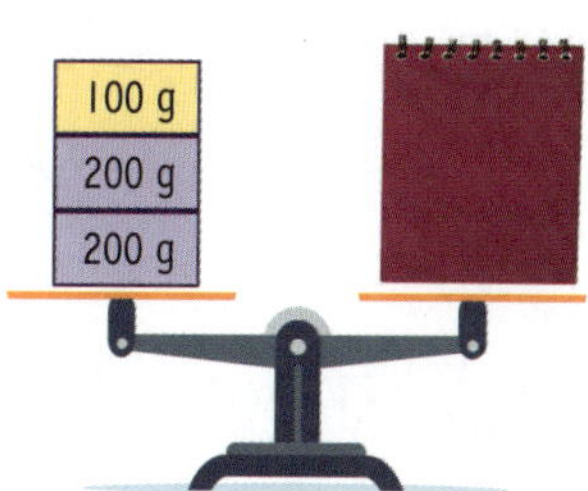

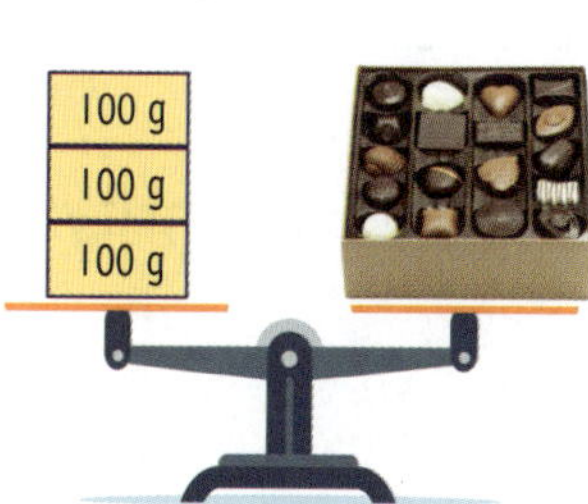

REVIEW

20 Read the scales and write the mass.

a ________ b ________ c ________ d ________

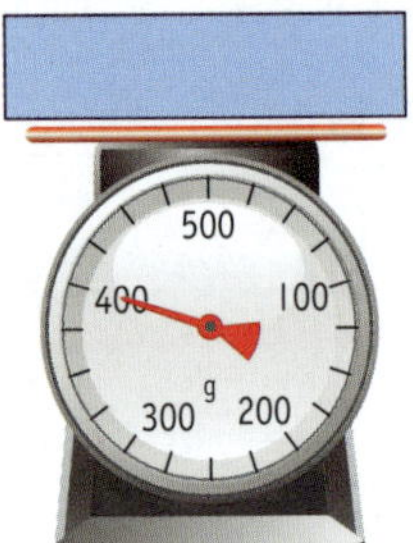

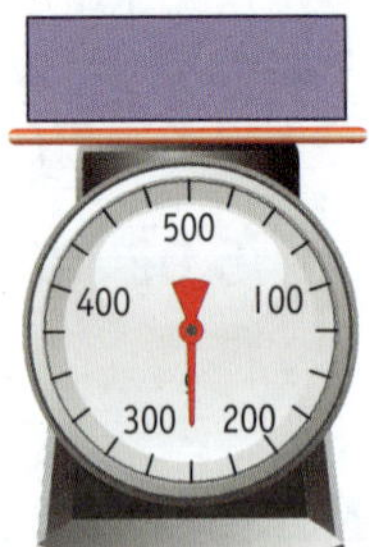

21 Complete the sentence.

A tonne is _____ kilograms.

22 Write the masses in short form.

a 6 tonnes _____

b 5 tonnes _____

c 24 tonnes _____

d 38 tonnes _____

e 135 tonnes _____

f 849 tonnes _____

23 How many kilograms?

a $\frac{1}{4}$ t ________

b $8\frac{3}{4}$ t ________

c $\frac{3}{4}$ t ________

d $8\frac{1}{4}$ t ________

e $4\frac{1}{2}$ t ________

f $\frac{1}{2}$ t ________

g $5\frac{1}{4}$ t ________

h 1 t ________

24 How many of each weight would equal one tonne?

a 50 kg Quantity: ___

b 200 kg Quantity: ___

c 1000 kg Quantity: ___

d 250 kg Quantity: ___

e 100 kg Quantity: ___

f 500 kg Quantity: ___

  ISBN: 9781925726190

25 Complete the tables.

	Grams	Kilograms	Tonnes
a	20 000		
b		60	
c			0.09
d	50 000		
e			0.045
f		85	

	Grams	Kilograms	Tonnes
g	15 500		
h		3000	
i			7
j	8 200 000		
k		6.350	
l			9.11

26 Fill in the missing information.

	Item	Net Mass	Packaging	Gross Mass
a	cookies	350 g	25 g	
b	cereal	925 g		970 g
c	olives		100 g	825 g
d	cake	550 g		585 g
e	tuna		35 g	525 g
f	box of chocolates	420 g	20 g	
g	bread	640 g		645 g
h	potato chips	175 g		185 g

27 Circle the best buy.

a	200 g for $2.20	cheese	1 kg for $10.00
b	250 g for $2.45	apples	1 kg for $10.00
c	500 g for $9.95	cookies	250 g for $5.00
d	100 g for $2.25	dog food	250 g for $5.50
e	100 g for $5.50	coffee	500 g for $28.00
f	500 g for $8.90	tea	250 g for $4.50
g	1 kg for $3.85	potato fries	750 g for $2.95
h	1 kg for $9.00	hot dogs	300 g for $2.50

O'CLOCK AND HALF PAST

Analogue clocks are divided into 12 parts.
Each part is equal to one hour passing.
There are two hands: a longer minute hand and a shorter hour hand.

It is 7 o'clock.

Minute hand (big hand) points to the 12.

Hour hand (little hand) points to the 7.

It is half past 7.

Hour hand is halfway between the 7 and the 8 because it is half past 7.

Minute hand points to the 6.

The minute hand pointing to the 12 tells you it is o'clock.

Imagine the clock face has been cut in half.

Digital clocks use numbers to show the time. They have no hands.

The numbers on this side show the hours. →

← The number on this side show the minutes past the hour.

It is seven o'clock.

7 hours

30 minutes past the hour

It is seven thirty.

Examples: Join the analogue clocks to the correct digital clock.

a

b

c

d

3:00

6:30

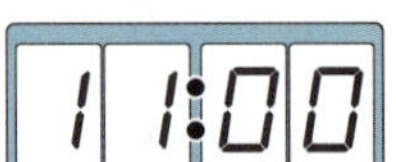

Check the answers on the video!

Your turn

Draw the missing minute hand to show the time.

half past 1

a 1 o'clock

b 9 o'clock

c half past 10

SELF CHECK Tick how you feel

Got it!	Need help...	I don't get it

Check your answers

How many did you get correct?

CATCH UP MATHS YEAR 6 BOOK B © PASCAL PRESS ISBN: 9781925726190

PRACTICE

Draw the missing hour hand to show the time.

4 o'clock | **a** half past 5 | **b** 10 o'clock | **c** half past 9

Show the times in Question 1 on the digital clocks below.

a

b

c

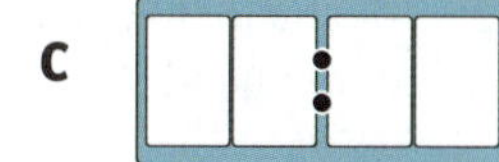

Use the same colour to colour the boxes that show the same time.

Analogue		**a**	**b**	**c**	**d**
How we read analogue time	six o'clock	two o'clock	half past 11	half past 4	five o'clock
Digital	4:30	6:00	2:00	11:30	5:00
How we read digital time	six o'clock	five o'clock	four thirty	two o'clock	eleven thirty

Complete the times.

1 o'clock

1 o'clock

a

half past 12

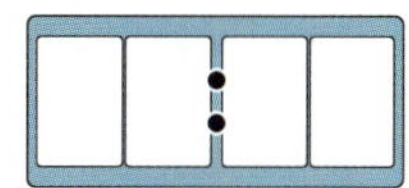

______

b

______

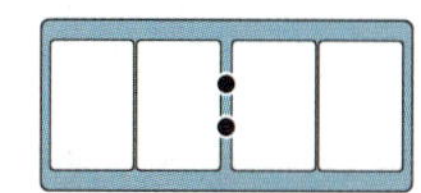

8 o'clock

c

______

______

QUARTER PAST AND QUARTER TO

Analogue times

It is a quarter past 1.

The minute hand (big hand) is one-quarter of the way around the clock. It points to the 3.

The hour hand (small hand) is one-quarter of the way between the 1 and the 2.

It is a quarter to 2.

The minute hand is three-quarters of the way around the clock. It points to the 9.

The hour hand is three-quarters of the way between the 1 and the 2.

Imagine the clock face has been cut into quarters.

$\frac{1}{4}$ hour = 15 minutes

Digital times

The numbers on this side show the hours. →

← The number on this side show the minutes past the hour.

It is one fifteen.

1 hour

45 minutes past the hour

It is one forty-five.

Examples: Join the analogue clocks to the correct digital clock.

a 

b 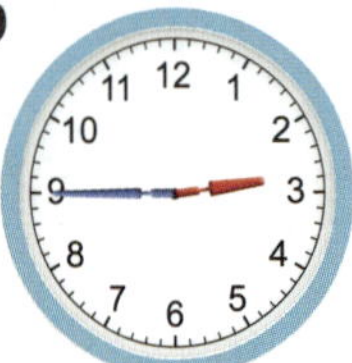

c 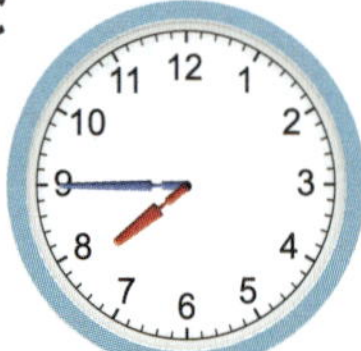

d 

Check the answers on the video!

Your turn

Draw the missing minute hand to show the time.

● $\frac{1}{4}$ past 9

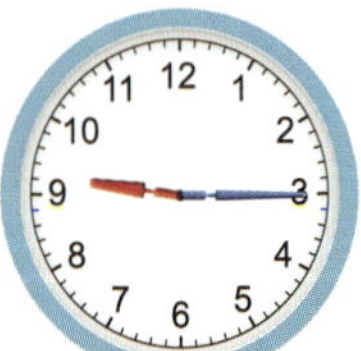

a $\frac{1}{4}$ to 6

b $\frac{1}{4}$ past 4

c $\frac{1}{4}$ to 7

SELF CHECK Tick how you feel

Got it!	Need help...	I don't get it

Check your answers

How many did you get correct?

CATCH UP MATHS YEAR 6 BOOK B © PASCAL PRESS ISBN: 9781925726190

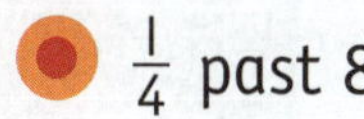

PRACTICE

1 Show the time on the clocks.

● $\frac{1}{4}$ past 8 | **a** $\frac{1}{4}$ to 4 | **b** $\frac{1}{4}$ to 5 | **c** $\frac{1}{4}$ past 7

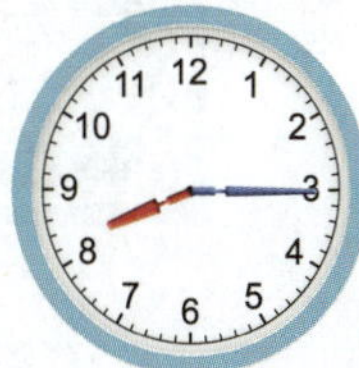   

2 Show the times in Question 1 on the digital clocks below. Then write how you would say the digital times.

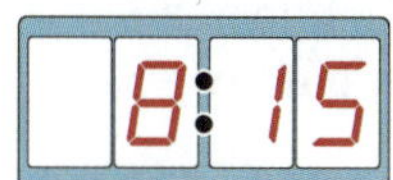

8 fifteen

a 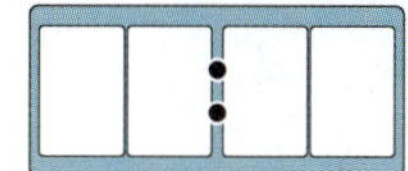 ______

b 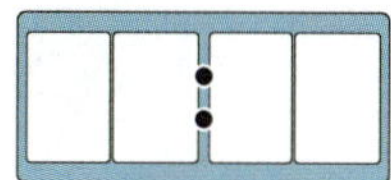 ______

c 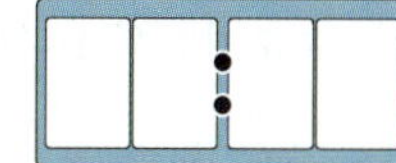 ______

3 Use the same colour to colour the boxes that show the same time.

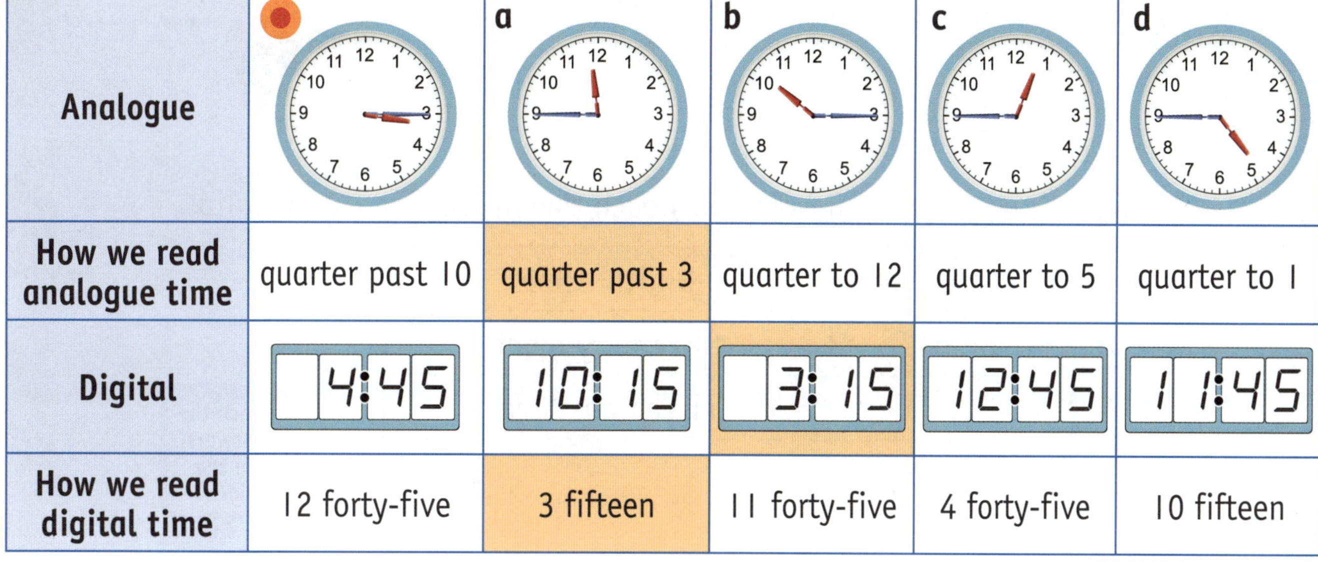

	●	a	b	c	d
Analogue					
How we read analogue time	quarter past 10	quarter past 3	quarter to 12	quarter to 5	quarter to 1
Digital	4:45	10:15	3:15	12:45	11:45
How we read digital time	12 forty-five	3 fifteen	11 forty-five	4 forty-five	10 fifteen

4 Write the time 15 minutes later. Use digital time.

● 1:00 1:15
a 2:15 ______
b $\frac{1}{4}$ past 10 ______
c $\frac{1}{4}$ to 12 ______
d $\frac{1}{4}$ to 3 ______
e 3:45 ______
f 6:15 ______
g $\frac{1}{4}$ past 11 ______
h 4:20 ______

5 Write the time 15 minutes before. Use digital time.

● 2:15 2:00
a $\frac{1}{4}$ to 3 ______
b $\frac{1}{4}$ to 12 ______
c 4:40 ______
d 6:35 ______
e 25 past 1 ______

INTERVALS

A time interval is the amount of time between two given points. For example, the time interval between two o'clock and six o'clock is four hours.

SCAN to watch video

The analogue time is 49 past 4 or 11 to 5.

The digital time would be 4 forty-nine.

If the big (minute) hand is on the 'to' side, count from the 12 to the **left**.

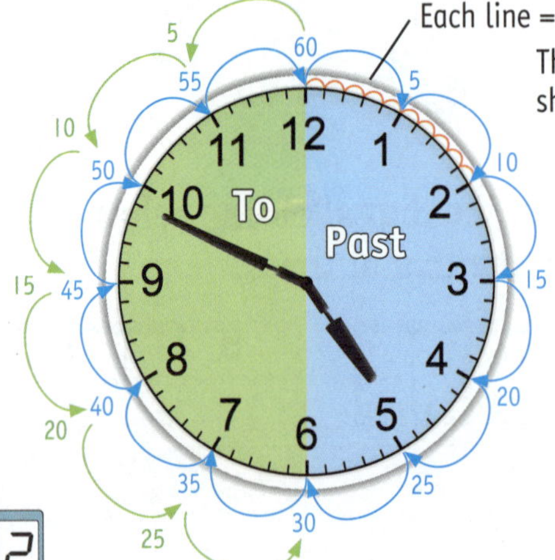

The black numbers show what hour it is.

There are 60 lines around the clock because there are 60 minutes in an hour.

If the big hand is on the 'past' side, count from the 12 to the **right**.

05:42

It is 18 to 6 or 42 past 5.

It is 24 past 10.

Important! When you convert to digital time, count from the 12 to the right, even if the big hand is on the 'to' side.

Examples: Write how you would read each time in analogue and digital.

	Analogue time	Digital time
a	12 to 4	three forty-eight
b		eleven-oh-three
c	3 to 11	
d	16 past 2	

Check the answers on the video!

Your turn

Show the times in the Examples on the clocks below.

a

3:48

b

c

d

SELF CHECK Tick how you feel

Got it!  Need help...  I don't get it 

Check your answers

How many did you get correct?

CATCH UP MATHS YEAR 6 BOOK B © PASCAL PRESS ISBN: 9781925726190

PRACTICE

1 Complete the times.

23 to 2

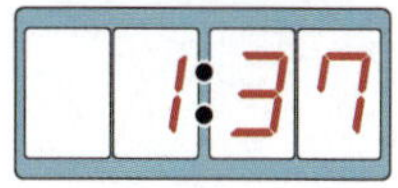

1 thirty-seven

a 

b 

c 

2 How would you read these times?

- 12:02 twelve-oh-two
- a 7:04 ________
- b 9:07 ________
- c 11:01 ________
- d 10:03 ________
- e 6:08 ________

3 Complete the table to help you better understand digital time.

	Digital time	How we read it	What it means
●	8:39	eight thirty-nine	21 minutes to 9
a	6:53		
b		eleven fourteen	
c			16 minutes to 5

4 Show the digital times on the analogue clocks.

12:06

a 8:57

b 2:33

c 4:24

5 Complete the table.

	Time	1 min before	5 mins before	5 mins after	13 mins after
●	10:14	10:13	10:09	10:19	10:27
a	11:45				
b	3:21				
c	5:05				

TIME CONVERSIONS

'Convert' means 'change into'. When we convert times, we change them – for example, from minutes to hours or minutes to seconds.

60 seconds = 1 minute

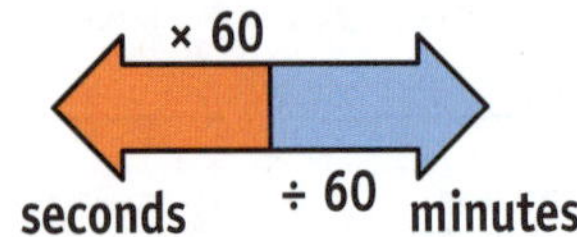

Seconds	Minutes
180	3
300	5

60 minutes = 1 hour

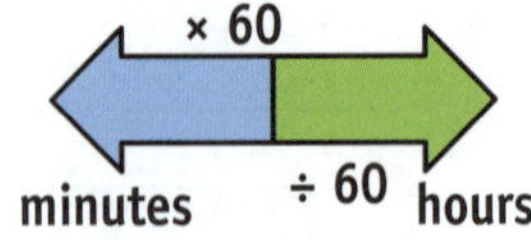

Minutes	Hours
240	4
60	1

Examples: Write the missing numbers.

a 2 minutes = 120 seconds (× 60)

b 8 minutes = ____ seconds (× 60)

c 300 seconds = ____ minutes (÷ 60)

d 120 seconds = ____ minutes (÷ 60)

e 3 hours = ____ minutes (× 60)

f 300 minutes = ____ hours (÷ 60)

g 720 minutes = ____ hours (÷ 60)

h 9 hours = ____ minutes (× 60)

Fill in the missing information.

● 1 minute = 60 seconds (× 60)

a 240 seconds = ____ minutes (__ 60)

b 5 minutes = ____ seconds (__ 60)

c 60 minutes = ___ hour (__ 60)

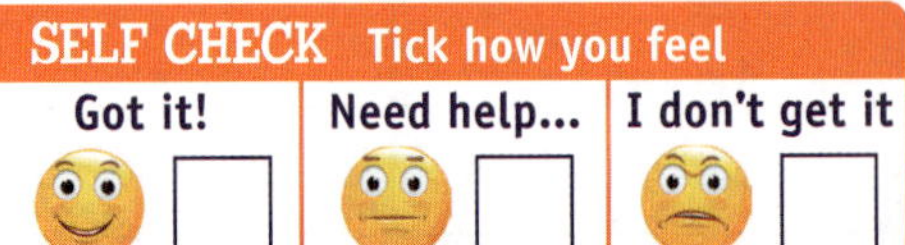

Check your answers
How many did you get correct?

CATCH UP MATHS YEAR 6 BOOK B © PASCAL PRESS ISBN: 9781925726190

PRACTICE

1 Complete the tables.

	Seconds	Minutes
	60	1
a	180	
b	390	
c	780	
d	600	
e	300	

	Minutes	Seconds
	1	60
f	7	
g	11	
h	$9\frac{1}{2}$	
i	12	
j	14	

2 Convert the minutes to hours.

- 60 minutes = __1__ hour
- a 420 minutes = _____ hours
- b 960 minutes = _____ hours
- c 660 minutes = _____ hours
- d 330 minutes = _____ hours
- e 450 minutes = _____ hours
- f 540 minutes = _____ hours
- g 210 minutes = _____ hours
- h 1320 minutes = _____ hours
- i 1080 minutes = _____ hours

3 Convert the hours to minutes.

- 1 hour = __60__ minutes
- a $2\frac{1}{2}$ hours = ________ minutes
- b 6 hours = ________ minutes
- c 10 hours = ________ minutes
- d $12\frac{1}{2}$ hours = ________ minutes
- e $8\frac{1}{2}$ hours = ________ minutes
- f 4 hours = ________ minutes
- g 17 hours = ________ minutes
- h 19 hours = ________ minutes
- i 24 hours = ________ minutes

4 Complete the conversion table. Hint: You will need a calculator.

	Seconds	Minutes	Hours
	180	3	0.05
a		5	
b	600		
c			0.25
d	1800		
e		45	

	Seconds	Minutes	Hours
f			1
g			3
h		480	
i			12
j	86 400		
k		2160	

24-HOUR TIME

There are 24 hours in one day. 12 hours are from midnight to midday and 12 hours are from midday to midnight.

12:00 midday or 00:00 midnight
01:00 or 13:00
02:00 or 14:00
03:00 or 15:00
04:00 or 16:00
05:00 or 17:00
06:00 or 18:00
19:00 or 07:00
20:00 or 08:00
21:00 or 09:00
22:00 or 10:00
23:00 or 11:00

The blue times are the 24-hour times you would read before midday (am).

The green times are the 24-hour times you would read after midday (pm).

Examples: Use blue to colour the times before midday and use green to colour the times after midday.

01:00	08:12	12:07	19:23	21:58
15:35	11:49	16:14	03:00	23:00

Write different times in 24-hour time.

Before midday	After midday
02:15	14:15

SELF CHECK Tick how you feel

Got it!	Need help...	I don't get it
☐	☐	☐

Check your answers
How many did you get correct? ☐

CATCH UP MATHS YEAR 6 BOOK B © PASCAL PRESS ISBN: 9781925726190

PRACTICE

1 Are these 24-hour times am or pm?

●	02:35	am	c	21:48	___	f	18:38	___	i	06:45	___
a	16:43	___	d	19:23	___	g	03:17	___	j	23:24	___
b	10:57	___	e	14:45	___	h	08:57	___	k	11:15	___

2 Convert these 24-hour times to am or pm times.

●	13:03	1:03 pm	c	22:56	________	f	11:48	________
a	06:16	________	d	07:12	________	g	15:44	________
b	18:17	________	e	04:55	________	h	08:27	________

3 Convert these times to 24-hour time.

●	5:01 am	05:01	c	9:10 am	______	f	8:48 pm	______
a	10:54 am	______	d	11:12 pm	______	g	3:07 pm	______
b	1:27 am	______	e	6:43 pm	______	h	7:26 pm	______

4 Use 24-hour time to complete the tables.

	1 hour before	Time	1 hour after
●	05:14	6:14 am	07:14
a		9:03 pm	
b		2:42 pm	
c		11:20 am	
d		3:51 pm	

	12 minutes before	Time	12 minutes after
●	01:07	1:19 am	01:31
e		7:47 am	
f		12:58 pm	
g		5:36 am	
h		10:45 pm	

5 Write T for true or F for false.

● [T] 03:25 = 3:25 am

a [ ] 13:50 is the same as 10 to 4.

b [ ] Midday is 12:00.

c [ ] 19:50 is in the evening.

d [ ] Dinner time could be 6:00.

e [ ] 11:15 = 11:15 am

f [ ] Some schools start at 21:00.

g [ ] Lunch time could be 12:30.

h [ ] Midnight is 12:00.

i [ ] 21:59 = 9:59 pm

THE CALENDAR

A calendar shows the twelve months of the year in order. It tells us what year, month, date and day of the week it is.

Hany's Calendar

April 2025						
S	**M**	**T**	**W**	**T**	**F**	**S**
		1 April Fool's Day	2	3	4	5
6	7	(8)	9	10 Lysa's B'day	11	12
13	14	15	16	17	18 Good Friday	19
20	21	22	23 World Book Day	24	25	26
27	28 Back to school	29 Tim's B'day	30			

The number 8 is circled on the calendar.
Read this date as Tuesday the 8th of April.

Days of the Week

S	Sunday
M	Monday
T	Tuesday
W	Wednesday
T	Thursday
F	Friday
S	Saturday

Months of the Year

Month	No. days
January	31
February	28 (29)
March	31
April	30
May	31
June	30
July	31
August	31
September	30
October	31
November	30
December	31

April is the 4th month of the year.

CATCH UP MATHS YEAR 6 BOOK B © PASCAL PRESS ISBN: 9781925726190

Example 1: How many days in these months?

a April 30 c August ___

b June ___ d October ___

Example 2: How are these days shortened on Hany's calendar?

a Friday F c Thursday __

b Monday __ d Sunday __

Example 3: What day of the week are these dates?

a 9th April 2025 Wednesday c 30th April 2025 ___________

b 25th April 2025 ___________ d 1st April 2025 ___________

Example 4: How many of each of these days are there in April 2025?

a Saturdays 4 c Tuesdays ___

b Sundays ___ d Thursdays ___

Check the answers on the video!

Your turn

Use Hany's calendar to answer the questions.

1 What day are these events?

- Lysa's birthday Thursday

a Tim's birthday ______________

b World Book Day ______________

c Good Friday ______________

2 Mark these dates on Hany's calendar.

- 1st April 2025: April Fool's Day

a 20th April 2025: Easter Sunday

b 12th April 2025: Lysa's Party

c 11th April 2025: last day of term

d 25th April 2025: Anzac Day

SELF CHECK Tick how you feel

Got it!	Need help...	I don't get it
☐	☐	☐

Check your answers

How many did you get correct? ☐

PRACTICE

1 This calendar is for February 2026. Fill in all the missing information.

___________ (month), ________ (year)						
S				T		S
1	2					
	Maths test					
						28

2 Mark these events on the calendar above.

- 9th February: Maths test
- **a** 14th February: Valentine's Day
- **b** 22nd February: Amy's birthday
- **c** 11th February: Jai's birthday
- **d** 19th February: Haircut 11:00 am
- **e** 27th February: Dentist at 4:00 pm
- **f** 13th February: Jai's party 7:00 pm
- **g** 28th Lunch out for Amy: 12 noon

3 Use the calendar above to complete the following.

- How many days are there in February 2026? 28
- **a** What is the next year after 2026 that February will have 29 days? ______
- **b** What was the day and date of the last day in January 2026?

  ____
- **c** What is the day and date of the first day in March 2026?

  ____
- **d** How many weekdays in February 2026? ______
- **e** How many weekend days in February 2026? ______

CATCH UP MATHS YEAR 6 BOOK B © PASCAL PRESS ISBN: 9781925726190

4 What season is February?

a in the northern hemisphere ______________

b in the southern hemisphere ______________

5 What is the date?

● one week from Wednesday, 10th February Wednesday, 17th February

a three weeks from Monday, 2nd February ______________

b one fortnight from Saturday, 14th February ______________

c one week after Wednesday, 11th February ______________

6 How many of these days are there in February 2026?

● Mondays 4

a Wednesdays ___

b Tuesdays ___

c Thursdays ___

d Fridays ___

e Saturdays ___

7 What day is it?

● 2nd February Monday

a 15th February ______________

b 26th February ______________

c 10th February ______________

d 20th February ______________

e 28th February ______________

f 16th February ______________

g 21st February ______________

h 17th February ______________

i 6th February ______________

8 How many weeks are there in February 2026? ______

9 What is the date one week later?

● 1st February 8th February

a 14th February ______________

b 21st February ______________

c 17th February ______________

10 What is the date one fortnight later?

● 5th February 19th February

a 9th February ______________

b 13th February ______________

c 11th February ______________

TIMETABLES

A timetable is a chart or schedule that tells you when something is due to happen.

Timetables are also called schedules or programs.

Sandiland Leisure Centre – Morning Program

Weekday	Mon	Tue	Wed	Thu	Fri	Weekend	Sat	Sun
5:45 am		**Pump** Julie			**Pump** Julie	7:00 am	**Yoga** Megan	
6:00 am				**Boxing** Boris				
8:00 am		**Master Pump** Deb		**Power Hour** Julie	**Rhythm Fit** Brock	8:00 am	**Rhythm Fit** Leonie	**Body Attack** Tanya
8:15 am	**Body Balance** Emily		**Pilates** Jude					
9:00 am					**Metafit** Brock	9:00 am	**Body Step** Leonie	**Pump** Tanya
9:15 am	**Power Hour** Leonie	**Body Step** Gordie	**Body Attack** Leonie	**Pump** Julie				
9:30 am					**Meta Power** Brock			
10:30 am	**Body Step** Leonie	**Pump** Gordie		**Rhythm Fit** Brock	**Pump** Lachlan	10:15 am	**Pump** Geri	
11:15 am			**Rhythm Fit** Glenys					
11:30 am				**Body Balance** Emily				
11:45 am		**Body Balance** Jude						
12:15 pm	**Pump** Julie		**Pump** Katie		**Body Balance** Jude			

All classes are 60 minutes except shaded classes, which are 55 minutes.

Many timetables use 24-hour time.

CATCH UP MATHS YEAR 6 BOOK B © PASCAL PRESS ISBN: 9781925726190

Example 1: Use the timetable to complete the following.

a How long is the Pump Class with Geri on Saturday? 55 minutes

b Where is this timetable from? ____________________

c How long are most classes? __________

d Do classes run every day of the week? ____

Example 2: What time does the earliest class start on these days?

a Monday 8:15 am

b Tuesday ______

c Saturday ______

d Sunday ______

Example 3: What time does the last class finish on these days?

a Wednesday 1:15 pm

b Thursday ______

c Saturday ______

d Sunday ______

e Tuesday ______

f Monday ______

Check the answers on the video!

Your turn

1 How many classes do these teachers have?

- Lachlan 1

a Deb ___

b Tanya ___

c Julie ___

d Glenys ___

e Brock ___

f Jude ___

g Gordie ___

2 How many of each of these sessions are there?

- Pump 9

a Body Balance ___

b Pilates ___

c Yoga ___

d Body Step ___

e Rhythm Fit ___

f Body Attack ___

g Meta Power ___

h Metafit ___

i Master Pump ___

SELF CHECK Tick how you feel

Got it!	Need help...	I don't get it
☐	☐	☐

Check your answers

How many did you get correct? ☐

PRACTICE

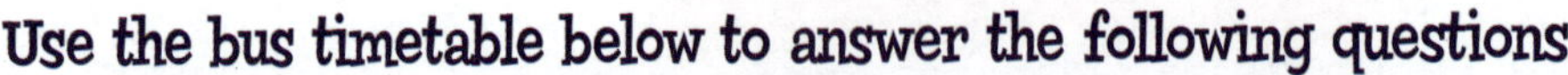

Use the bus timetable below to answer the following questions.

Minanda to Main			Monday–Friday			
Minanda			09:08	10:08	11:08	12:08
Westfield			09:11	10:11	11:11	12:11
Baring	08:00	08:35	09:16	10:16	11:16	12:16
Gannons				10:24	11:24	12:24
Pilli Lilli	08:04	08:39	09:20	10:27	11:27	12:27
Southland	08:07		09:23			12:30
Crescent	08:11	08:41	09:29	10:29	11:29	12:34
Fielder	08:13	08:43	09:31	10:31	11:31	12:36
Dolans	08:16	08:46	09:34	10:34	11:34	12:39
Gow	08:19	08:49	09:37	10:37	11:37	12:42
South	08:29	09:00	09:46	10:46	11:46	12:51
East		09:04	09:50	10:50	11:50	12:55
Main		09:09	09:55	10:55	11:55	13:00

1 How many stops do these buses have?

- 09:08 bus from Minanda 11
- **b** 10:08 bus from Minanda ___
- **a** 08:00 bus from Baring ___
- **c** 08:35 bus from Baring ___

2 If I boarded the 11:08 bus at each location, how many stops to reach Fielder?

- Crescent 1
- **b** Baring ___
- **d** Minanda ___
- **a** Gannons ___
- **c** Westfield ___
- **e** Pilli Lilli ___

3 If I need to get to Main by each of these times, what bus should I catch from Baring?

- 10 am 09:16
- **b** 12 pm ______
- **a** 11 am ______
- **c** 1 pm ______

4 How many times does this bus stop at these locations?

- Westfield 4
- **b** Southland ___
- **d** Main ___
- **a** Gannons ___
- **c** Baring ___
- **e** South ___

5 How many times before 1 pm does this bus stop at each location?

- South 6
- **b** Main ___
- **a** Gannons ___
- **c** Gow ___

CATCH UP MATHS YEAR 6 BOOK B © PASCAL PRESS ISBN: 9781925726190

Kirra's High School Timetable

Term 1

Monday		Tuesday		Wednesday		Thursday		Friday	
Period 0	8:00–8:55			Period 0	8:00–8:55	Period 0	8:00–8:55	Period 0	8:00–8:55
Period 1	9:00–10:05	Period 1	9:00–10:05	Period 1	9:00–9:50	Period 1	9:00–10:00	Period 1	9:00–10:05
Period 2	10:05–11:05	Recess	10:05–10:25	Period 2	9:50–10:35	Period 2	10:00–10:55	Period 2	10:05–11:05
						Assembly	10:55–11:10		
Recess	11:05–11:25	Period 2	10:25–11:25	Recess	10:35–10:55	Recess	11:10–11:30	Recess	11:05–11:25
Period 3	11:25–12:25	Period 3	11:25–12:25	Period 3	10:55–11:40	Period 3	11:30–12:25	Period 3	11:25–12:25
Period 4	12:25–1:25	Lunch 1	12:25–12:45	Period 4	11:40–12:25	Period 4	12:25–1:25	Period 4	12:25–1:25
Lunch 1	1:25–1:45	Lunch 2	12:45–1:05	Lunch 1	12:25–12:45	Lunch 1	1:25–1:45	Lunch 1	1:25–1:45
Lunch 2	1:45–2:05	Period 4	1:05–2:05	Lunch 2	12:45–1:05	Lunch 2	1:45–2:05	Lunch 2	1:45–2:05
Period 5	2:05–3:05	Teacher planning session		Sport		Period 5	2:05–3:05	Period 5	2:05–3:05

Use the highschool timetable to answer these questions.

6 How many days does this timetable run for? ______

7 How many minutes does each of these periods last?

- ● Period 1 on Monday __65__
- a Period 0 on Monday ______
- b Period 3 on Wednesday ______
- c Period 4 on Thursday ______
- d Period 5 on Friday ______
- e Period 1 on Thursday ______

8 What is the day and time?

- a Assembly ____________________
- b Sport ____________________
- c Teacher planning session ____________________

9 What time does lunch start?

- a Monday ________
- b Tuesday ________
- c Thursday ________

10 How long does lunch go for?

- a Wednesday ________
- b Tuesday ________
- c Friday ________

11 What time does recess start?

- a Tuesday ________
- b Thursday ________
- c Friday ________

TV Guide: 24 January 2025

Time	ATV	BTV	CTV	6	6two	6mate	Vie	Yeah!
2 pm	2:00 pm Grand line 2:30 pm Waste War	2:06 pm Dora the Builder 2:18 pm Get Dirty TV 2:30 pm Dash 2:42 pm Numberhouse 2:48 pm Little D and Big Bro	2:05 pm Danger Dog 2:16 pm Thundercats 2:39 pm Deadly Snakes		2:00 pm All Things 2:30 pm Great Adventures	2:00 pm Storage Solutions 2:30 pm Storage Solutions	2:00 pm House Finders 2:30 pm Flip N Move	2:29 pm Dance Dads
3 pm	3:31 pm Everyone's a Star	3:03 pm Curious Cate 3:31 pm Play Story 3:59 Bananas in Queensland	3:07 pm Teenage Surfers 3:35 pm World Histories		3:30 pm Escape to the City	3:00 pm Night Lightning	3:32 pm Fix my House	3:27 pm Movie: Mart Patrol
4 pm	4:00 pm Football W League	4:12 pm Ready, Steady, Go! 4:25 pm Dana and Charlie 4:36 pm Kiddies 4:48 pm Play Story 4:57 pm Gals	4:05 pm Operation Wreck 4:18 pm Afternoon Adventure 4:30 pm Almost Always	4:00 pm Amazing Backyards	4:30 pm Escape to the City	4:00 pm Bushfire Battle 4:30 pm Graveyard Homes	4:32 Homes Australia	
5 pm		5:10 pm Dinosaur Bus 5:25 pm Firegirl Ally 5:35 pm Jenny Rabbit 5:51 pm Penny Pig 5:57 pm Sally and Ben's Castle	5:00 pm The Headmaster 5:30 pm School of Charm 5:52 pm Sadie Sylvester	5:00 pm Six News at 5 5:30 pm Sydney Getaways	5:30 pm Escape to the City	5:30 pm Counting Cows	5:33 pm House Finders	5:25 pm Love Heart

12 Answer the following about the TV Guide.

a What is the date of this TV Guide? ______________________

b How many channels does this TV Guide show? _____

c What time does the TV Guide start? ________

13 What time does each of these shows start?

- Dinosaur Bus 5:10 pm
- **a** Fix my House ________
- **b** Love Heart ________
- **c** Get Dirty TV ________
- **d** Grand line ________
- **e** Counting Cows ________

CATCH UP MATHS YEAR 6 BOOK B © PASCAL PRESS ISBN: 9781925726190

14 What channel are each of the shows in Question 13?

- ● BTV
- a ____________
- b ____________
- c ____________
- d ____________
- e ____________

15 How long does each of these shows run for?

- ● Teenage Surfers 18 mins
- a Dance Dads ________
- b Football W League ________
- c Curious Cate ________
- d Penny Pig ________
- e Graveyard Homes ________

Use the timetable below to answer the following questions.

6K's Class Timetable

	Monday	Tuesday	Wednesday	Thursday	Friday
9:15	Literacy	Literacy	Language	Testing Spelling and Maths	Sport
10:15			Writing		
11:15	Lunch				
12:05	Maths	Maths	Maths	Maths	Assembly
1:05	Writing		Art	Library	Wellbeing
1:50	Fruit break				
2:10	Technology	Nature	Literacy	Science	Language
3:15					

16 What time does school start and finish?

Start: ______ Finish: ______

17 How many sessions per week in each of these subjects?

- ● Art 1
- a Literacy ___
- b Maths ___
- c Language ___
- d Technology ___
- e Science ___

18 How many minutes do each of these sessions go for?

- ● Library 45
- a Maths on Tuesday ____
- b Wellbeing ____
- c Literacy on Monday ____

TIMELINES

A timeline is a number line that shows special dates in order, starting from the earliest date.

SCAN to watch video

Aria's Timeline

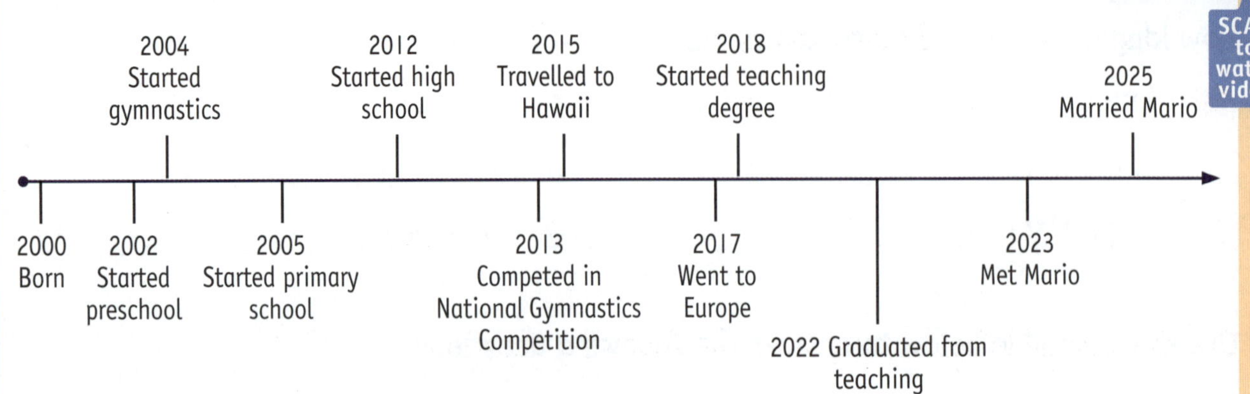

Examples: In what year did these events happen?

a Started preschool 2002

b Travelled to Hawaii _____

c Met Mario _____

d Started primary school _____

1 What special event did Aria record for each of these years?

● 2000 Born

a 2005 ______________________

b 2025 ______________________

c 2013 ______________________

2 Add these events to Aria's timeline.

a 2009: Moved house

b 2019: Moved out of home

c 2024: Bought first house

d 2026: Had first baby

Check your answers
How many did you get correct?

CATCH UP MATHS YEAR 6 BOOK B © PASCAL PRESS ISBN: 9781925726190

PRACTICE

The History of Qantas

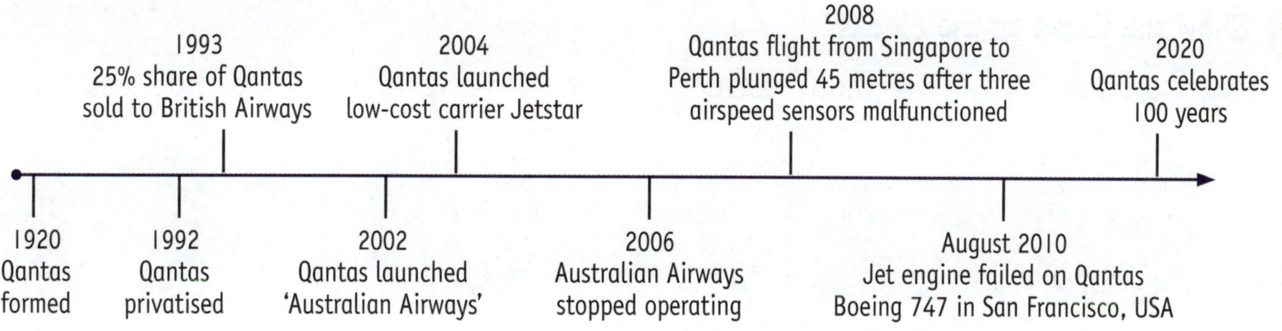

Source: sbs.com.au

Use the History of Qantas timeline above to answer the following.

1 **In what year did these events occur?**

- Qantas formed 1920
- **a** Qantas privatised ______
- **b** Jetstar launched ______
- **c** Qantas 100 years old ______

2 **What event happened at Qantas in these years?**

- 1993 25% share sold to British Airways
- **a** 2002 ______
- **b** 2006 ______
- **c** 2008 ______

3 **Draw arrows from the label to the correct spot on Mirri's Timeline**

Mirri's Timeline

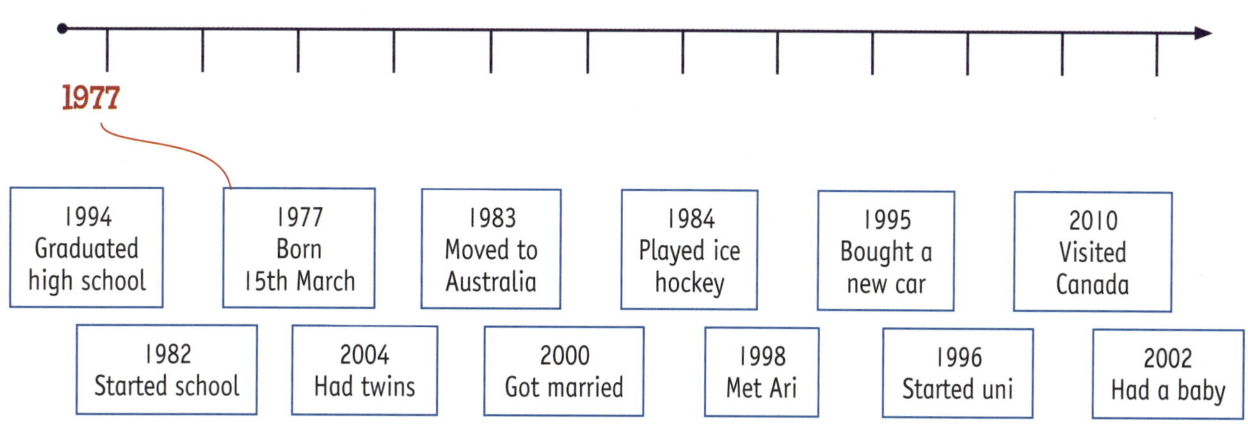

4 **Mark at least five special events in your life on the timeline below.**

TIME REVIEW

Show the times on the clocks.

a 3 o'clock

b half past 6

c 5 o'clock

d half past 12

Show the times in Question 1 on the digital clocks below.

a 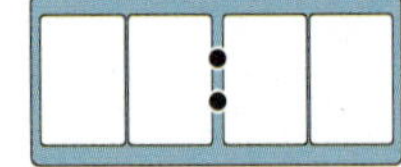

b 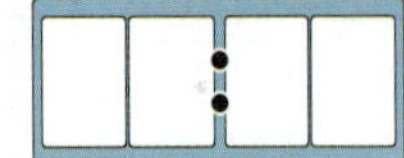

c 

d 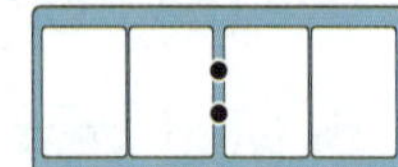

Use the same colour to colour the boxes that show the same time.

Analogue	a	b	c	d
How we read analogue time	quarter past 6	quarter past 2	quarter to 11	quarter to 12
Digital	10:45	11:45	6:15	2:15
How we read digital time	eleven forty-five	six fifteen	two fifteen	ten forty-five

Write the time 15 minutes after. Use digital time.

a 2:00 ______

b 10:45 ______

c $\frac{1}{4}$ past 9 ______

d $\frac{1}{4}$ to 10 ______

e $\frac{1}{4}$ to 12 ______

f 9:45 ______

Show the time on the digital clocks.

a

b

c

d

CATCH UP MATHS YEAR 6 BOOK B © PASCAL PRESS ISBN: 9781925726190

6 How would you read the time?

a ________ 9:32

b ________ 11:13

c ________ 12:58

d ________

7:45

7 Complete the table.

	Digital time	How we read it	What it means
a	9:27		
b		ten sixteen	
c			14 minutes to 5
d	7:23		
e			26 minutes past 1
f		twelve thirty-three	

8 What was the time 5 minutes before?

a 10:12 ______ c 3:36 ______ e 7:22 ______ g 9:43 ______

b 5:02 ______ d 4:41 ______ f 6:58 ______ h 11:09 ______

9 What time is 14 minutes after?

a 11:52 ______ c 6:29 ______ e 2:43 ______ g 3:45 ______

b 5:06 ______ d 9:18 ______ f 4:37 ______ h 10:01 ______

10 Complete the times.

a

half past 11

____

b

____

three forty-five

c

____

8:15

____

d

24 to 5

____

REVIEW

11 How would you say these times?

a ____________

b ____________

12:06

c ____________

d ____________

2:02

e ____________

f ____________

6:09

g ____________

h ____________

8:07

12 Complete the tables.

	Seconds	Minutes	Hours
a		5	
b		8	
c	540		
d			0.75
e		60	
f	900		
g		30	
h			1.5
i		10	

	Seconds	Minutes	Hours
j			2
k		300	
l			11
m	86 400		
n			48
o			3.5
p		390	
q			2.5
r	45 000		

13 Are these 24-hour times am or pm?

a 01:36 ____
b 11:47 ____
c 07:49 ____
d 21:25 ____
e 14:14 ____
f 16:10 ____
g 17:03 ____
h 12:33 ____
i 06:34 ____
j 22:36 ____
k 10:43 ____
l 23:47 ____

14 Convert these times to 24-hour time.

a 5:02 am ______
b 9:14 am ______
c 12:52 pm ______
d 2:40 pm ______
e 6:53 pm ______
f 11:31 am ______
g 4:23 pm ______
h 3:46 pm ______
i 1:01 am ______

CATCH UP MATHS YEAR 6 BOOK B © PASCAL PRESS ISBN: 9781925726190

Convert these 24-hour times to am or pm times.

a	12:16	______	d	22:11	______	g	11:58	______
b	05:06	______	e	08:28	______	h	03:39	______
c	10:40	______	f	14:27	______	i	17:09	______

Use 24-hour time to complete the tables.

	1 hour before	Time	1 hour after
a		7:11 am	
b		12:03 pm	
c		6:48 am	
d		8:59 pm	
e		1:32 pm	
f		10:34 am	

	13 mins before	Time	13 mins after
g		11:04 am	
h		8:53 pm	
i		2:42 pm	
j		9:31 am	
k		4:52 am	
l		6:41 pm	

This is a blank calendar for June 2024. Fill in the missing information.

____________ (month), ________ (year)						
M	T					
					1	2
						30

REVIEW

18 **Mark these events on the calendar in Question 17.**

a 9th June: Party 11 am

b 17th June: Haircut 3 pm

c 21st June: Lunch out 12 pm

d 5th June: Gran's birthday

e 10th June: Dentist 11:30 am

f 14th June: Maths test

g 11th June: Excursion

h 19th June: Sports carnival

19 **Write the day of the week for each event in Question 18.**

a ______________

b ______________

c ______________

d ______________

e ______________

f ______________

g ______________

h ______________

20 **Use the calendar above to complete the following.**

a What was the day and date of the last day in May 2024? ____________________

b What was the day and date of the first day of July 2024? ____________________

c What season is June in the northern hemisphere? ____________

d What season is June in the southern hemisphere? ____________

e How many weekdays are there in June 2024? _____

21 **What day is it?**

a 11th June ______________

b 22nd June ______________

c 29th June ______________

d 2nd June ______________

e 16th June ______________

f 1st June ______________

22 **What is the day and date?**

a one week after 14th June ________________________

b one week from 9th June ________________________

c one fortnight after 7th June ________________________

d a fortnight from 17th June ________________________

e three weeks from 3rd June ________________________

f three weeks before 30th June ________________________

  ISBN: 9781925726190

Use the bus timetable to answer the following.

Tomar to James				Monday–Friday			
Route	938	937	937	938	938	938	938
Tomar	13:08	14:08	15:08		15:08		15:48
Manly	13:16	14:16		15:30	15:16	16:00	15:56
Strand	13:24			15:38	15:24	16:12	
Clucky		14:28		15:44	15:30	16:20	
Els	13:37	14:35	15:23	15:51	15:37	16:27	
James	13:46	14:44	15:32	16:00	15:46	16:36	16:03

23 How many stops do these buses have?

a 938, 13:08 from Tomar _____

b 937, 14:08 from Manly _____

c 938, 15:48 from Tomar _____

d 937, 15:08 from Tomar _____

24 If I board the 15:30 938 bus at the following locations, how many stops would I have to James?

a Manly _____

b Clucky _____

c Els _____

d Strand _____

25 Which bus from Tomar Station should I catch to arrive at James by these times?

a 2:00 pm _______________

b 3:30 pm _______________

c 4:00 pm _______________

d 3:00 pm _______________

26 How many times before 4:00 pm does the 938 stop at these locations?

a Strand _____

b Manly _____

c Els _____

d James _____

e Clucky _____

27 How often does the 938 bus:

a leave from Tomar? _____

b travel to Clucky? _____

c travel to Strand? _____

d leave at 15:08? _____

28 How many stops on these routes?

a 938 at 15:30 _____

b 937 at 15:08 _____

c 938 at 15:08 _____

d 937 at 14:08 _____

  ISBN: 9781925726190

REVIEW

Use this spin studio timetable to complete the following.

Weekday	Mon	Tue	Wed	Thu	Fri	Weekend	Sat	Sun
5:45 am	Spin Deb		Gentle Spin Leoni		Spin Christi			
6:00 am		Spin Karen		Spin Emma				
7:00 am	Spin Jan							
8:15 am	Gentle Spin Bob		Spin Ray		Gentle Spin Ray	9:00 am		Gentle Spin Tanya
9:15 am	Spin Jeni	Spin Chris	Gentle Spin Deb	Spin Bernice	Spin Emma	10:00 am	Gentle Spin Mel	
12:15 pm		Gentle Spin Deb						
4:30 pm	Spin Emma					5:00 pm	Spin Tatiana	
5:15 pm			Gentle Spin Jeni	Spin Bob				
5:45 pm	Spin Leoni	Spin Mel						
6:30 pm								

29 How many classes run on these days?

a Monday _____

b Tuesday _____

c Friday _____

d the weekend _____

30 What time does the earliest class start on these days?

a Wednesday ________

b Thursday ________

c Sunday ________

d Saturday ________

e Tuesday ________

f Monday ________

31 What time does the latest class start on these days?

a Tuesday ________

b Friday ________

c Saturday ________

d Monday ________

e Wednesday ________

f Thursday ________

32 How many classes do each of these teachers lead?

a Deb _____

b Leoni _____

c Emma _____

d Mel _____

e Ray _____

f Bernice _____

g Bob _____

h Tanya _____

33 Which class is on at this time?

a Monday 5:45 pm ______________

b Thursday 9:15 am ______________

CATCH UP MATHS YEAR 6 BOOK B © PASCAL PRESS ISBN: 9781925726190

Olivia Newton-John

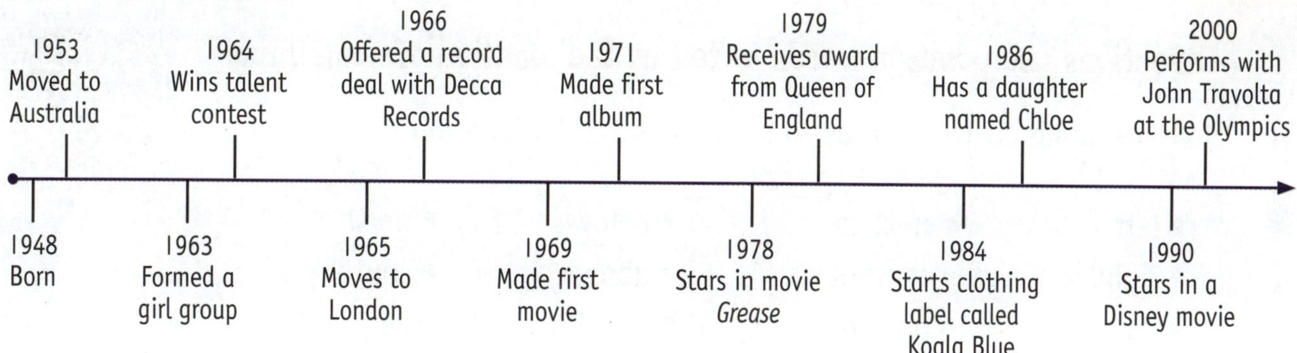

Use Olivia Newton-John's timeline above to answer the following.

34 **In what year did these events happen?**

a Moved to Australia _____

b Made her first movie _____

c Made her first album _____

d Performed at the Olympics _____

e Had a baby _____

35 **What event in Olivia's life happened in these years?**

a 1948 ______________________________

b 1963 ______________________________

c 1978 ______________________________

d 1984 ______________________________

e 1990 ______________________________

Construct a timeline for someone you know. Include at least five significant events.

DIRECTIONS

Directions use position words to tell us the location of something.

Here are some position words to help you describe location.

• left	• next to	• below	• north
• right	• in front of	• above	• south
• up	• on top of	• beside	• east
• down	• behind	• between	• west

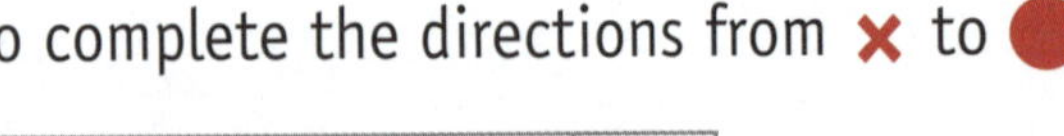

Example: Write the words **up**, **down**, **right** and **left** to complete the directions from × to ●.

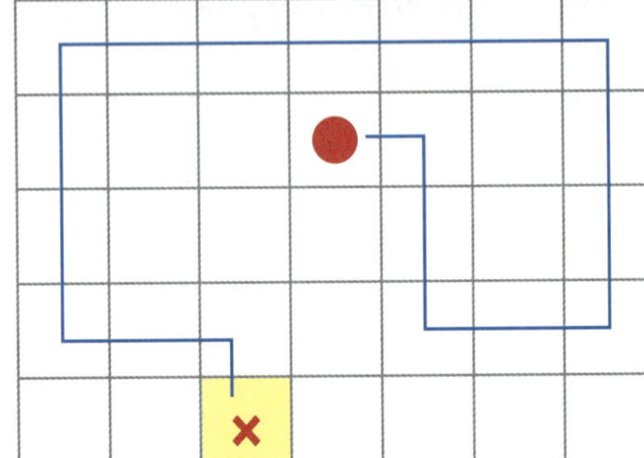

Start at ×.

Move:

a 1 space <u>up</u>

b 2 spaces ______

c 3 spaces ______

d 6 spaces ______

e 3 spaces ______

f 2 spaces ______

g 2 spaces ______

h 1 space ______

Check the answers on the video!

Your turn

Describe the moves to get from × to ●.

1

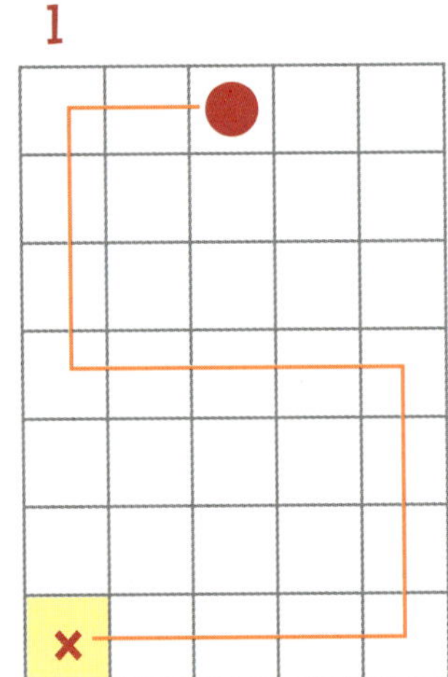

● <u>4 right</u>

a ______

b ______

c ______

d ______

2

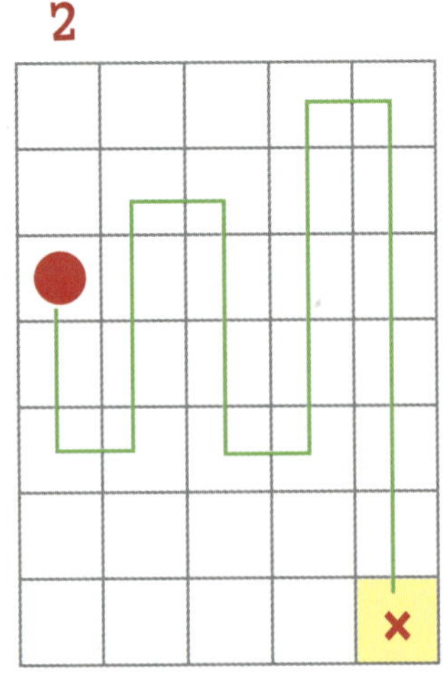

● <u>6 up</u>

a ______

b ______

c ______

d ______

e ______

f ______

g ______

h ______

SELF CHECK Tick how you feel

Got it!	Need help...	I don't get it
☐	☐	☐

Check your answers

How many did you get correct? ☐

CATCH UP MATHS YEAR 6 BOOK B © PASCAL PRESS ISBN: 9781925726190

PRACTICE

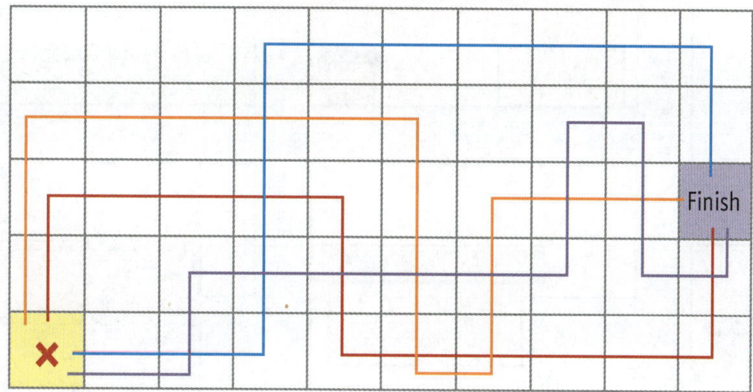

1 **Write directions for following each path from × to Finish.**

- red path Start at ×. 2 up, 4 right, 2 down, 5 right, 2 up
- **a** blue path ________
- **b** orange path ________
- **c** purple path ________

Use the teachers' message trays below to answer the following questions.

Mr Jennings	Ms Dillon	Miss Pontifex	Mrs Toman	Mrs Reid
Mrs Brandman	Miss Graham	Ms Young	Mr Hoff	Mrs Evans
Mrs Lennon	Miss Dowe	Mr Thornton	Miss Willow	Ms Ward
Mr Knight	Ms Rositano	Miss Saka	Mrs Katsi	Mr Callow

2 **Write the name on the message tray that matches each location.**

- Below Miss Graham's Miss Dowe
- **a** To the right of Ms Young's ________
- **b** Above Mr Hoff's ________
- **c** In the bottom right corner ________

3 **Describe the position of these teachers' message trays.**

- Miss Saka's below Mr Thornton's
- **a** Mrs Reid's ________
- **b** Miss Pontifex's ________
- **c** Mrs Lennon's ________
- **d** Ms Ward's ________

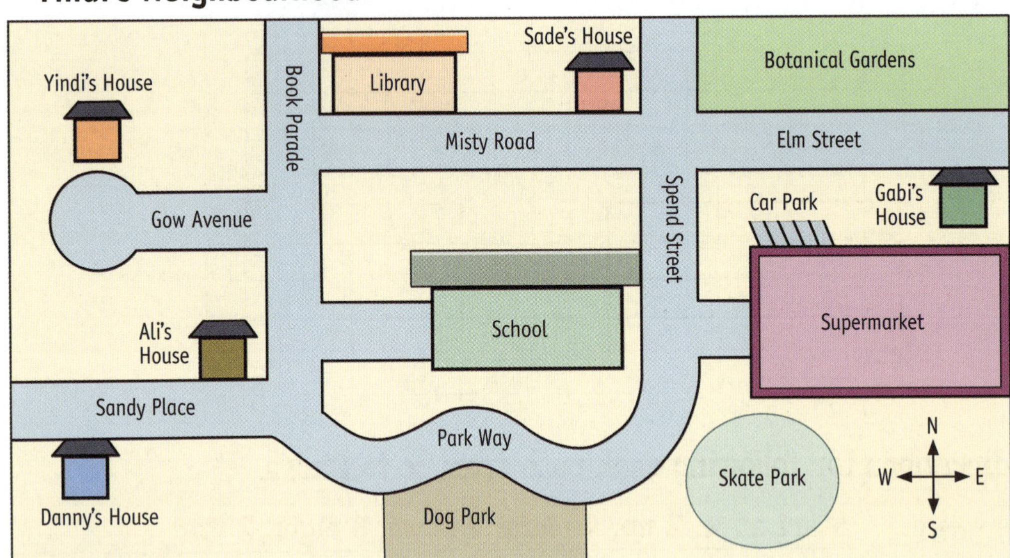

4 Where is Yindi going?

● Yindi leaves her house and walks east on Gow Ave. She turns left and heads north on Book Parade. She turns right into Misty Road, crosses over Spend Street onto Elm Street, and goes past the supermarket carpark.

Where is Yindi? Gabi's house

a Yindi leaves her house and walks east on Gow Avenue. She turns right and heads south on Book Parade. She turns left on Park Way, passing the dog park on her right, and stops before she reaches the supermarket.

Where is Yindi? ____________

5 Write directions from Yindi's house to these locations.

a Supermarket ____________

____

b School ____________

____

c Danny's house ____________

____

6 Write directions from these locations to Danny's house.

a Library ____________

____

b Sade's house ____________

____

CATCH UP MATHS YEAR 6 BOOK B © PASCAL PRESS ISBN: 9781925726190

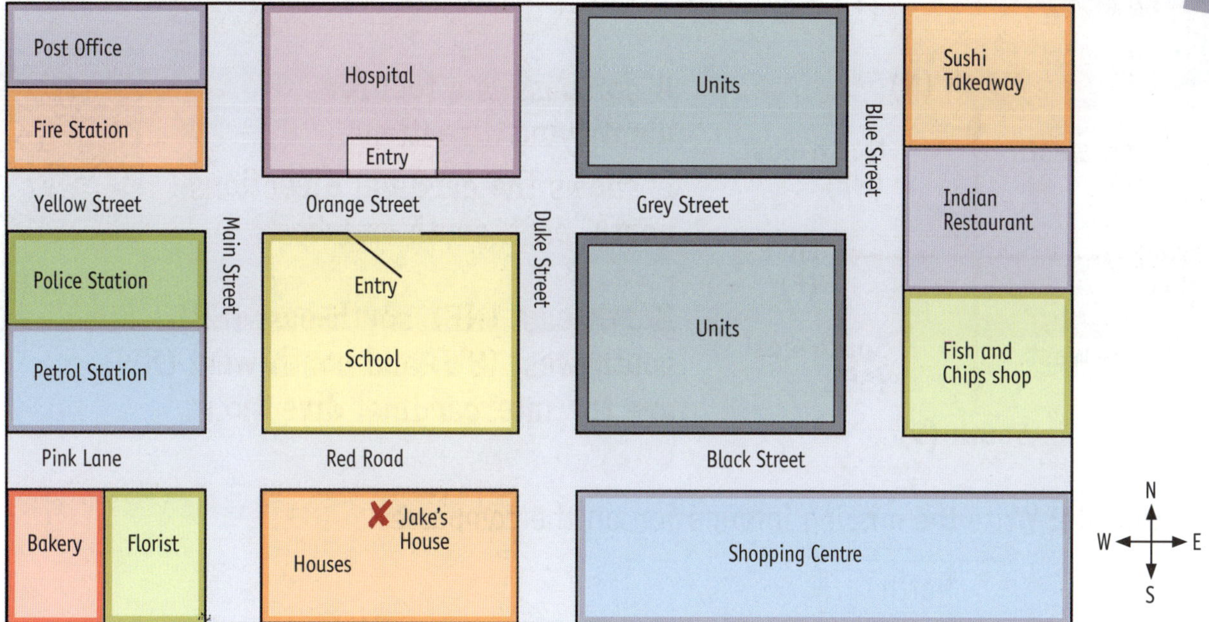

7 Where is Jake going?

- Jake leaves his house, turns left and walks to the end of Red Road. He turns right onto Main Street and walks north, then turns right onto Orange Street.
  He turns right and walks into the school.

a Jake leaves his house and heads east on Red Road.
He crosses Duke Street and walks east on Black Street.
He turns right and walks into the __________.

b Jake leaves his house and heads west on Red Road.
He turns right into Main Street then heads north past the Police Station.
He crosses Yellow Street and continues past the Fire Station.
He turns left into the __________.

8 Write directions from Jake's house to these locations.

a Hospital __________

b Fish and Chips Shop __________

c Bakery __________

d Sushi Takeaway __________

THE COMPASS

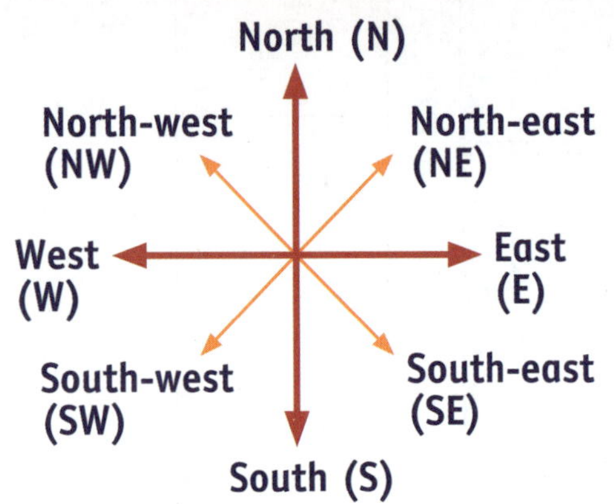

A compass is a tool for identifying directions.
It shows the cardinal directions: north, east, south and west.

North-east (NE), south-east (SE), south-west (SW) and north-west (NW) are the intercardinal directions.

SCAN to watch video

Example: Write the missing information on the compasses.

a North, West, South-east, South

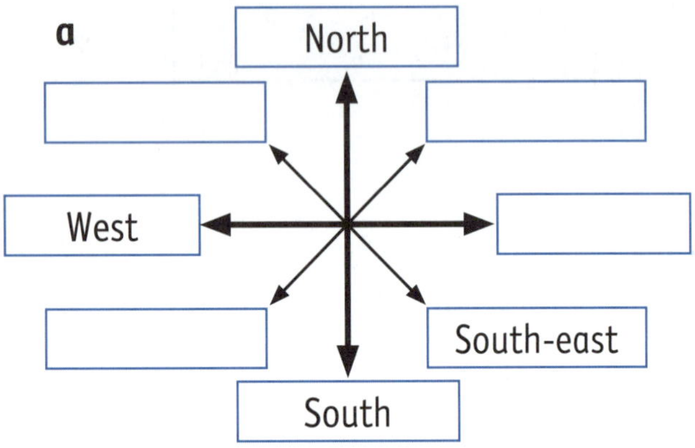

b North, North-east, West, East

c North-east, East, South-west, South

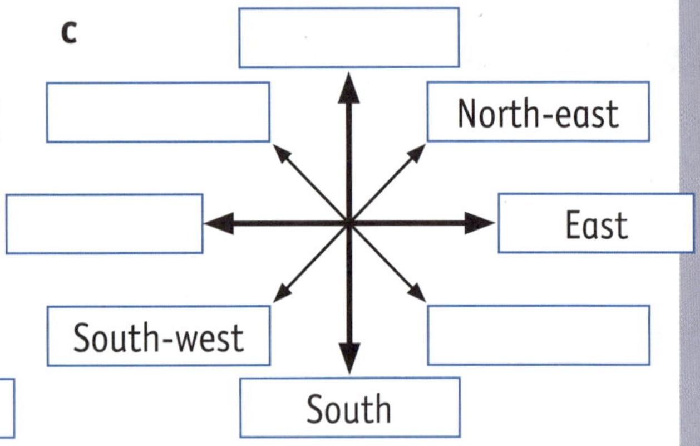

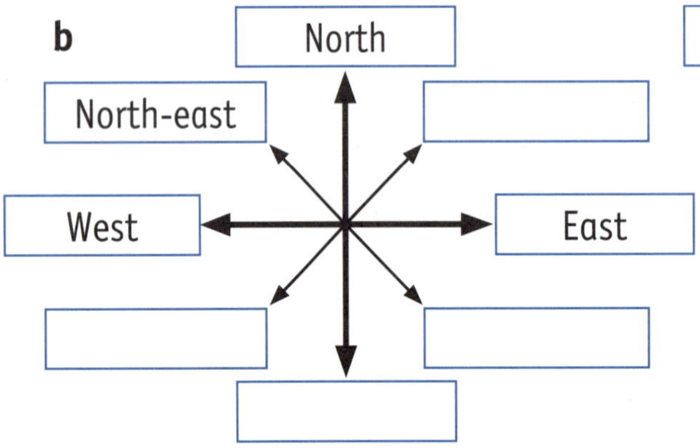

Check the answers on the video!

Your turn

Which way is the compass pointing?

a south-west

a ____________

b ____________

c ____________

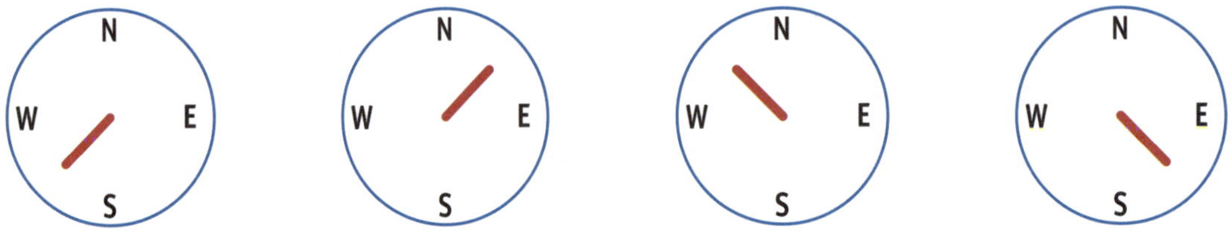

N W E S

SELF CHECK Tick how you feel

Got it!	Need help...	I don't get it
☐	☐	☐

Check your answers
How many did you get correct?

PRACTICE

Show the direction on the compass.

a south	a west	b north-east	c north-west
	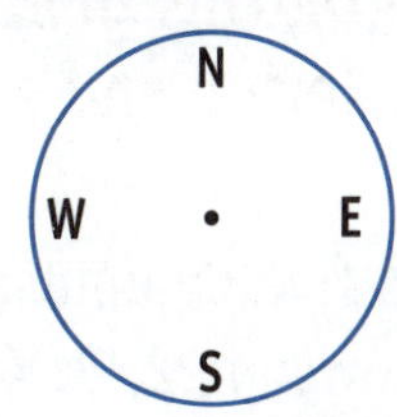	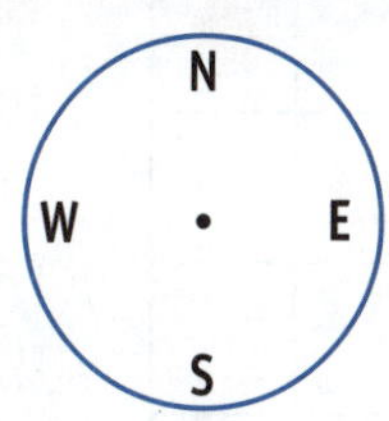	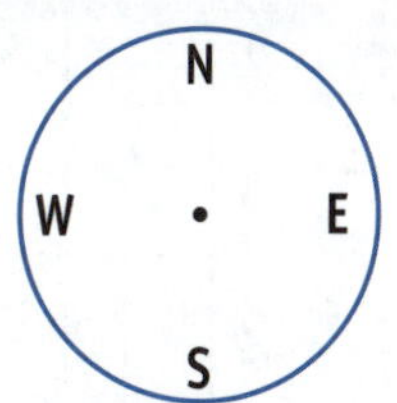

Use the grid at the right to complete the following.

Z		B
	X	A
	D	

From X, what letter is in each of these directions?

● south D a east ___ b north-west ___

Write the letters in the given location.

● B: NE of X a S: SW of X b L: W of X c E: N of X d K: SE of X

Using the grid at the right, write directions from ● to each shape.

● ⏢ North of ●
a ■ ____________
b ⬬ ____________
c ⬢ ____________
d ▲ ____________
e ▬ ____________
f ⬟ ____________
g ⯃ ____________

▬	⏢	⬢
▲	●	■
⯃	⬟	⬬

Write the places on the map of Magpie Island.

● Canary: S of Eagle Point
a Flamingo: W of Hawk Lookout
b Pigeon: N of Hawk Lookout
c Finch: NE of Budgie Point
d Magpie: S of Canary
e Kookaburra: S of Flamingo
f Cockatoo: NW of Kookaburra
g Galah: SE of Flamingo
h Emu: SE of Finch

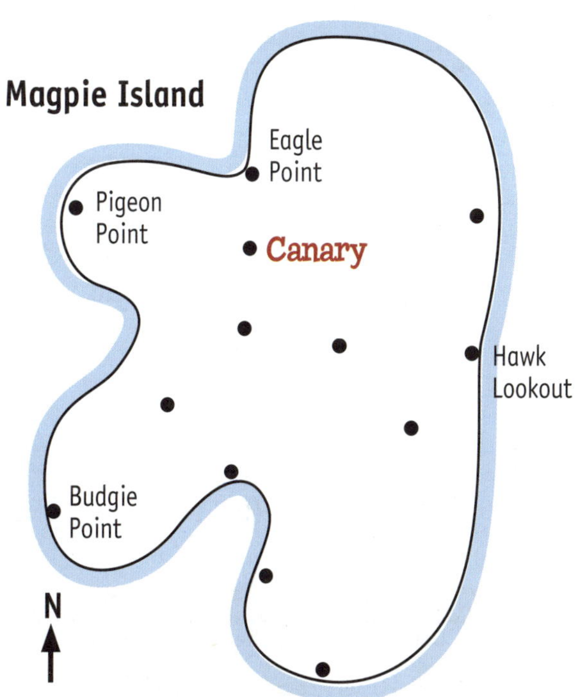

GRID REFERENCES

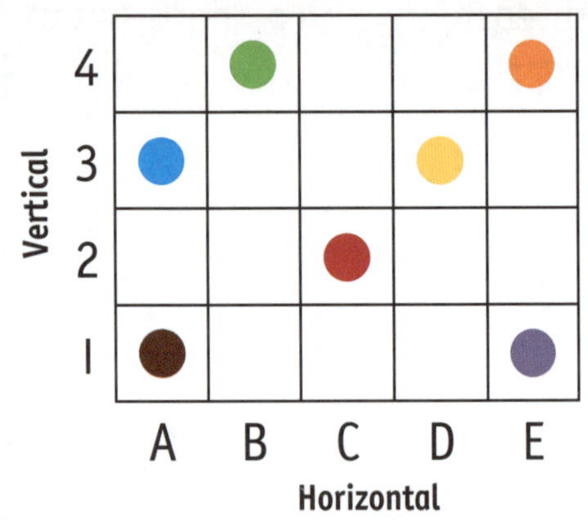

Each space on this grid is named by a letter and a number.

This grid has five columns: A, B, C, D, E.
It has four rows: 1, 2, 3, 4.

The ● has the grid reference (A,3).

The ● has the grid reference (E,4).

SCAN to watch video

When you give grid references, say the horizontal grid label first and then the vertical grid label.

Put grid references in brackets with a comma between them.

Examples: Complete the grid references using the grid above.

a ● has the grid reference (A,1).

b ● has the grid reference _____.

c ● has the grid reference _____.

d ● has the grid reference _____.

1 Using the grid below right, write the grid references for these.

● × (A,4)

a × _____

b × _____

c × _____

4	×			
3			×	
2		×		×
1		×		
	A	B	C	D

2 Draw the crosses at these grid references in the grid at the right.

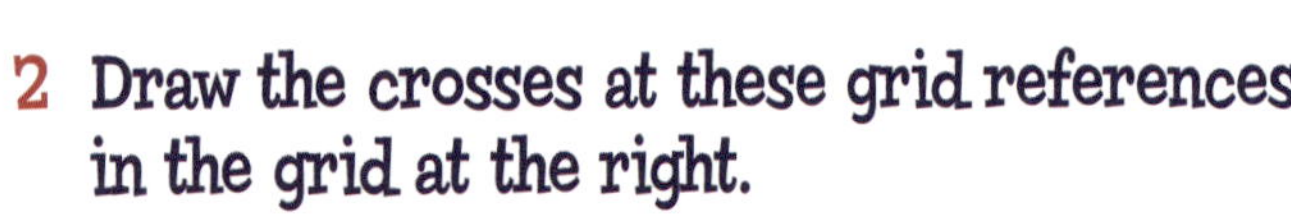

● × (B,2)

a × (A,1)

b × (C,1)

c × (B,4)

d × (A,3)

e × (D,1)

SELF CHECK Tick how you feel

Got it!	Need help...	I don't get it
☐	☐	☐

Check your answers

How many did you get correct? ☐

CATCH UP MATHS YEAR 6 BOOK B © PASCAL PRESS ISBN: 9781925726190

PRACTICE

Write the grid references.

- × (A,1)
- a × ______
- b × ______
- c × ______
- d × ______
- e × ______
- f × ______
- g × ______

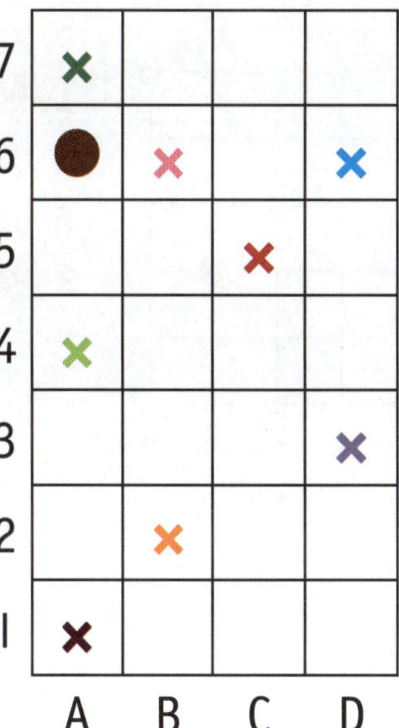

Draw the coloured circles at these grid references.

- (A,6)
- a (A,3)
- b (C,4)
- c (B,5)
- d (D,1)
- e (C,7)
- f (D,7)
- g (D,5)

Sam's teacher gave him the following grid references to plot on the grid. Tick or cross each grid reference to show whether Sam drew the circles in the correct places.

- (2,4) ✗
- a (5,8) ☐
- b (2,1) ☐
- c (3,8) ☐
- d (4,10) ☐
- e (3,9) ☐
- f (7,3) ☐
- g (6,6) ☐
- h (1,2) ☐
- i (5,5) ☐
- j (7,1) ☐

COORDINATES

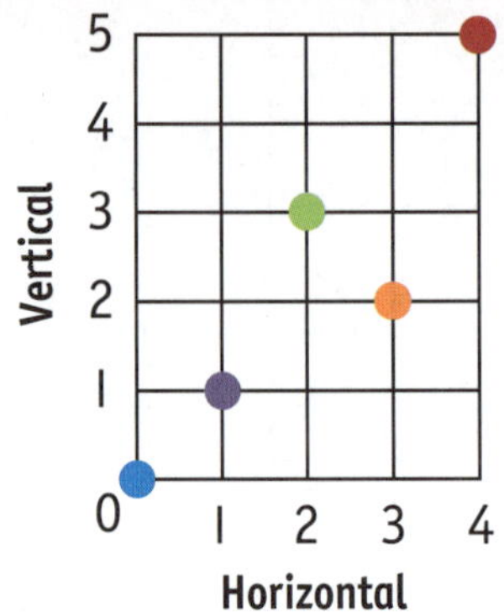

This is a grid where each line is named by a number.
A pair of coordinates names a point where lines meet.

SCAN to watch video

This grid has 5 vertical lines: 0, 1, 2, 3, 4.
It has 6 horizontal lines: 0, 1, 2, 3, 4, 5.
The ● has the coordinates (3,2).
The ● has the coordinates (0,0).

When you give coordinates, say the horizontal grid label first and then the vertical grid label.

Put coordinates in brackets with a comma between them.

Examples:

Complete the coordinates using the grid above.

a ● has the coordinates (2,3).

b ● has the coordinates (__,__).

c ● has the coordinates (__,__).

1 Draw the crosses at these coordinates in the grid below.

● × (0,D)
a × (4,F)
b × (5,A)
c × (1,B)
d × (2,E)
e × (6,D)

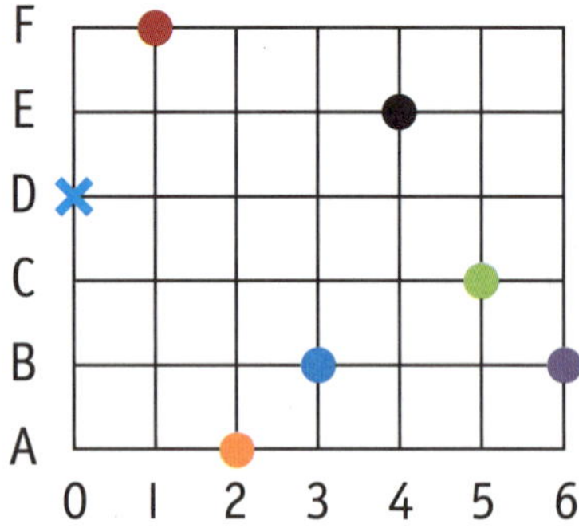

2 Write the coordinates for these.

● ● (3,B)
a ● _____
b ● _____
c ● _____
d ● _____
e ● _____

SELF CHECK Tick how you feel

Got it! | Need help... | I don't get it

Check your answers
How many did you get correct?

CATCH UP MATHS YEAR 6 BOOK B © PASCAL PRESS ISBN: 9781925726190

PRACTICE

1 List the coordinates of each point on the letters.

- V (E,F), (F,D), (G,F)

a D ______

b E ______

c X ______

d Y ______

e Z ______

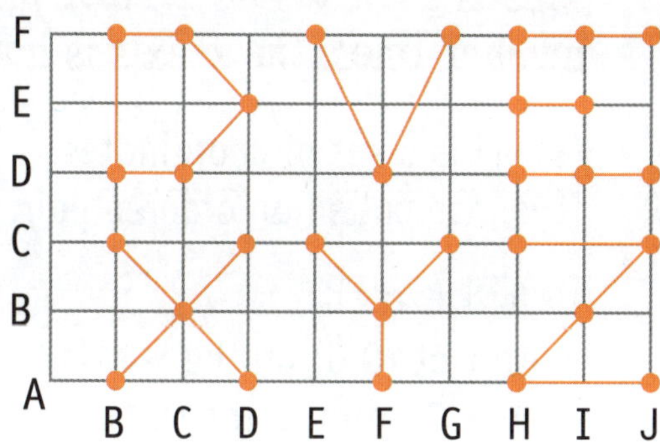

2 Plot each set of points.
Join the points and then name the shape you made.

- (0,5), (1,7), (2,5) triangle

a (3,5), (3,7), (5,5), (5,7) ______

b (6,3), (6,7), (8,7), (8,3) ______

c (8,2), (7,0), (5,0), (6,2) ______

d (2,1), (2,4), (0,4), (0,2) ______

e (3,1), (3,3), (4,4), (5,3), (5,1) ______

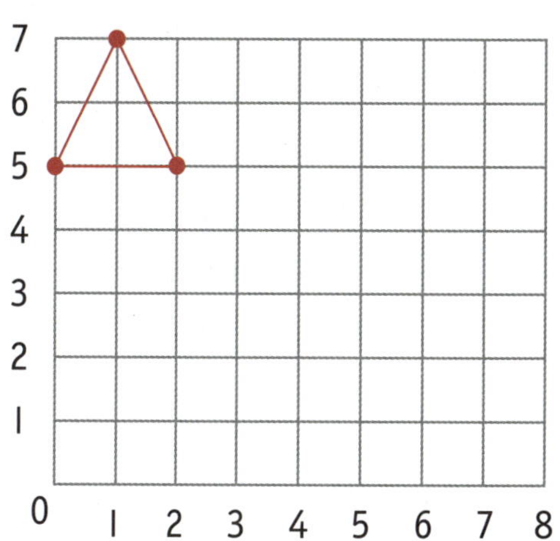

3 Using the map at the right, write the coordinates of these mountains.

- Mount Hilda (4,4)

a Mount Arron ______

b Mount Killey ______

c Mount Philip ______

4 Which mountain is at the given coordinates?

- (5,8) Mount Louis

a (3,2) ______

b (6,3) ______

c (2,4) ______

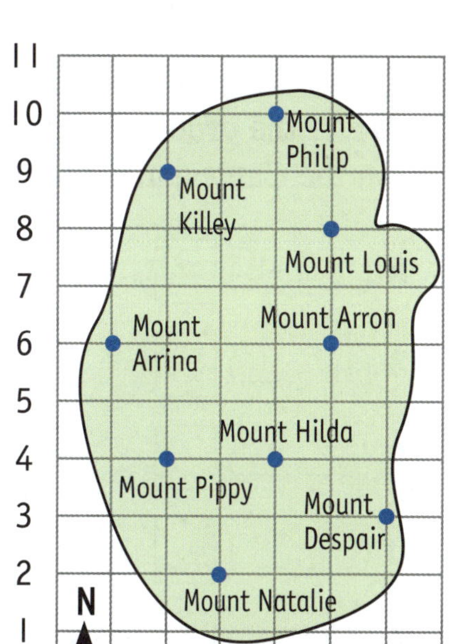

THE CARTESIAN NUMBER PLANE

SCAN to watch video

This is a Cartesian number plane. It is made up of two perpendicular number lines: the x-axis is horizontal and the y-axis is vertical.

(3,–4) is a set of coordinates. It is also called an ordered pair.

To plot (3,–4):

- Start at (0,0) on the *x*–axis. Move 3 units to the right.
- Then move 4 units down the *y*–axis.

(3,–4) is in Quadrant 4.

You could join the points (–5,–2), (–3,–2), (–3,–4) and (–5,–4) to make a square.

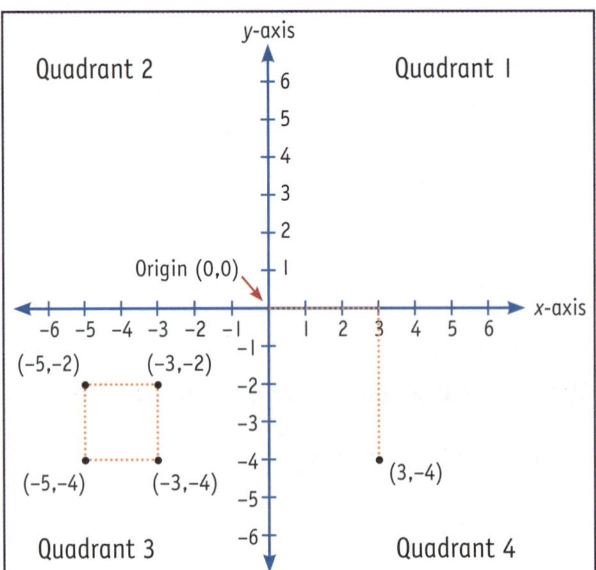

Pay attention to the negative signs!

Examples:

a Circle the ordered pair that is not in Quadrant 1: (2,2) (4,6) (5,–3) (5,6)

b Circle the coordinates that are not in Quadrant 3: (–3,–2) (–4,–5) (–1,–6) (–3,1)

c Circle the ordered pairs that are in Quadrant 2: (–6,4) (2,–3) (4,–4) (–2,3)

Your turn

Write the ordered pair and quadrant for each point on the Cartesian number plane.

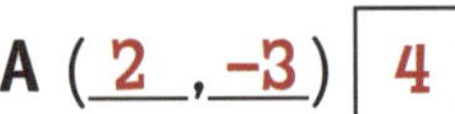

B (__,__)	☐	F (__,__)	☐
C (__,__)	☐	G (__,__)	☐
D (__,__)	☐	H (__,__)	☐
E (__,__)	☐	I (__,__)	☐
		J (__,__)	☐

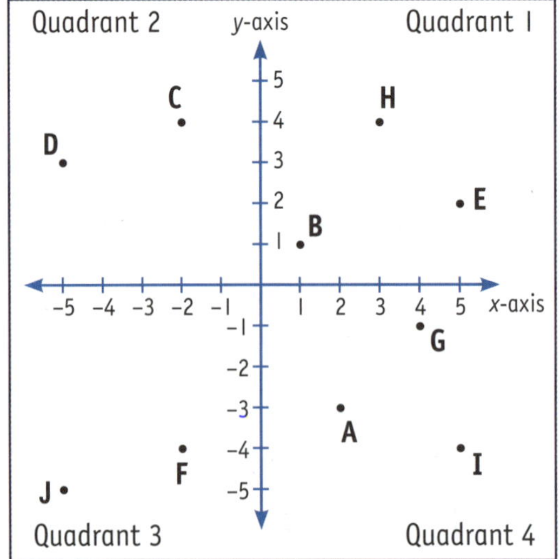

SELF CHECK Tick how you feel

Got it!	Need help...	I don't get it
☐	☐	☐

Check your answers

How many did you get correct? ☐

CATCH UP MATHS YEAR 6 BOOK B © PASCAL PRESS ISBN: 9781925726190

PRACTICE

1 Plot these ordered pairs on the Cartesian plane.

- A (−1,−4)
- B (3,6)
- C (1,−2)
- D (−2,4)
- E (−4,−3)
- F (2,−6)
- G (1,2)
- H (−2,−4)
- I (4,5)
- J (5,4)

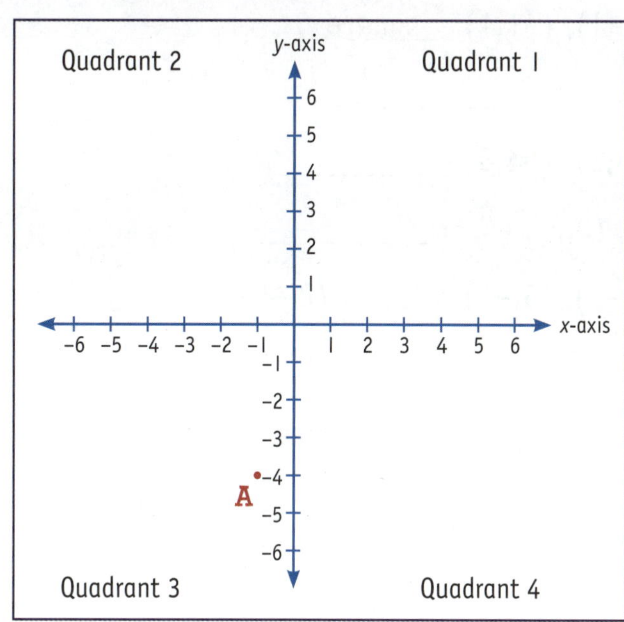

2 Write the quadrant for each ordered pair from Question 1.

A Quadrant 3   D ______   G ______   I ______

B ______   E ______   H ______   J ______

C ______   F ______

3 Use the Cartesian plane to complete the table.

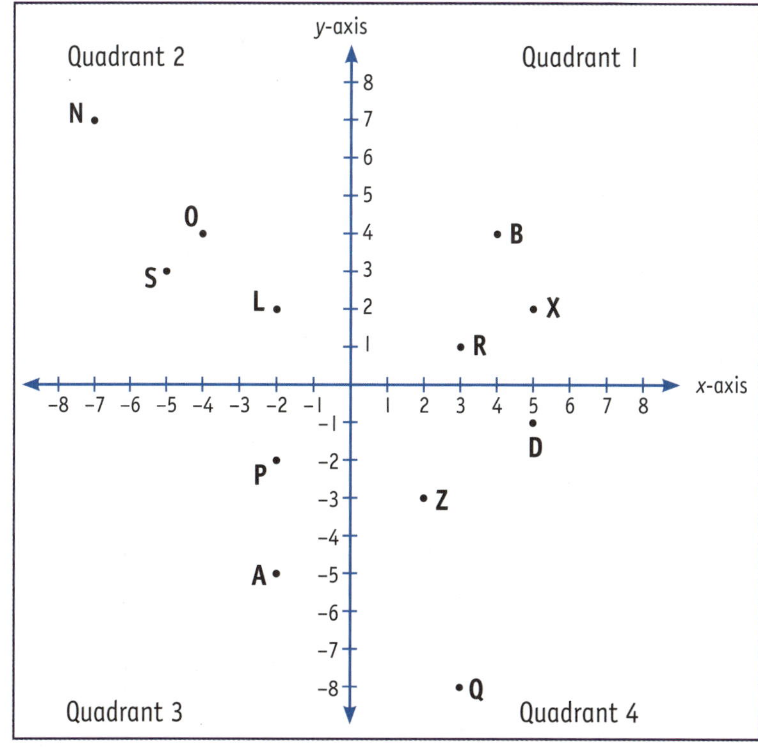

	Ordered pair	Quadrant
B		
Z		
	(−2,−2)	
D		
O		
X		
N		
	(−2,2)	
Q		
R		
S		
A		

Plot the ordered pairs on the Cartesian plane. Then colour and name the 2D shape you drew.

- (1,1), (1,–1), (–1,–1), (–1,1) square
- **a** (–6,0), (–3,0), (–3,–5) ________
- **b** (–4,6), (–6,6), (–6,3), (–4,3) ________
- **c** (3,6), (1,4), (5,4), (2,2), (4,2) ________
- **d** (3,–3), (5,–3), (6,–5), (5,–7), (3,–7), (2,–5) ________
- **e** (7,0), (5,0), (7,6) ________

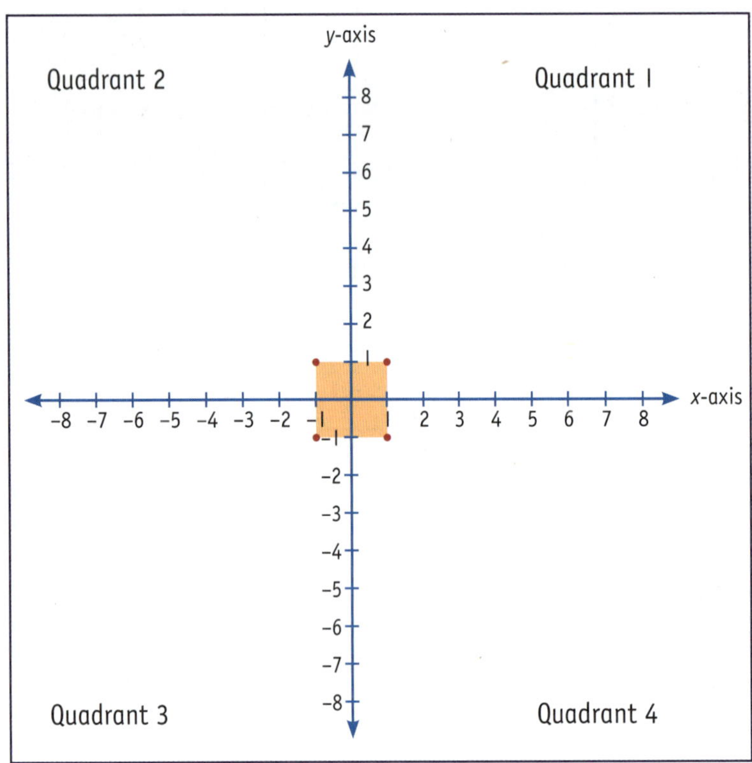

Translate the shapes 3 units to the right and 2 units down.

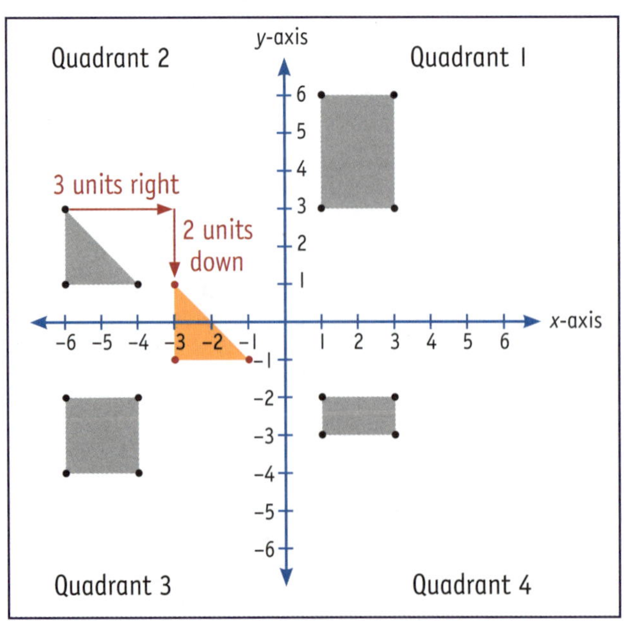

CATCH UP MATHS YEAR 6 BOOK B © PASCAL PRESS ISBN: 9781925726190

LEGENDS

Legends show where important places or landmarks are on a map. Legends are also called keys.

Here are some symbols you may see in the legend of a map.

hospital

parking

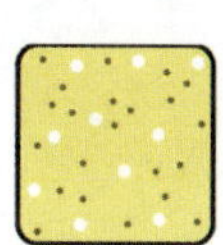

sand

Examples: Join each symbol with the correct label.

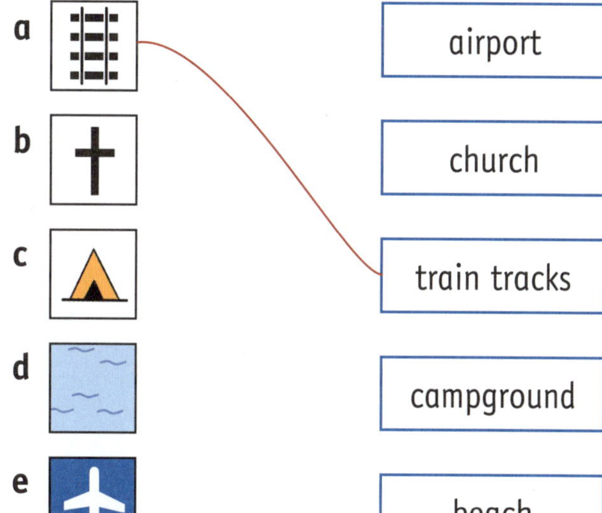

Your turn

Draw a symbol in the box that could represent these things.

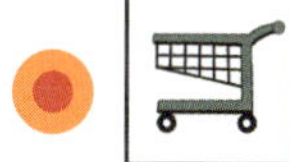 shop

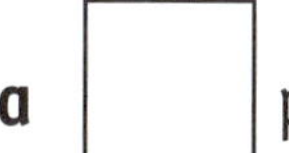

a post office

b library

c  road

d 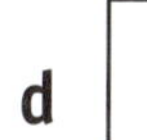 national park

e  no parking

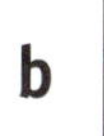

SELF CHECK Tick how you feel

Got it!	Need help...	I don't get it

Check your answers

How many did you get correct?

  ISBN: 9781925726190

PRACTICE

Stayaway Camping Ground

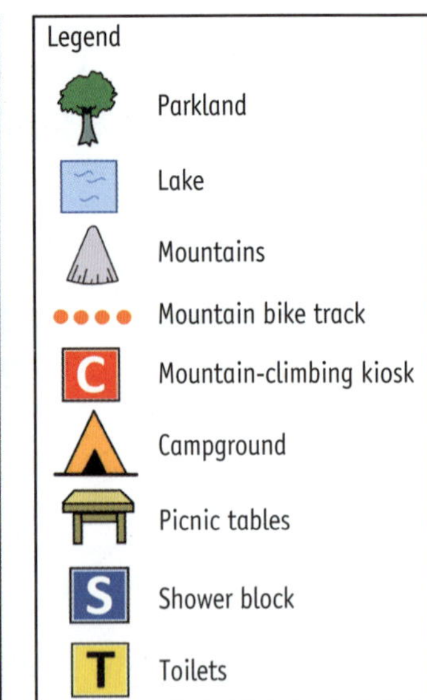

1 What do these symbols represent?

a ____________ c ____________ f ____________

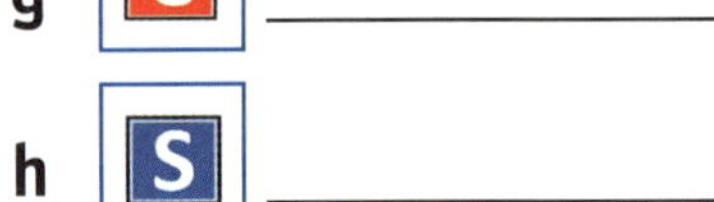

b ____________ d ____________ g ____________

e ____________ h ____________

2 How many are at Stayaway Camping Ground?

- Toilets 1
- a Shower blocks __
- b Campgrounds __
- c Lakes __
- d Picnic tables __

3 Draw the symbols on the right in appropriate places on the map.

Harry's Recreation Ground

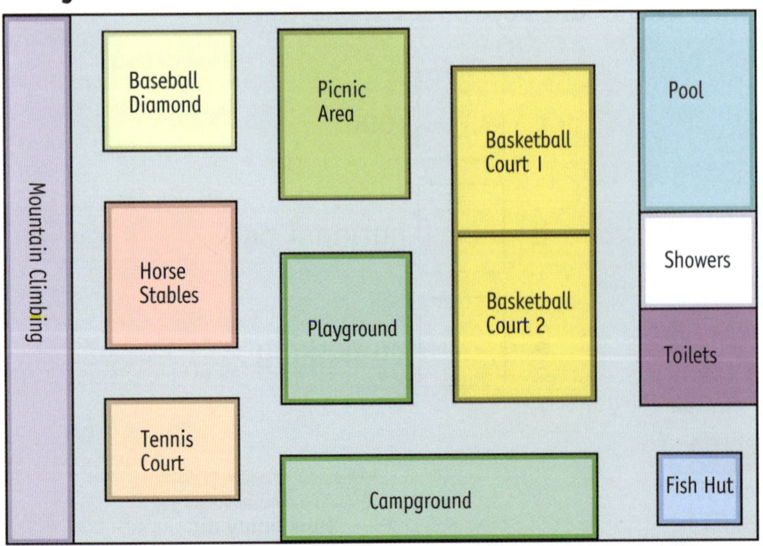

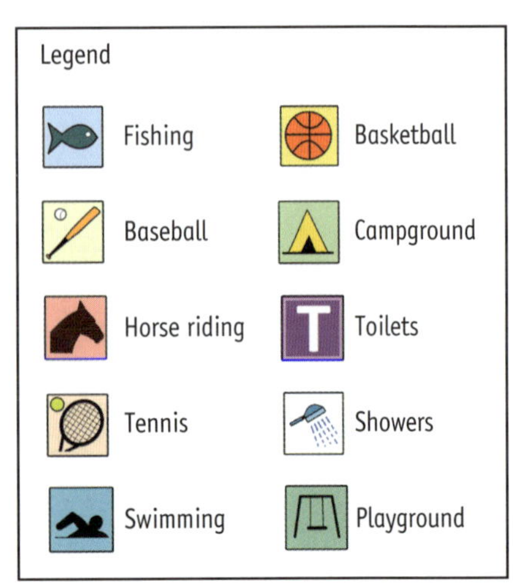

CATCH UP MATHS YEAR 6 BOOK B © PASCAL PRESS ISBN: 9781925726190

POSITION REVIEW

1 Describe the moves to get from the blue box to the red circle.

a

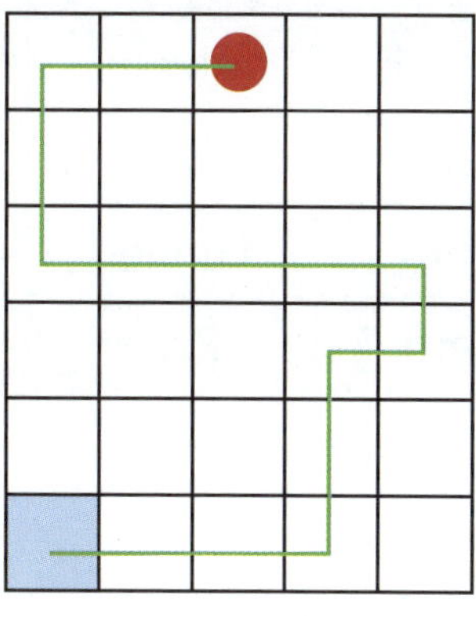

b

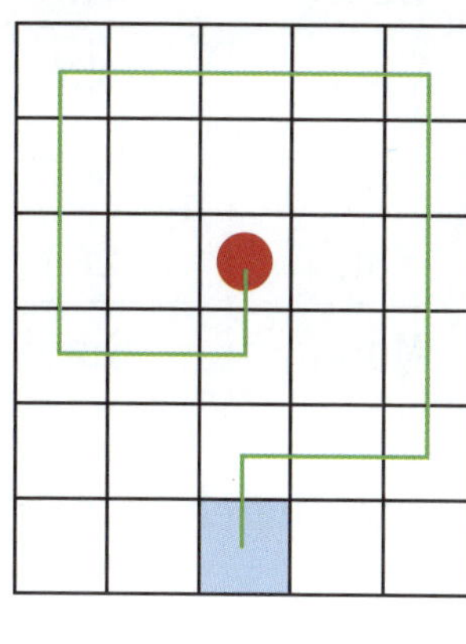

c

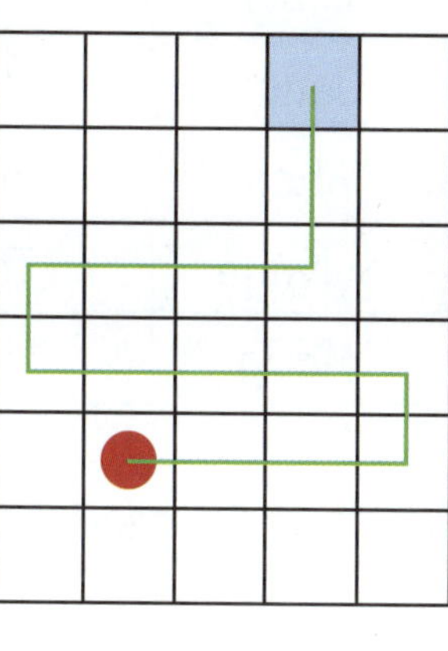

2 Follow the directions to draw the paths. Start at ×.

a Blue path: up 3, left 4, down 2, right 2, down 4, left 5, up 3

b Black path: down 2, left 3, up 5, left 1, down 3, left 3

c Orange path: left 3, down 1, left 3, up 4, left 2, down 3, right 1

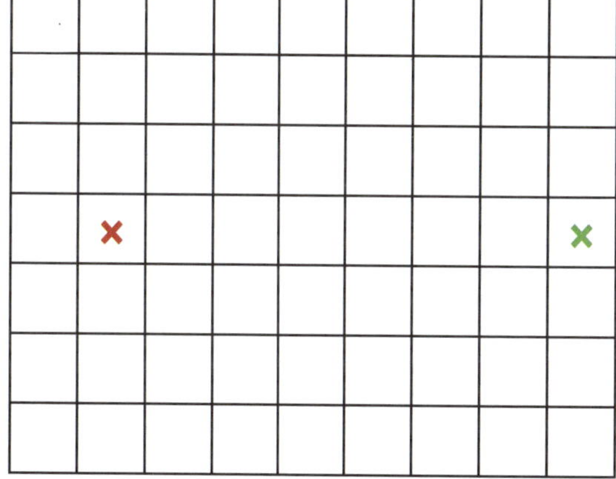

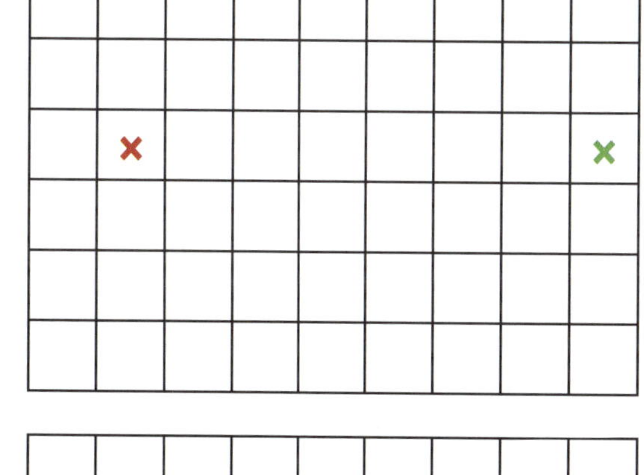

3 Draw an X on the grid and your own path that has 8 steps. Then write the directions.

Start at X.

1 ____________ 5 ____________

2 ____________ 6 ____________

3 ____________ 7 ____________

4 ____________ 8 ____________

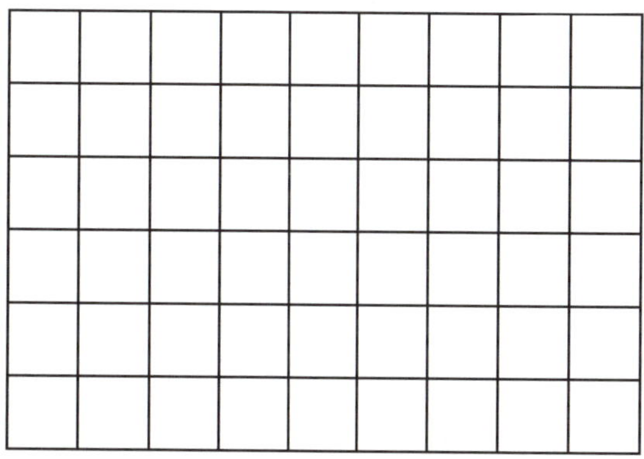

REVIEW

Here are the tote trays for 6 Gold. Use them to complete the following.

Emily	Jake	Beni	Kylie	Amir	Tiana
Jenai	Chen	Annie	Nia	Deborah	Alyssa
Allan	Danny	Lisa	Matt	Shuri	Rob
Sylvana	Tony	Kali	Len	Nina	Carmen
Mario	Liz	Mark	Bindi	Gaeto	Rose

4 Write the name on the tote tray that matches each location.

a Below Amir's ______________

b Above Allan's ______________

c To the left of Nia's ______________

d To the right of Sylvana's ______________

e At the top left corner ______________

f At the bottom right corner ______________

g In the top row, third from the right ______________

h In the bottom row, fourth from the left ______________

i In the middle row, second from the right ______________

5 Describe the position of these tote trays.

a Jenai's ____

b Jake's ____

c Mario's ____

d Chen's ____

e Carmen's ____

f Danny's ____

g Bindi's ____

h Annie's ____

i Tony's ____

j Rose's ____

k Amir's ____

CATCH UP MATHS YEAR 6 BOOK B © PASCAL PRESS ISBN: 9781925726190

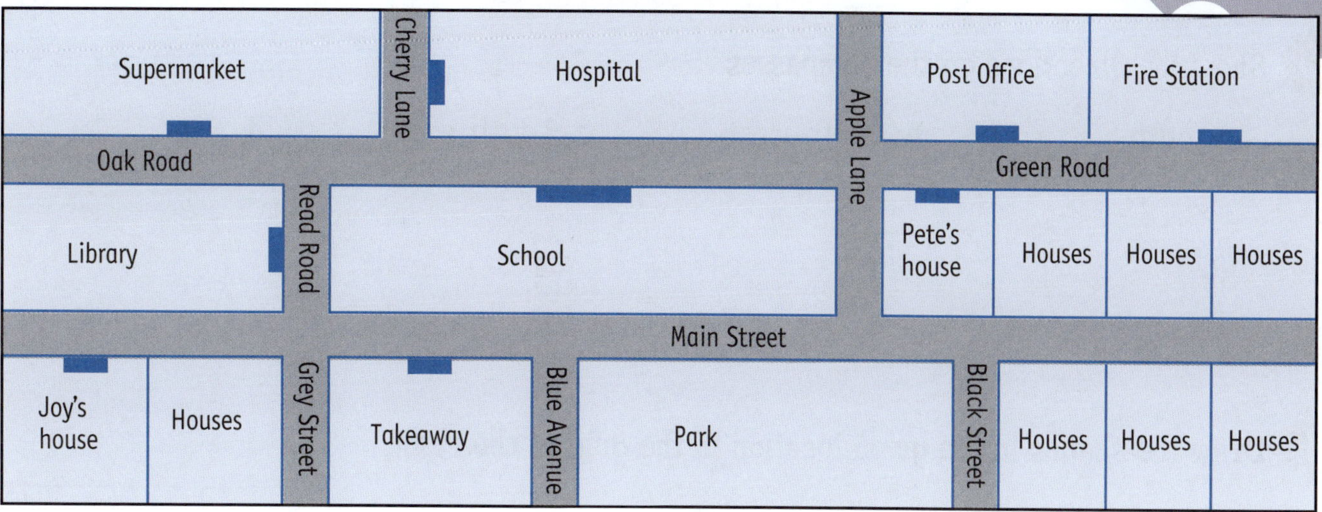

6 Follow the directions, mark the paths on the map and write the destination.

a Joy leaves her house and walks east on Main Street. She turns left on Read Rd, walks past the library and crosses Oak Road. She walks past the supermarket, turns left on Cherry Lane and then crosses the road to ______________.

b Pete leaves his house, heads west on Green Rd, turns left on Apple Lane, turns right on Main Street, walks past the park, and then crosses Blue Avenue to get to ______________.

7 Write directions.

a From Joy's house to the Fire Station ________________________________

____

____

b From Pete's house to Joy's house ________________________________

____

____

8 Complete the sentences.

a A compass is a tool for identifying ______________.

b It shows north, east, south and west. These are called ____________ points. North-east, south-east, south-west and north-west are called ______________ points.

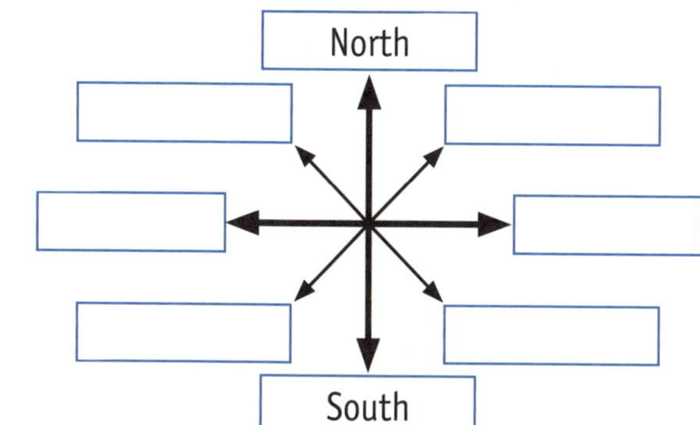

9 Write the missing points on the compass.

REVIEW

Show the directions on the compasses.

a south

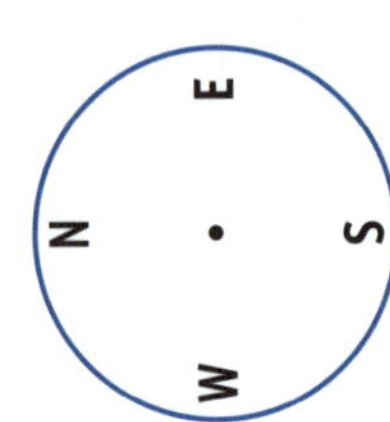

b north-west

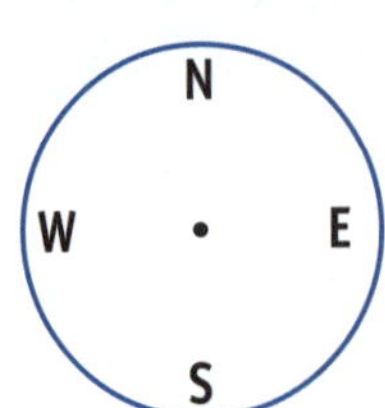

c south-east

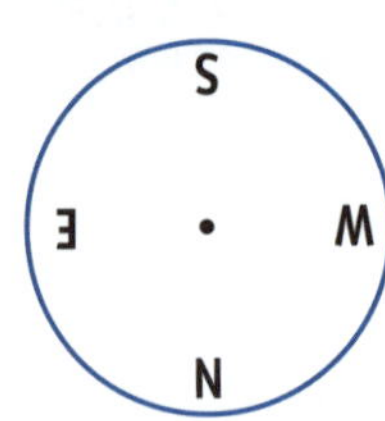

d west

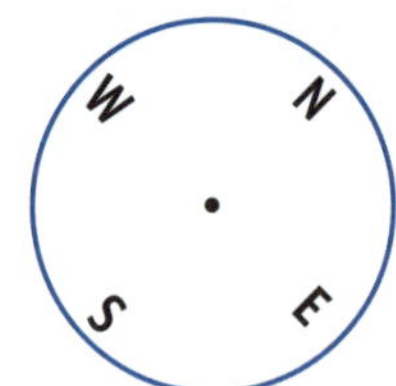

Draw the shapes in the given location in the grid at the right.

a ● south of ×

b ♥ north-west of ×

c ▲ east of ×

d ★ south-west of ×

e ■ north of ×

f ⬢ south-east of ×

g trapezium north-east of ×

h ▮ west of ×

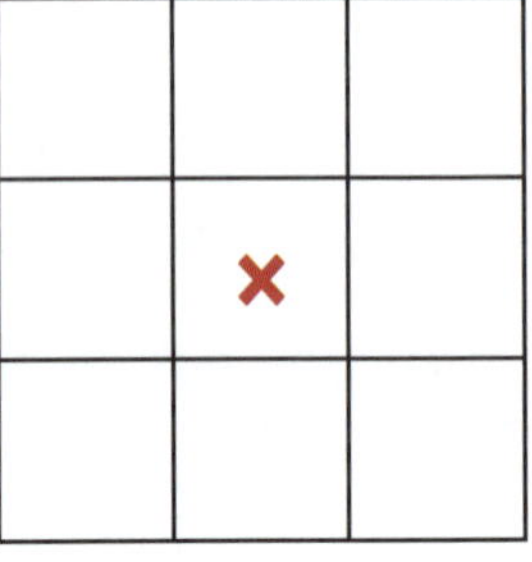

Write the grid references.

a ● ______

b ● ______

c ● ______

d ● ______

e ● ______

f ● ______

g ● ______

h ● ______

i ● ______

j ● ______

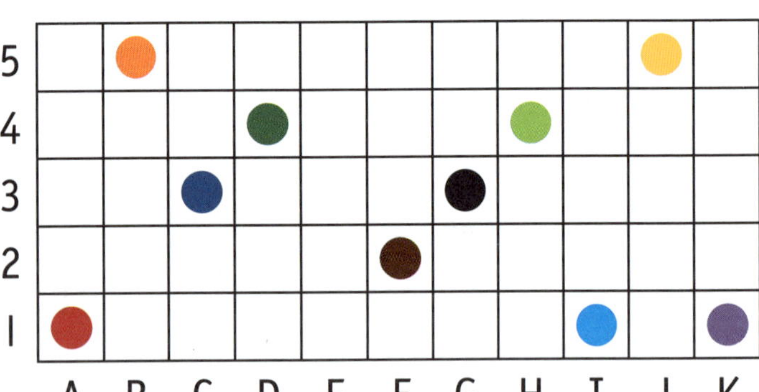

Draw the coloured crosses at these grid references in the grid above.

a × (B,2)

b × (E,3)

c × (H,2)

d × (K,2)

e × (A,3)

f × (C,1)

g × (G,5)

h × (J,2)

i × (F,4)

j × (I,4)

14 Write the coordinates.

a ● ______

b ● ______

c ● ______

d ● ______

e ● ______

f ● ______

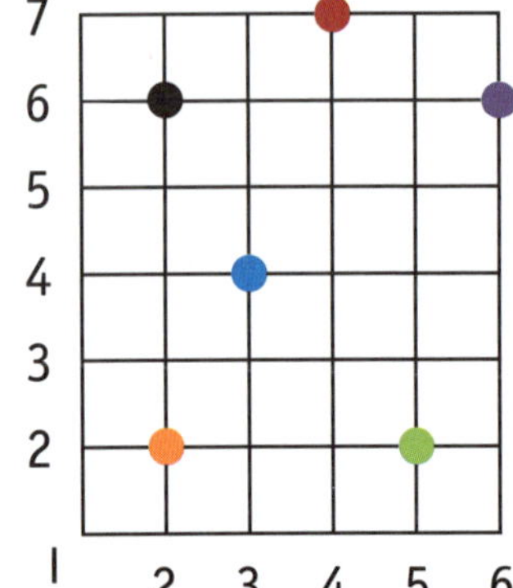

15 Draw the circles at these coordinates.

a ● (6,4)

b ● (1,4)

c ● (4,1)

d ● (4,6)

CATCH UP MATHS YEAR 6 BOOK B © PASCAL PRESS ISBN: 9781925726190

16 Fill in the missing information on the Cartesian number plane.

17 Write the coordinates for each point.

A ________

B ________

C ________

D ________

E ________

F ________

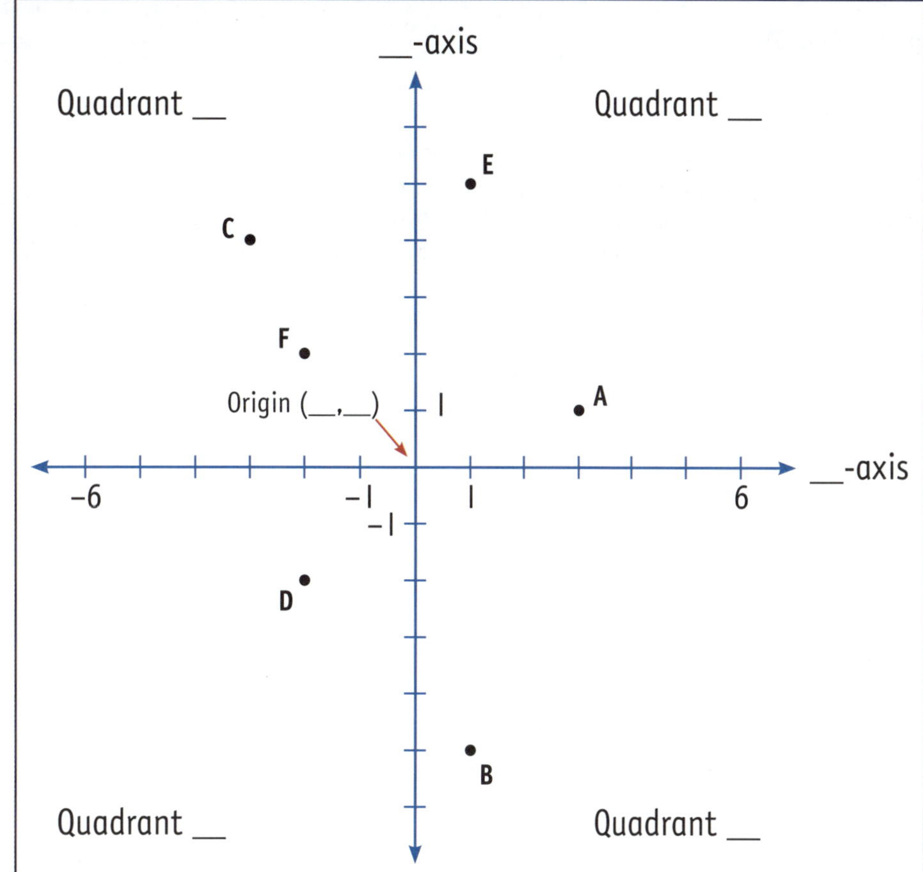

18 Plot the ordered pairs on the Cartesian number plane.

G (6,0)

H (−5,2)

I (1,1)

J (−6,−3)

K (6,−3)

19 Draw the symbols in the correct boxes to show where each fruit can be found in Ada's Orchard.

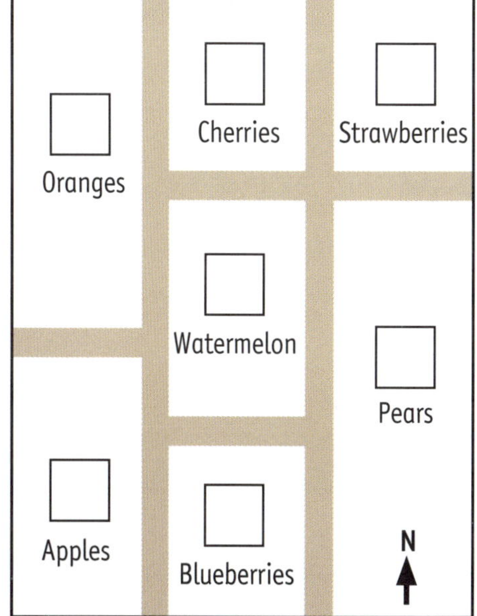

20 What fruit is in each of these directions?

a south of the oranges ________________

b north-east of the watermelon ________________

c south-west of the watermelon ________________

d north of the blueberries ________________

e east of the cherries ________________

f south of the strawberries ________________

ANSWERS

1. LENGTH

Metres

Page 1 – Your Turn

Adult to check

Page 2 – Practice

1 a 327 cm d 454 cm g 709 cm
  b 864 cm e 992 cm h 1204 cm
  c 642 cm f 538 cm i 1528 cm

2 a 1 m 43 cm d 2 m 4 cm g 67 m 51 cm
  b 1 m 56 cm e 2 m 67 cm h 28 m 95 cm
  c 4 m 81 cm f 9 m 2 cm i 30 m 86 cm

3 236 cm, 327 cm, 454 cm, 538 cm, 642 cm, 709 cm, 864 cm, 992 cm, 1204 cm, 1528 cm

4 6751 cm, 3086 cm, 2895 cm, 902 cm, 526 cm, 481 cm, 267 cm, 204 cm, 156 cm, 143 cm

5

a	438 cm	4 m 38 cm	4.38 m
b	906 cm	9 m 6 cm	9.06 m
c	810 cm	8 m 10 cm	8.10 m
d	649 cm	6 m 49 cm	6.49 m
e	470 cm	4 m 70 cm	4.70 m
f	851 cm	8 m 51 cm	8.51 m
g	560 cm	5 m 60 cm	5.60 m
h	103 cm	1 m 3 cm	1.03 m
i	215 cm	2 m 15 cm	2.15 m
j	772 cm	7 m 72 cm	7.72 m
k	1042 cm	10 m 42 cm	10.42 m
l	2449 cm	24 m 49 cm	24.49 m
m	1503 cm	15 m 3 cm	15.03 m

Centimetres

Page 3 – Your Turn

a 9 cm b 10 cm c 8 cm d 13 cm

Page 4 – Practice

1 a $6\frac{1}{2}$ cm d $8\frac{1}{2}$ cm g 5 cm
  b 7 cm e $5\frac{1}{2}$ cm h 4 cm
  c 2 cm f $9\frac{1}{2}$ cm i 3 cm

2 Adult to check

3 a 0.37 m d 0.30 m g 1.43 m
  b 4 cm e 10 cm h 2.08 m
  c 0.15 m f 8 cm i 350 cm

Millimetres

Page 5 – Your Turn

1 a 36 mm b 58 mm c 138 mm

2

E F G H

0 CM 1 2 3 4 5 6 7 8 9 10 11 12 13 14 15

Page 6 – Practice

1 a mm b mm c mm d cm e cm

2 Adult to check

3 a 67 m b 86 mm c 103 mm d 111 mm e 125 mm

4

	mm	cm	m
a	50 mm	5 cm	0.05 m
b	30 mm	3 cm	0.03 m
c	200 mm	20 cm	0.2 m
d	600 mm	60 cm	0.6 m
e	700 mm	70 cm	0.7 m
f	800 mm	80 cm	0.8 m
g	3000 mm	300 cm	3 m
h	2500 mm	250 cm	2.5 m
i	4350 mm	435 cm	4.35 m
j	7250 mm	725 cm	7.25 m
k	9500 mm	950 cm	9.5 m
l	8900 mm	890 cm	8.9 m
m	6820 mm	682 cm	6.82 m

Kilometres

Page 7 – Your Turn

1 a 7000 m c 28 000 m e 274 000 m
  b 15 000 m d 44 000 m

2 a 6 km b 13 km c 37 km d 24 km e 62 km

Page 8 – Practice

1 a 5420 d 15 210 g 624 025
  b 6062 e 72 043 h 580 010
  c 8008 f 437 003 i 804 001

2 a 2 km 515 m d 6 km 494 m g 10 km 780 m
  b 5 km 381 m e 7 km 730 m h 125 km 680 m
  c 8 km 620 m f 96 km 3 m i 654 km 964 m

3 a 0.940 d 7.314 g 8.175
  b 0.032 e 5.606 h 65.075
  c 0.105 f 1.140 i 416.410

4

	cm	m	km
a	825	8.25	0.008 25
b	400 000	4000	4
c	700	7	0.007
d	1500	15	0.015
e	625 000	6250	6.25
f	613	6.13	0.00613
g	500 000	5000	5
h	200	2	0.002
i	900 000	9000	9
j	112 500	1125	1.125
k	515 m	5.15 m	0.005 15
l	5 437 000	54 370	54.37
m	100	1	0.001

Perimeter

Page 9 – Your Turn

a P = 20 cm

Page 10 – Practice

1 a P = 30 m b P = 17 cm c P = 18 m

2 Adult to check

3 a 5 m b 9 m c 58 cm d 18 m e 152 mm f 9 m g 17 m h 10.5 cm i 18.5 mm j 36 m k 8 cm l 4 mm m 1 m n 70 cm o 15 mm p 8 cm q 84 m

4 a P = 28 m b P = 25 cm c P = 36 m d P = 20 m e P = 51 m

Temperature

Page 12 – Your Turn

Page 13 – Practice

1 a ten degrees Celsius
b forty-six degrees Celsius
c one hundred and five degrees Celsius
d four hundred and ninety degrees Celsius

2 a 63 °C b 70 °C c 19 °C d 302 °C

3 a 73 °C b 100 °C c 5 °C

4 Adult to check

Length Review Page 14

1 a 326 cm b 144 cm c 202 cm d 920 cm e 1020 cm f 667 cm g 405 cm h 545 cm i 8414 cm

2 a 2 m 49 cm b 4 m 80 cm c 8 m 2 cm d 7 m 34 cm e 1 m 75 cm f 56 m 2 cm g 30 m 23 cm h 94 m 14 cm i 0 m 65 cm j 43 m 84 cm

3

	cm	m and cm	Decimal m
a	529 cm	5 m 29 cm	5.29 m
b	648 cm	6 m 48 cm	6.48 m
c	735 cm	7 m 35 cm	7.35 m
d	620 cm	6 m 20 cm	6.2 m
e	415 cm	4 m 15 cm	4.15 m
f	108 cm	1 m 8 cm	1.08 m
g	805 cm	8 m 5 cm	8.05 cm
h	342 cm	3 m 42 cm	3.42 m
i	106 cm	1 m 6 cm	1.06 m
j	585 cm	5 m 8 cm	5.85 cm
k	266 cm	2 m 66 cm	2.66 m
l	689 cm	6 m 89 cm	6.89 m
m	1127 cm	11 m 27 cm	11.27 m
n	1838 cm	18 m 38 cm	18.38 m

4 a $4\frac{1}{2}$ cm b 6 cm c $5\frac{1}{2}$ cm d $2\frac{1}{2}$ cm e 3 cm f $7\frac{1}{2}$ cm g 1 cm h 4 cm i 5 cm j 2 cm

5 Adult to check

6

cm	32	78	5	1000	300	890	422	803	950
m	0.32	0.78	0.05	10	3	8.9	4.22	8.03	9.5

7 one

8 a 40 mm b 120 mm c 280 mm d 490 mm e 980 mm f 80 mm g 160 mm h 360 mm i 580 mm j 640 mm

9 a 3 mm b 34 mm c 71 mm d 116 mm e 152 mm f 175 mm

10 G H I J K L
0 1 2 3 4 5 6 7 8 9 10 11 12 13 14 15 16 17 18

11 a 24 mm b 76 mm c 101 mm d 122 mm e 143 mm

12

	mm	cm	m
a	120	12	0.12
b	150	15	0.15
c	50	5	0.05
d	700	70	0.70
e	360	36	0.36
f	4000	400	4
g	1250	125	1.25
h	5980	598	5.98
i	6350	635	6.35
j	9900	990	9.9

13 1000

14 a 291 000 m b 601 000 m c 137 000 m d 512 000 m e 8000 m f 415 000 m g 359 000 m h 774 000 m

15 a 4 km b 32 km c 49 km d 805 km e 1.5 km f 83 km g 70 km h 99 km

16 a 10 258 m b 4059 m c 7034 m d 3010 m e 19 345 m f 25 004 m g 8246 m h 3753 m

17 a 3 km 565 m b 4 km 858 m c 3 km 2 m d 7 km 130 m e 9 km 267 m f 3 km 749 m g 5 km 931 m h 2 km 478 m

18 a 0.039 km b 0.215 km c 0.437 km d 0.608 km e 0.590 km f 8.269 km g 1.401 km h 7.268 km

19

	cm	m	km
a	400	4	0.004
b	300 000	3000	3
c	625	6.25	0.006 25
d	200	2	0.002
e	875 000	8750	8.75
f	512	5.12	0.0512
g	600 000	6000	6
h	400 000	4000	4
i	2300	23	0.023
j	950 000	9500	9.5

1. LENGTH CONTINUED

20 a P = 10 m b P = 32 cm c P = 44 m d P = 24.2 m

21 a 5 cm b 30 mm c 19 m d 290 cm e 23 mm f 144 m g 204 cm h 7.5 mm i 30 m j 4.2 cm k 60 mm l 75 m

22 Adult to check

23 a 21 °C b 44 °C c 63 °C d 72 °C e 99 °C

24 Adult to check

2. ANGLES

Right Angles and Complementary Angles

Page 20 – Your Turn

Adult to check

Page 21 – Practice

1 Circle: a b c d e f g

2 a 17° b 80° c 40° d 50° e 68° f 8° g 60° h 82° i 45° j 18° k 11°

Acute and Obtuse Angles

Page 22 – Your Turn

Adult to check

Page 23 – Practice

1 a irregular pentagon
b irregular hexagon
c parallelogram (irregular quadrilateral)
d rhombus (irregular quadrilateral)
e regular pentagon
f regular hexagon
g irregular hexagon

2 a ∠PQR or ∠RQP, obtuse b ∠BCD or ∠DCB, acute

3 a acute

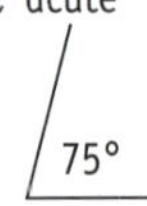

b obtuse

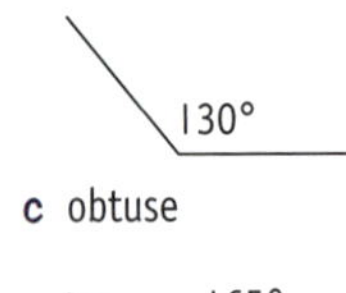

c obtuse

165°

d acute

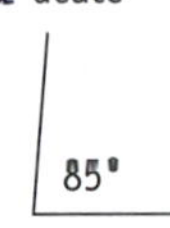

e acute

10°

f obtuse

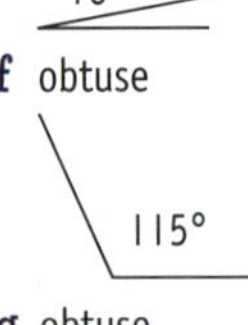

g obtuse

170°

Using a protractor

Page 24 – Your Turn

a Inside b Outside c Inside

Page 25 – Practice

1 a 125° b 160° c 30° d 93° e 155°

2 Adult to check

3 a 95° b 20° c 120°

2. ANGLES CONTINUED

Supplementary Angles

Page 26 – Your Turn

a 158° b 10° c 136° d 85° e 124° f 65° g 53°

Page 27 – Practice

1 a 159° b 77° c 9° d 137° e 88° f 30° g 65° h 170° i 94° j 110° k 33° l 146° m 85° n 48° o 178°

2 a 142° b 22° c 61° d 164° e 60° f 49° g 31° h 37° i 78° j 111° k 70°

3 a 103° 77° b 142° 38° c 134° 46°

Straight, Reflex and Revolution Angles

Page 28 – Your Turn

Adult to check

Page 29 – Practice

1 a revolution b reflex c straight d straight e reflex f reflex g straight

2 a C b E c B d A

3 Adult to check

4 Adult to check

5 Adult to check

Vertically Opposite Angles

Page 30 – Your Turn

a $x = 60°$, $y = 50°$

b $x = 35°$, $y = 145°$

Page 31 – Practice

1 a ∠DOE or ∠EOD b ∠AOB or ∠BOA c ∠BOF or ∠FOB d ∠COF or ∠FOC e ∠AOE or ∠EOA

2

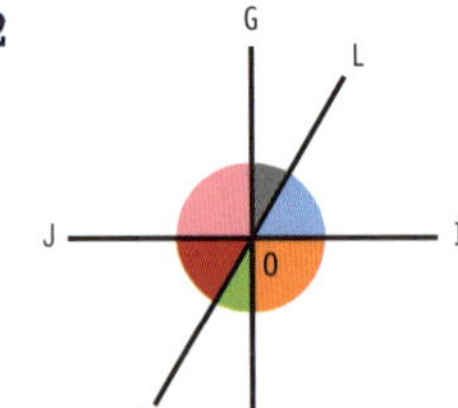

3 a $x = 90°$, $y = 90°$
b $x = 95°$, $y = 45°$
c $x = 35°$, $y = 95°$, $z = 50°$
d $x = 50°$, $y = 65°$
e $x = 64°$, $y = 55°$, $z = 61°$

Angle Sum of a Triangle

Page 32 – Your Turn

a $x = 100°$, scalene b $x = 75°$, isosceles c $x = 60°$, equilateral

Page 33 – Practice

1 a 180° ✔ b 180° ✔ c 180° ✔ d 185° ✘ e 170° ✘

2 a 30° b 99° c 45° d 65° e 40° f 60° g 30° h 75°

  ISBN: 9781925726190

Angle Sum of Quadrilaterals

Page 34 – Your Turn

a $x = 90°$ b $x = 130°$ c $x = 70°$

Page 35 – Practice

1 a $y = 70°$ b $y = 55°$ c $y = 95°$ d $y = 55°$ e $y = 100°$

2 a $50° + 145° + 110° + 55° = 360°$
b $90° + 90° + 115° + 65° = 360°$
c $50° + 195° + 55° + 60° = 360°$

Angles Review Page 36

1 90

2 Adult to check

3 a 55° b 75° c 66° d 9° e 48°

4 a less b greater/more

5 a 115° (blue) b 25° (red) c 140° (blue) d 55° (red) e 175° (blue)

6 Adult to check angles a acute b obtuse c obtuse d acute

7 a 180° b straight

8 a 116° b 148° c 79° d 13° e 130° f 31° g 177° h 169°

9 a 130°, 50° b 50°, 130° c 25°, 155° d 60°, 120°

10 a 114° 66°
b 33° 147°

c 104° 76°
d 82° 98°

11 Straight, 180°
Reflex, 180°, 360°
Revolution, 360°

12 Adult to check

13 equal

14 a $x = 50°, y = 130°$
b $x = 40°, y = 140°$
c $x = 32°, y = 148°$
d $x = 127°, y = 53°$
e $x = 138°, y = 42°, z = 138°$
f $x = 122°, y = 58°, z = 122°$
g $x = 154°, y = 26°, z = 26°$
h $x = 60°, y = 80°, z = 40°$
i $x = 159°, y = 21°, z = 159°$
j $x = 120°, y = 30°, z = 30°$
k $x = 85°, y = 47°, z = 48°$
l $w = 44°, x = 36°, y = 70°, z = 30°$
m $x = 25°, y = 54°, z = 101°$
n $w = 47°, x = 69°, y = 32°, z = 32°$

15 180°

16 a 97° b 60° c 20° d 81° e 63° f 44° g 42° h 60° i 60° j 50° k 53° l 47°

17 a scalene b equilateral c scalene d scalene e isosceles f right angled g scalene h right angled i equilateral j isosceles k scalene l scalene

18 360°

19 a 60° b 123° c 48° d 63°

3. 2D SHAPES

2D Shapes

Page 40 – Your Turn

a irregular quadrilateral (rectangle)
b regular hexagon
c regular triangle (equilateral)
d regular quadrilateral (square)
e irregular hexagon
f irregular triangle

Page 41 – Practice

1 a irregular pentagon
b regular heptagon
c regular dodecagon
d irregular heptagon
e regular triangle
f parallelogram (irregular quadrilateral)
g irregular octagon
h trapezium (irregular quadrilateral)
i regular octagon
j irregular triangle
k regular pentagon
l regular decagon
m irregular dodecagon
n regular hexagon
o regular nonagon
p irregular octagon
q irregular heptagon
r rhombus (irregular quadrilateral)
s irregular nonagon

2 Adult to check

3 Adult to check

4 Adult to check

5 Adult to check

6

	Name	Letters that help identify shape	No. angles	No. sides	No. vertices
a	heptagon	hepta	7	7	7
b	pentagon	penta	5	5	5
c	decagon	deca	10	10	10
d	hexagon	hexa	6	6	6
e	dodecagon	dodeca	12	12	12
f	nonagon	nona	9	9	9
g	octagon	octa	8	8	8

Types of Lines

Page 44 – Your Turn

Adult to check

Page 45 – Practice

1 Adult to check

2 a b c

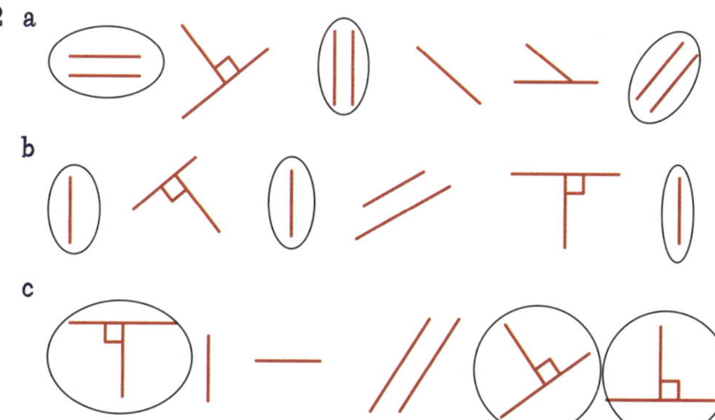

3. 2D SHAPES CONTINUED

Circles

Page 46 – Your Turn

1 a egg shape b oval c semicircle

Page 47 – Practice

1 Adult to check

2 a centre b arc c sector d diameter e chord f segment g radius h tangent

3 Adult to check

Triangles

Page 48 – Your Turn

1 a equilateral triangle
b right-angled triangle
c right angled triangle
d equilateral triangle
e isosceles triangle
f scalene triangle
g equilateral triangle
h right-angled triangle
i scalene triangle
j isosceles triangle

Page 49 – Practice

1 Adult to check

2 a isosceles triangle
b right-angled triangle
c scalene triangle
d scalene triangle
e right-angled triangle
f equilateral triangle
g isosceles triangle

Transformations

Page 50 – Your Turn

a reflection b translaton c rotation

Page 51 – Practice

1 a translation b reflection c rotation d translation

2 a 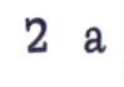    

b    

c   

d   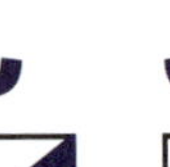  

3 a 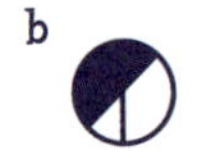   

b    

c    

4 a  

b  

c  

Tessellations

Page 52 – Your Turn

Adult to check

Page 53 – Practice

a ✗ b ✓ c ✓ d ✗ e ✓ f ✓ g ✗

2 a b

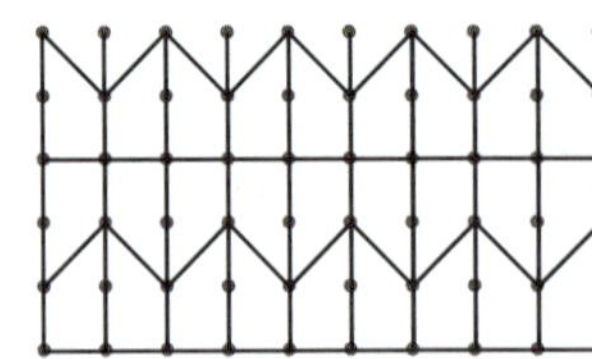

3 Adult to check

Symmetry

Page 54 – Your Turn

a 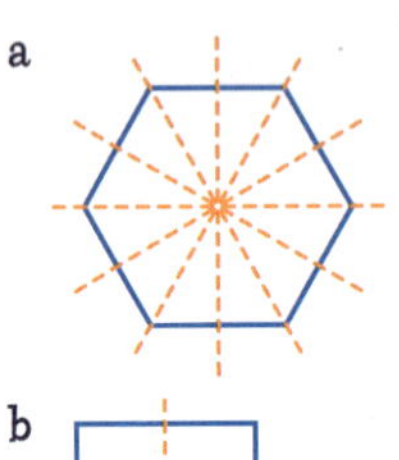

b 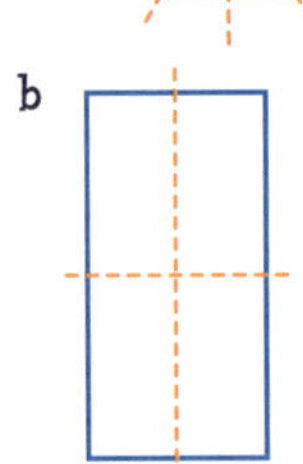

c 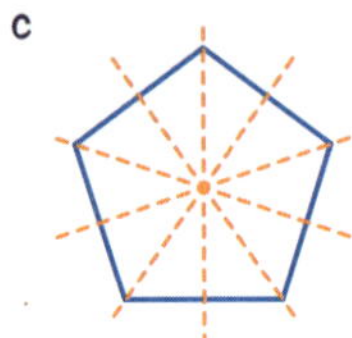

d 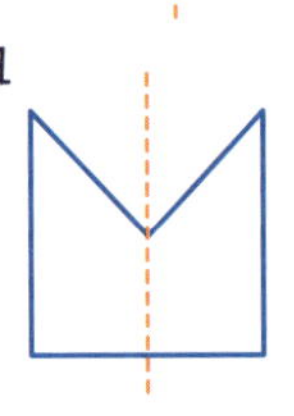

Page 55 – Practice

1 Adult to check

2 a b c

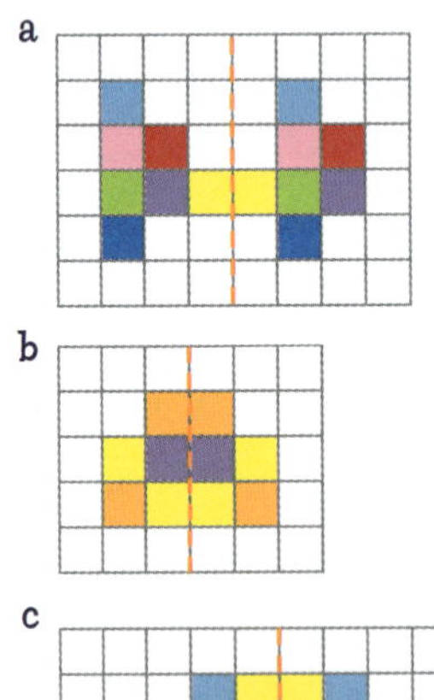

d 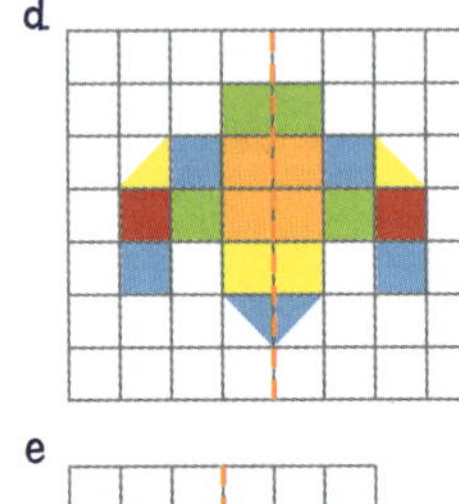

e 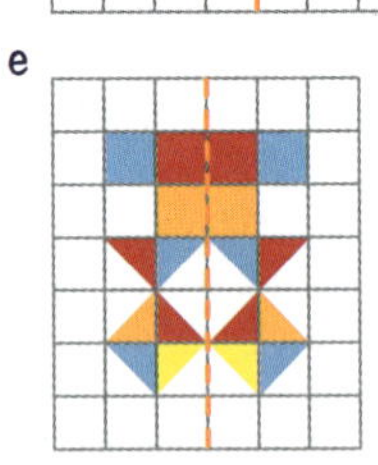

CATCH UP MATHS YEAR 6 BOOK B © PASCAL PRESS ISBN: 9781925726190

3

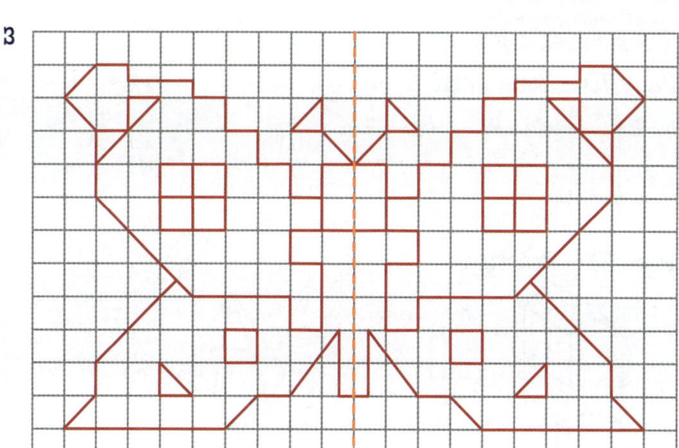

2D Shapes Review Page 56

1 dimensions, length, flat

2

	Name	Letters that help identify shape	No. angles	No. sides	No. vertices
a	heptagon	hepta	7	7	7
b	pentagon	penta	5	5	5
c	decagon	deca	10	10	10
d	hexagon	hexa	6	6	6
e	quadrilateral	quad	4	4	4
f	nonagon	nona	9	9	9
g	triangle	tri	3	3	3
h	octagon	octa	8	8	8
i	dodecagon	dodeca	12	12	12

3 Adult to check

4 Adult to check

5 Adult to check

6 Adult to check

7 a circle (regular) b egg shape (irregular) c oval (irregular) d semicircle (irregular)

8 a right-angled triangle
b isosceles triangle
c equilateral triangle
d scalene triangle

9 Adult to check

10 a right-angled triangle
b scalene triangle
c isosceles triangle
d equilateral triangle
e scalene triangle
f isosceles triangle
g equilateral triangle
h right-angled triangle

11 a slide b turn c flip

12 a reflection b translation c rotation

13 a reflection b rotation c translation

14 a rotation
b translation
c reflection
d translation
e rotation

15 a

b

c

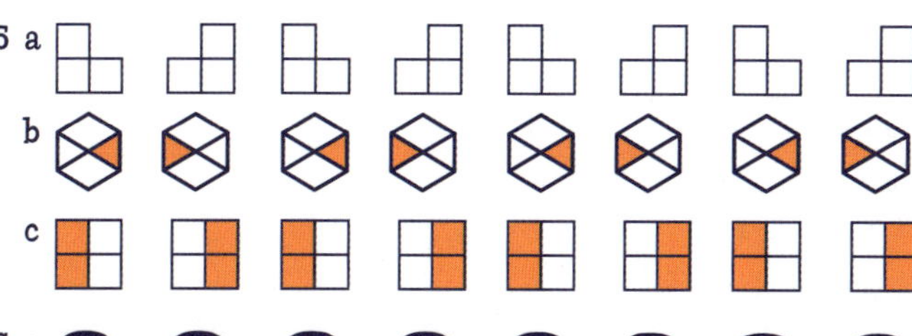

16 a

b

c

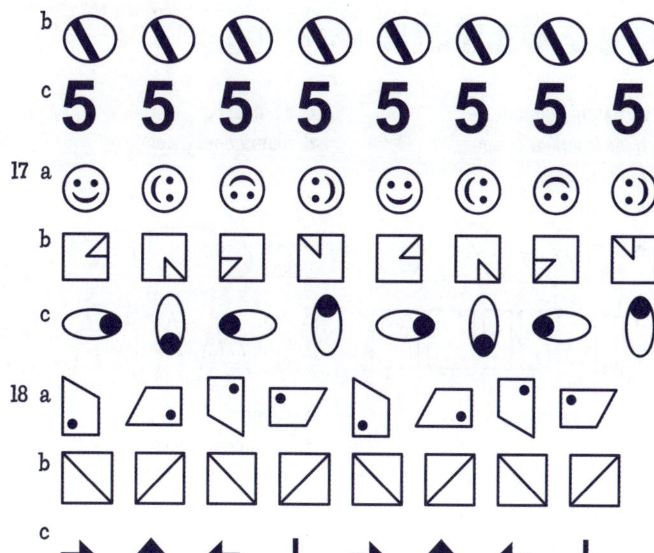

17 a

b

c

18 a

b

c

19 For example: square, rectangle, equilateral triangle, right-angled triangle, hexagon

20 No

21 Adult to check

22 Adult to check

23 Tick a b c d e f

24 a

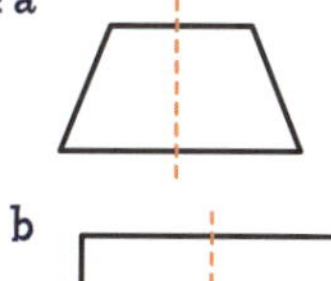

b

c

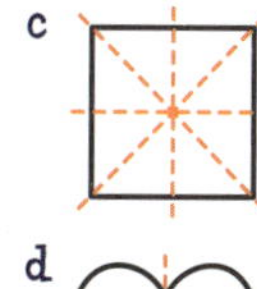

d

e

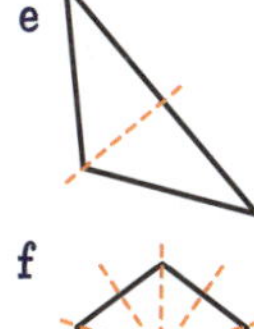

f

25 a

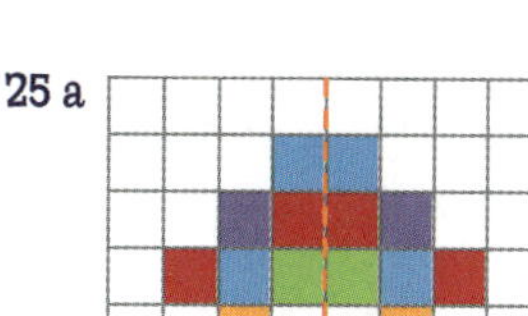

c

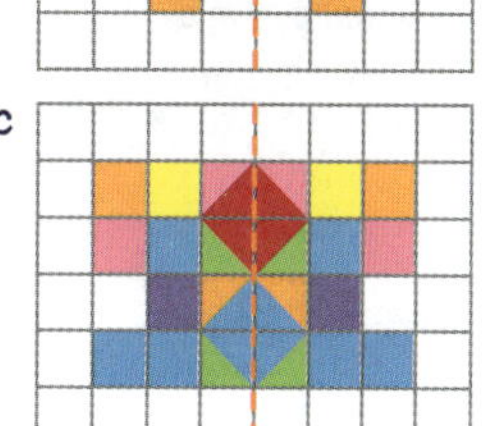

e

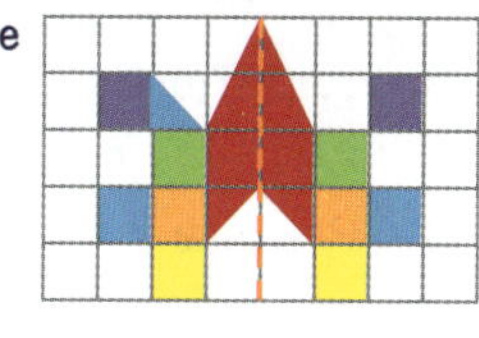

4. 3D OBJECTS

Prisms and Pyramids

Page 62 – Your Turn

Adult to check

Page 63 – Practice

1 a hexagonal prism
b cone
c sphere
d pentagonal prism
e triangular prism
f cube
g triangular pyramid
h cylinder
i octagonal prism
j pentagonal pyramid
k octagonal pyramid

2 Circle blue: A, C, D
Circle red: B, E

4. 3D OBJECTS CONTINUED

3 a square pyramid
b rectangular prism
c cylinder
d pentagonal pyramid

4 Adult to check

5 Adult to check

6 a 

b 

c

d 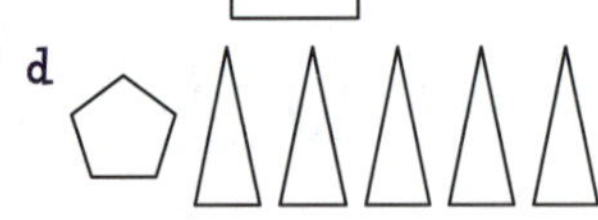

7 a hexagonal prism
b cone
c triangular pyramid
d rectangular pyramid
e cylinder
f octagonal prism
g sphere

8 Colour a c d i

9 a triangular prism: 2 triangular bases, 3 rectangular faces
b octagonal prism: 2 octagonal bases, 8 rectangular faces
c pentagonal prism: 2 pentagonal bases, 5 rectangular faces
d octagonal pyramid: 1 octagonal base, 8 triangular faces
e hexagonal pyramid: 1 hexagonal base, 6 triangular faces

Drawing 3D Objects

Page 66 – Your Turn

Adult to check

Page 67 – Practice

1 Adult to check
2 Adult to check
3 Adult to check

Faces, Edges and Corners

Page 68 – Your Turn

a 7, 15, 10 b 4, 6, 4 c 7, 12, 7

Page 69 – Practice

1 a cone
b square pyramid
c hexagonal prism
d cylinder
e hexagonal pyramid
f pentagonal prism
g rectangular pyramid

2

	Name	Object	Vertices	Edges	Faces
a	sphere		0	0	1
b	hexagonal prism		12	8	8
c	cone		1	0	1
d	pentagonal prism		10	15	7
e	square pyramid		5	8	5
f	cylinder		0	0	2

Cross-sections

Page 70 – Your Turn

a 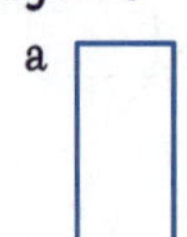

b 

c 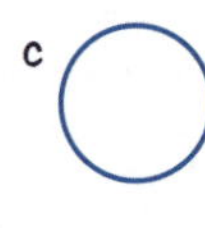

Page 71 – Practice

1 a 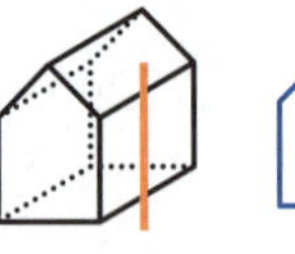

d 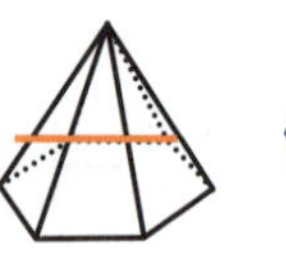

b 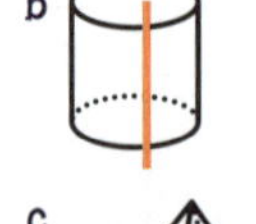

c 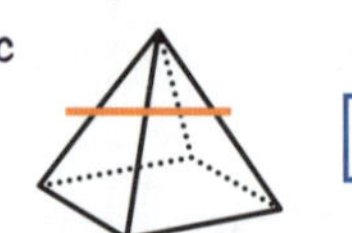

e 

2 a triangle b rectangle c triangle

3 a

b 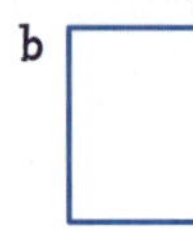

c 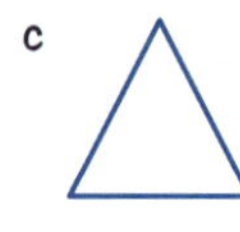

4 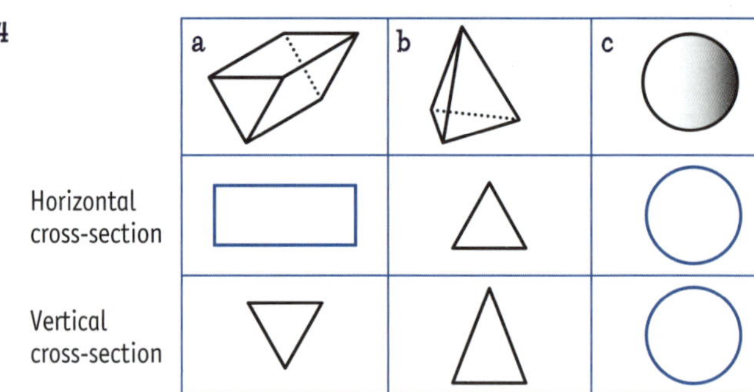

Nets

Page 72 – Your Turn

a 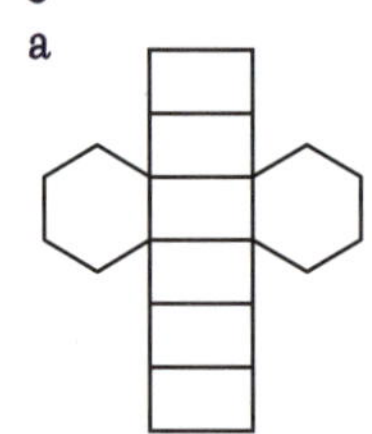

b 

c 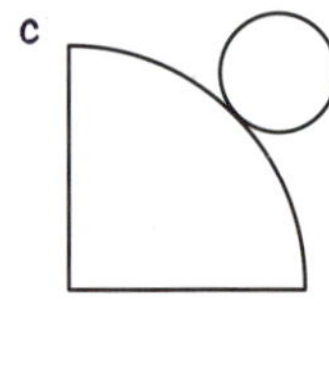

Page 73 – Practice

1 A and D, B and F, C and I, E and G, H and J

2 Sample answer:

3 a ✕ b ● c ♥

4 a B b C c A

  ISBN: 9781925726190

5

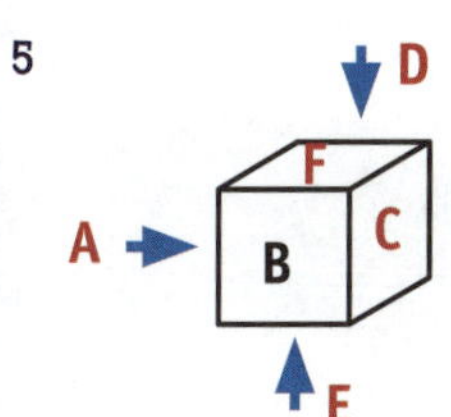

Views

Page 74 – Your Turn

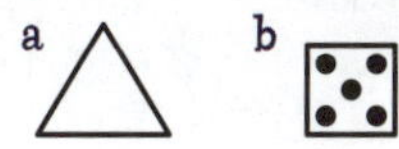

Page 75 – Practice

1 Adult to check

2 a

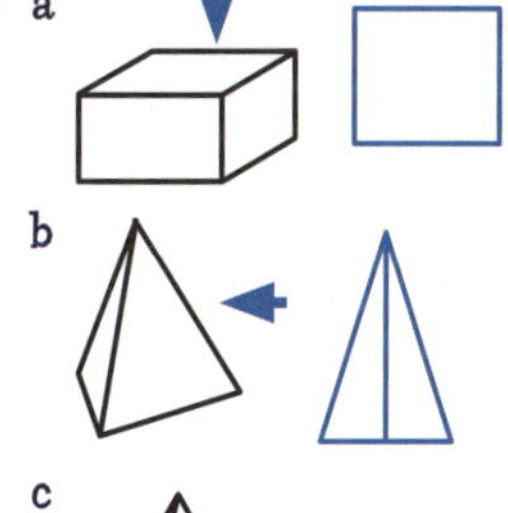

d

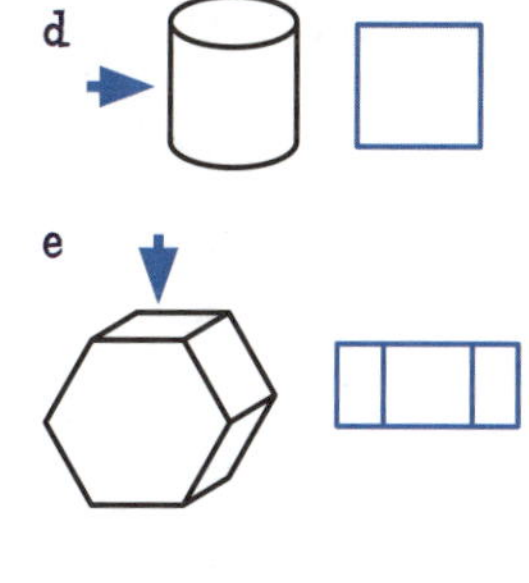

b

e

c

2

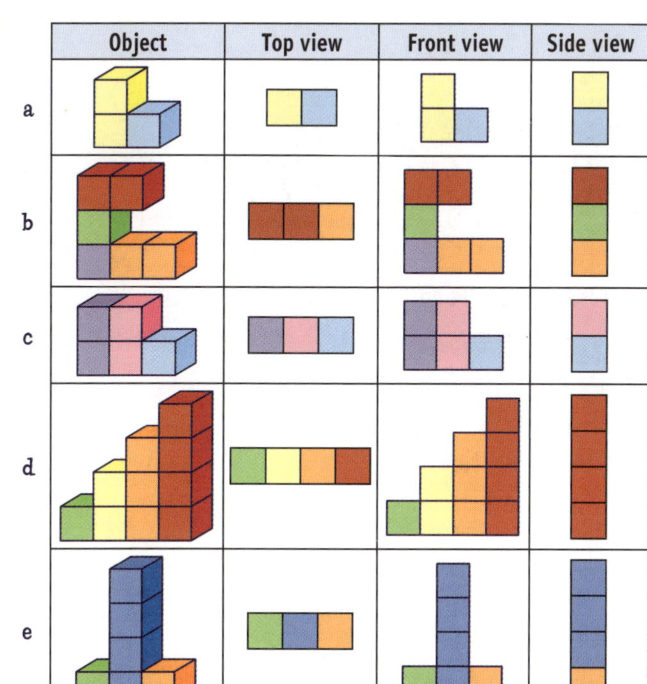

3D Objects Review Page 76

1 a width
b cones, cylinders, spheres
c bases, bases, rectangles
d base, triangles, apex

2 a cube
b triangular pyramid
c cone
d rectangular prism
e sphere
f rectangular pyramid
g octagonal prism
h hexagonal pyramid
i triangular prism
j cylinder
k pentagonal pyramid
l square pyramid
m pentagonal prism
n octagonal pyramid

3 Adult to check

4 Adult to check

5 Adult to check

6 flat, 3D, edge, vertex

7

	Name	Vertices	Edges	Faces
a	cube	8	12	6
b	triangular pyramid	4	6	4
c	octagonal prism	16	24	10
d	cylinder	0	0	2
e	hexagonal prism	12	18	8
f	rectangular prism	8	12	6
g	cone	1	0	1
h	sphere	0	0	0

8 Adult to check

9 Adult to check

10 a ● b A c ✕

11 a rectangular pyramid
b cylinder/ sphere
c cone
d hexagonal prism
e triangular prism
f pentagonal pyramid

12 a cube
b rectangular prism
c triangular prism
d pentagonal prism
e sphere
f hexagonal prism

13

	Object	Top view	Front view	Side view
a				
b				
c				
d				
e				
f				

5. AREA

The Square Centimetre

Page 82 – Your Turn

a A b B and C c 31 cm^2

Page 83 – Practice

1 a 5 cm^2 b 2 cm^2 c 16 cm^2 d 6 cm^2 e 7 cm^2 f 9 cm^2 g 9 cm^2

2 a irregular pentagon b triangle c irregular hexagon d rectangle e irregular octagon f triangle g square

3 a b b c c f and g

4 a 9 cm^2 b 1 cm^2 c 7 cm^2

5 Adult to check

The Square Metre

Page 84 – Your Turn

1 a Blue: shape c
 b Red: shape d
 c Green: shapes a and b

2 a 16 m^2 b 19 m^2 c 18 m^2 d 17 m^2

Page 85 – Practice

1 Tick: b g h

2 a 9 m^2 b 58 m^2 c 40 m^2 d 110 m^2 e 363 m^2

3 a m^2 b m^2 c m^2 d cm^2 e cm^2

4 a 12 m^2 b 10 m^2 c 9 m^2 d 9 m^2 e 6 m^2 f 10 m^2 g 4 m^2 h 2 m^2 i 1 m^2

5 108 m^2

6 a entry b toys c books d baby clothes e men's clothes and food

Hectares

Page 86 – Your Turn

1 Colour: b d e i

Page 87 – Practice

1 a 49 hectares b 79 hectares c 66 hectares d 352 hectares e 210 hectares f 149 hectares g 3764 hectares h 4935 hectares i 3980 hectares j 6002 hectares k 5963 hectares

2 a 2.4 b 7.1 c 14.9 d 87.2 e 20.9 f 14.2 g 90 000 h 460 000 i 3 690 000 j 4 810 000 k 1 560 000 l 6 390 000

3 a 20 ha b 15 ha c 15 ha d 10 ha e 24 ha

4 90 ha

The Square Kilometre

Page 88 – Your Turn

a 40 km^2 b 9 km^2 c 700 km^2 d 1634 km^2 e 4902 km^2 f 8076 km^2

Page 89 – Practice

1 1 Russia 2 Canada 3 USA 4 China 5 Brazil 6 Australia 7 India 8 Argentina 9 Kazakhstan 10 Algeria

2 a 9 357 022 km^2 b 562 363 km^2 c 6 816 560 km^2

3 a 2 330 000 km^2 b 450 000 km^2 c 620 000 km^2 d 492 000 km^2

4 a 8 100 000 km^2 b 675 000 km^2

Using Multiplication to Calculate Area – Squares and Rectangles

Page 90 – Your Turn

a length × width
 = 6 m × 6 m
 = 36 m^2

Page 91 – Practice

1 a 49 m^2 b 20 m^2 c 8 cm^2 d 100 cm^2 e 42 cm^2 f 4 m^2 g 96 m^2

2 a Area 1 = 36 m^2, Area 2 = 4 m^2, Area = 32 m^2
 b Area 1 = 80 m^2, Area 2 = 4 m^2, Area 3 = 1 m^2, Area 4 = 2 m^2, Area = 73 m^2

Using Multiplication to Calculate Area – Triangles

Page 93 – Your Turn

a $\frac{1}{2}$ × length × width
 = $\frac{1}{2}$ × 11 m × 4 m
 = 22 m^2

Page 94 – Practice

1 a 24 cm^2 b 30 m^2 c 12 cm^2 d 42 m^2 e 5 cm^2 f 25 cm^2 g 24 m^2

Perimeter

Page 95 – Your Turn

a P = 2 cm + 2 cm + 2 cm + 2 cm + 2 cm + 2 cm + 2 cm + 2 cm
 = 16 cm
b P = 4 m + 4 m + 4 m + 4 m + 4 m
 = 20 m

Page 96 – Practice

1 a 13 cm b 12 m c 8 m d 16 m e 16 cm

2 a 36 m b 32 m

3 a 42 cm b 44 cm c 34 cm d 60 m e 66 m

Area Review Page 98

1 a $4\frac{1}{2}$ cm^2 b 8 cm^2 c 6 cm^2 d 9 cm^2 e 8 cm^2 f $6\frac{1}{2}$ cm^2 g 7 cm^2 h 7 cm^2 i 5 cm^2 j 4 cm^2 k 12 cm^2 l $4\frac{1}{2}$ cm^2

2 a triangle b rectangle c octagon d square e triangle f heptagon g octagon h pentagon i octagon j hexagon k rectangle l pentagon

3 Adult to check

4 a 6 m^2 b 4 m^2 c 8 m^2 d 10 m^2 e 6 m^2 f 6 m^2

5 a indoor plants
 b trees

6 56 m^2

  ISBN: 9781925726190

7 a 2 m^2 b 2 m^2 c 2 m^2

8 1 hectare = 10 000 m^2
100 m × 100 m

9 a 70 000 m^2 b 50 000 m^2 c 10 000 m^2 d 100 000 m^2

10 Adult to check

11 a 2.9 b 4.6 c 8.2 d 1 e 7.7 f 11.4

12 Adult to check

13 a 27 km^2 b 64 km^2 c 100 km^2 d 602 km^2 e 2894 km^2

14 a 9 m^2 b 25 cm^2 c 4 cm^2 d 49 m^2

15 a 12 m^2 b 8 m^2 c 168 cm^2 d 60 m^2 e 15 cm^2 f 300 m^2

16 a 36 cm^2 – 3 cm^2 = 33 cm^2
b 80 m^2 – (6 m^2 + 6 m^2) = 68 m^2

17 a 10 cm^2 b 30 cm^2 c 20 cm^2 d 60 m^2 e 36 cm^2 f 48 m^2

18 a 24 cm b 16 m c 6 m d 21 cm e 28 cm f 44 m g 22 m h 54 cm i 30 m j 28 cm k 36 m

6. VOLUME & CAPACITY

Millitres and Litres

Page 104 – Your Turn

a 15 mL b 500 mL c 250 mL d 2 L e 10 L

Page 105 – Practice

1 a 6 L b 39 L c 88 mL d 100 mL e 380 mL f 8 L g 52 L h 2000 mL i 5000 mL

2 a 250 mL b 500 mL c 750 mL

3 a 2 L b $2\frac{1}{4}$ L c $1\frac{3}{4}$ L d $8\frac{1}{4}$ L e $4\frac{1}{2}$ L f $10\frac{1}{2}$ L g $6\frac{3}{4}$ L h 5 L i $7\frac{3}{4}$ L

4 a 3250 mL b 4000 mL c 6250 mL d 9500 mL e 7750 mL f 5500 mL g 8750 mL h 12 250 mL i 10 500 mL

5 a 0 L 490 mL b 1 L 810 mL c 3 L 590 mL d 2 L 660 mL e 8 L 750 mL f 6 L 480 mL g 5 L 370 mL h 7 L 230 mL i 4 L 930 mL

6 a 9 L 250 mL b 12 L 500 mL c 15 L 440 mL d 5 L 250 mL e 12 L 190 mL

7 a 1500 mL b 3.575 L c 7425 mL d 2.65 L e 3200 mL f 4.9 L g 8750 mL h 9325 mL i 6600 mL j 5.85 L k 2.7 L

8 a 4375 mL b 2750 mL c 7125 mL d 5450 mL e 6900 mL f 9575 mL g 15 600 mL h 11 820 mL i 12 665 mL

9 a 3.125 L b 7.52 L c 5.57 L d 2.32 L e 4.645 L f 6.45 L g 8.28 L h 9.795 L i 2.59 L

10 a 3, 1, 4, 2, 5 b 5, 3, 4, 2, 1

11 a 300 mL b 2250 mL c 500 mL d 250 mL e 150 mL f 4500 mL g 150 mL

6. VOLUME & CAPACITY CONTINUED

12 a

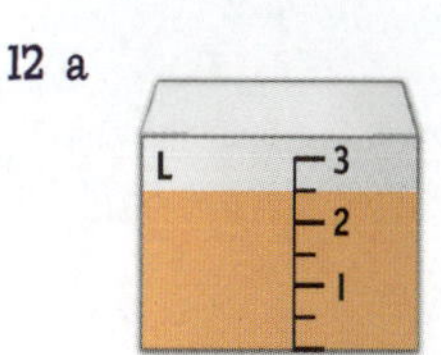

b

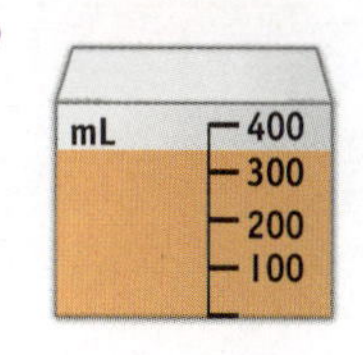

c

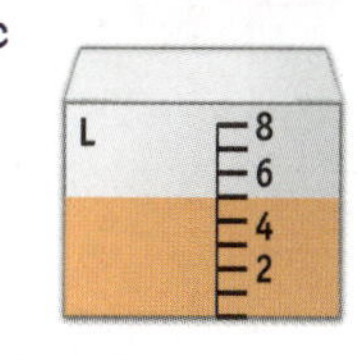

d

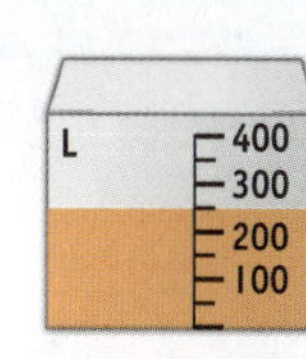

e

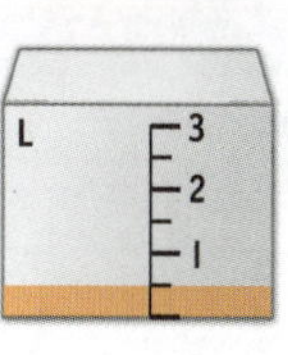

f

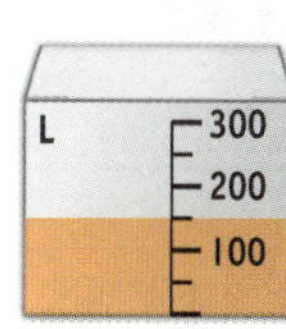

g

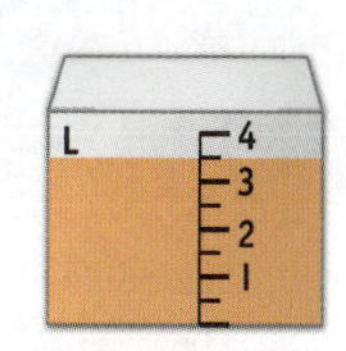

Capacity

Page 108 – Your Turn

4, 3, 2, 6, 1, 5

Page 109 – Practice

1 Adult to check

2 Adult to check

3 a L b L c L d mL e L f L g L

4 a 1 L b 1.3 L c 35 L d 9 L e 1.55 L

Volume

Page 110 – Your Turn

Colour B, D, F

Page 111 – Practice

1 cm^3

2 a 18 b 14

3 a 1 b 2 c 2, 3, 1 d 4 cm^3 e 8 cm^3 f 4 cm^3 g 42 cm^3

4 Adult to check

Cubic Metres

Page 112 – Your Turn

Adult to check

Page 113 – Practice

1 1 cm × 1 cm × 1 cm = 1 cm^3
1 m × 1 m × 1 m = 1 m^3

2 a 15 m^3 b 24 m^3 c 38 m^3 d 157 m^3 e 295 m^3 f 1495 m^3 g 6258 m^3

3 a 10 cubic metres b 19 cubic metres c 26 cubic metres d 392 cubic metres e 748 cubic metres f 1264 cubic metres g 8976 cubic metres

4 a yellow box 12 m^3 b green box 5 m^3 c pink box 7 m^3 d blue box 10 m^3 e orange box 14 m^3

6. VOLUME & CAPACITY CONTINUED

Displacement

Page 114 – Your Turn

1 a 2, 3, 1
  b 200, 600, 400

Page 115 – Practice

1 a 200 mL, 200 cm^3
  b 150 mL, 150 cm^3
  c 250 mL, 250 cm^3
  d 1500 mL, 1500 cm^3
  e 200 mL, 200 cm^3

2

	Water Level		Displacement	Volume of object
	Before	After		
a	2500 mL	4000 mL	1500 mL	1500 cm^3
b	3200 mL	4800 mL	1600 mL	1600 cm^3
c	150 mL	230 mL	80 mL	80 cm^3
d	500 mL	1000 mL	500 mL	500 cm^3
e	4250 mL	5 L	750 mL	750 cm^3
f	6 L	8.250 L	2250 mL	2250 cm^3
g	450 mL	610 mL	160 mL	160 cm^3

Measuring Volume

Page 116 – Your Turn

a = length × width × height
= l × w × h
= 9 cm × 2 cm × 4 cm
= 72 cm^3

Page 117 – Practice

1 a 96 cm^3 b 540 cm^3 c 52 m^3
2 a 24 cm^3 b 4 cm c 252 m^3 d 12 cm e 75 m^3 f 7 cm g 3.5 m h 9.5 m i 36 m^3 j 8 cm

Volume and Capacity Review Page 118

1 a 8 mL b 6 L c 12 L d 24 mL e 97 mL f 105 L g 254 L h 19 L i 430 mL j 380 mL
2 a $2\frac{1}{2}$ L b 3 L c $1\frac{1}{4}$ L d $6\frac{3}{4}$ L e $2\frac{1}{4}$ L f 6 L
3 a 1000 mL b 750 mL c 500 mL d 250 mL
4 a 8500 mL b 5750 mL c 2250 mL d 4000 mL e 7500 mL f 1750 mL
5 a 6 L 200 mL b 3 L 490 mL c 1 L 810 mL d 2 L 730 mL e 10 L 365 mL f 23 L 428 mL
6 a 2.805 L b 6.23 L c 8.46 L d 4.375 L e 3.71 L f 12.436 L
7 a 10 325 mL b 4984 mL c 6485 mL d 9610 mL e 23 120 mL f 8600 mL
8 a 4, 2, 5, 1, 3 b 3, 5, 4, 1, 2 c 3, 1, 4, 2, 5
9 a 3 L b 1.5 L c 500 mL d 6 L

10 a

b 2 L, 1

c mL 600, 400, 200

d 8 L, 6, 4, 2

11 liquid, hold
12 a left container b right container c left container
13 Adult to check
14 a mL b L c L d mL
15 a 2100 mL b 1125 mL c 18 250 mL d 11 500 mL
16 cm^3
17 a 32 b 60 c 24 d 72
18 a 40 cm^3 b 8 cm^3 c 12 cm^3 d 36 cm^3 e 48 cm^3 f 28 cm^3
19 188 cm^3
20 1 m^3
21 Adult to check
22 a 9 m^3 b 374 m^3 c 73 m^3 d 1428 m^3 e 25 m^3 f 47 m^3 g 169 m^3 h 862 m^3
23 a 2 cubic metres b 63 cubic metres c 12 cubic metres d 128 cubic metres e 8 cubic metres f 25 cubic metres g 57 cubic metres h 198 cubic metres

24

	Water level		Displacement	Volume of object
	Before	After		
a	300 mL	650 mL	350 mL	350 cm^3
b	500 mL	1000 mL	500 mL	500 cm^3
c	2000 mL	2130 mL	130 mL	130 cm^3
d	1000 mL	3200 mL	2200 mL	2200 cm^3
e	400 mL	1850 mL	1450 mL	450 cm^3
f	3250 mL	4930 mL	1680 mL	1680 cm^3
g	4650 mL	5045 mL	395 mL	395 cm^3
h	1275 mL	1895 mL	620 mL	620 cm^3

25 a 30 m^3 b 360 cm^3 c 16 560 mm^3 d 30 514 cm^3
26 a 6 cm b 3 m c 48 mm^3 d 8 cm e 4 m f 960 m^3 g 480 cm^3 h 9 mm i 3.4 m j 6.2 m

7. MASS

Kilograms

Page 124 – Your Turn

1 Circle: b, e, h, i, k
2 a 9 kg b 16 kg c 20 kg d 73 kg e 105 kg f 250 kg g 748 kg

Page 125 – Practice

1 a 2 kg b 5 kg c 1 kg
2 a a child b a chicken c shopping bag

  ISBN: 9781925726190

7. MASS CONTINUED

3 a 3250 g b 6750 g c 4000 g d 5250 g e 7500 g

4 a 20 b 5 c 4 d 2 e 1

5 a 4, 3, 5, 1, 2 b 5, 2, 4, 1, 3 c 4, 2, 5, 1, 3

6 a

kg	grams
$\frac{1}{2}$ kg	500 g
5 kg	5000 g
2 kg	2000 g
$3\frac{3}{4}$ kg	3750 g
$7\frac{1}{2}$ kg	7500 g

b

kg	grams
0.25 kg	250 g
1.68 kg	1680 g
2.75 kg	2750 g
0.75 kg	750 g
0.3 kg	300 g

c

kg, g	grams
$2\frac{3}{4}$ kg	2750 g
2 kg 300 g	2300 g
5 kg 625 g	5625 g
4 kg 750 g	4750 g
$4\frac{1}{4}$ kg	4250 g

d

kg	grams
12 kg	12 000 g
$14\frac{1}{4}$ kg	14 250 g
$13\frac{1}{2}$ kg	13 500 g
$12\frac{3}{4}$ kg	12 750 g
$11\frac{3}{4}$ kg	11 750 g

e

kg	grams
8.125 kg	8125 g
8.25 kg	8250 g
8.3 kg	8300 g
8.75 kg	8750 g
8.98 kg	8980 g

f

kg, g	grams
4 kg 4 g	4004 g
4 kg 8 g	4008 g
4 kg 400 g	4400 g
4 kg 425 g	4425 g
4 kg 800 g	4800 g

7 a 8050, 8.05 b 4427, 4.427 c 3060, 3.06 d 205, 2.05

8 a 726 g = 0.726 kg
b 85.2 g = 0.0852 kg
c 1510 g = 1.51 kg
d 4125 g = 4.125 kg
e 80.6 g = 0.0806 kg

9 a 3120 g, 3.12 kg
b 1575 g, 1.575 kg
c 675 g, 0.675 kg
d 1200 g, 1.2 kg
e 4800 g, 4.8 kg
f 2250 g, 2.25 g
g 213 g, 0.213 kg
h 420 g, 0.420 kg

10 a 3120 g b 5650 g c 2460 g d 2750 g e 4350 g

11 a 10 kg b 28.33 kg c 15.65 kg d 8.11 kg e 14.35 kg f 10.22 kg g 15.58 kg h 5.21 kg

12 boots, brick, lunchbox, ladder

Grams

Page 128 – Your Turn

1 a 24 g b 37 g c 99 g d 170 g e 250 g f 375 g g 420 g h 750 g

2 Adult to check

Page 129 – Practice

1 a ✗ b ✗ c ✓ d ✗ e ✗ f ✗ g ✗ h ✓ i ✓

2 1, 3, 5, 6, 2, 4, 7

3 a 875 g b 900 g c 500 g d 245 g e 755 g f 975 g g 925 g

4 a 175 g = 0.175 kg
b 500 g = 0.5 kg
c 490 g = 0.49 kg
d 3775 g = 3.775 kg
e 750 g = 0.75 kg

5 a 156 g b 2735 g c 6700 g d 3070 g e 5803 g f 10 420 g g 200 g h 530 g

Reading Scales

Page 130 – Your Turn

a 4 × 100 g, 1 × 25 g
b 3 × 100 g, 1 × 50 g

Page 131 – Practice

1 a 2 kg b 1.5 kg c 7 kg

2 a 2000 g b 1500 g c 7000 g

3 orange, brown, green, blue

4 a

b

c

d

e

f

g

5 a 2500 g = 2.5 kg
b 200 g = 0.2 kg
c 250 g = 0.25 kg
d 550 g = 0.55 kg
e 750 g = 0.75 kg
f 1200 g = 1.2 kg
g 1050 g = 1.05 kg

The Tonne

Page 132 – Your Turn

1 Colour: a, d, g

2 a 500 kg b 750 kg c 4500 kg d 3750 kg e 5250 kg

7. MASS CONTINUED

Page 133 – Practice

1

	Grams	Kilograms	Tonnes
a	50 000	50	0.05
b	700 000	700	0.7
c	90 000	90	0.09
d	400 000	400	0.4
e	100 000	100	0.1
f	1 000 000	1000	1
g	6 000 000	6000	6
h	7 000 000	7000	7
i	4 500 000	4500	4.5
j	5 750 000	5750	5.75
k	13 250 000	13 250	13.25
l	9 500 000	9500	9.5
m	3 000 000	3000	3
n	4 000 000	4000	4
o	20 000	20	0.02
p	8 250 000	8250	8.25
q	9 750 000	9750	9.75

2 a 5 b 10 c 20 d 1 e 40

3 a False b True c True d True e True f True g True h False i False j False k False l True m False n True o True

Gross Mass and Net Mass

Page 134 – Your Turn

a 95 g, 435 g, 530 g, 435 g

Page 135 – Practice

1 a 100 g b 5 g c 310 g d 575 g e 15 g f 20 g g 195 g h 25 g

2 a 280 g b 650 g c 775 g d 120 g e 750 g f 255 g g 570 g h 215 g

Best Buys

Page 136 – Your Turn

a 5 kg for $24.00 b 2 kg for $11.50 c 1 kg for $4.14

Page 137 – Practice

1 a 250 g for 70c b 500 g for $10.30 c 4 kg for $5.00 d 100 g for $1.55 e 600 g for $8.00 f 1 kg for $5.90 g 1 kg for $9.00 h 1 kg for $2.95 i 5 kg for $12.00 j 100 g for $2.50 k 500 g for $9.50 l 750 g for $4.20 m 500 g for $5.50

Mass Review Page 138

1 a ✗ b ✓ c ✗ d ✓ e ✗ f ✓ g ✓ h ✗ i ✓ j ✓

2 a 3 kg b 74 kg c 56 kg d 189 kg e 725 kg

3 a 80 kg b 2 kg c 5 kg d 8 kg

4 a 3500 g b 6250 g c 2750 g d 5000 g e 15 250 g f 12 000 g

5

25 g	50 g	100 g	125 g	200 g	250 g	500 g	1 kg
975 g	950 g	900 g	875 g	800 g	750 g	500 g	0 g

6 a 3, 1, 5, 2, 4 b 1, 3, 4, 2, 5 c 2, 1, 4, 3, 5

7 a

kg	grams
$\frac{1}{4}$ kg	250 g
$3\frac{3}{4}$ kg	3750 g
$\frac{1}{2}$ kg	500 g
5 kg	5000 g
$7\frac{1}{4}$ kg	7250 g
$8\frac{1}{2}$ kg	8500 g

b

kg	grams
0.75 kg	750 g
2.68 kg	2680 g
0.4 kg	400 g
1.34 kg	1340 g
4 kg	4000 g
3.62 kg	3620 g

c

kg	grams
$3\frac{1}{4}$ kg	3250 g
3.5 kg	3500 g
3 kg 200 g	3200 g
3 kg 750 g	3750 g
3 kg	3000 g
0.3 kg	300 g

8 a 3370 g = 3.37 kg b 6700 g = 6.7 kg c 5827 g = 5.827 kg d 1680 g = 1.68 kg e 2.6 g = 0.0026 kg f 12 870 g = 12.87 kg g 38 500 g = 38.5 kg h 6100 g = 6.1 kg

9 a 1240 g = 1.24 kg b 1440 g = 1.44 kg c 2125 g = 2.125 kg d 1050 g = 1.05 kg e 510 g = 0.51 kg f 256 g = 0.256 kg g 2544 g = 2.544 kg h 882 g = 0.882 kg

10 a 8 g b 12 g c 26 g d 59 g e 173 g f 245 g

11 Adult to check

12 a 760 g b 275 g c 670 g d 5 g e 475 g f 390 g g 102 g h 900 g i 545 g j 657 g k 92 g l 212 g

13

	Item	Weight	Grams to make 1 kg
a	packet of chips	90 g	910 g
b	tin of tuna	425 g	575 g
c	protein bar	25 g	975 g
d	bag of lollies	245 g	755 g
e	jar of jam	340 g	660 g

14 a 2125 g = 2.125 kg b 720 g = 0.72 kg c 1470 g = 1.47 kg d 225 g = 0.225 kg e 1360 g = 1.36 kg

ANSWERS

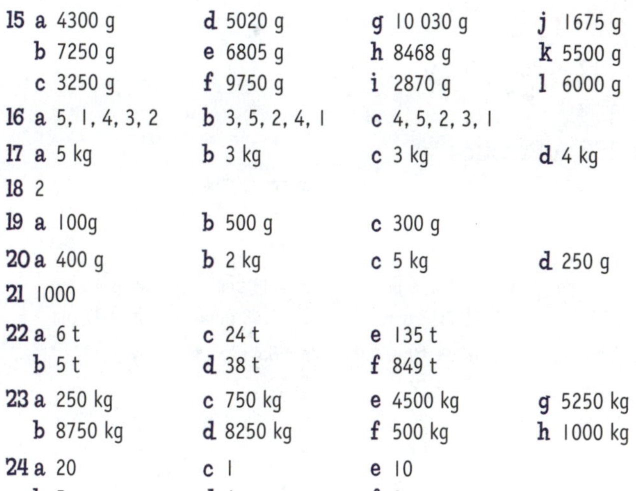

15 a 4300 g
b 7250 g
c 3250 g
d 5020 g
e 6805 g
f 9750 g
g 10 030 g
h 8468 g
i 2870 g
j 1675 g
k 5500 g
l 6000 g

16 a 5, 1, 4, 3, 2 b 3, 5, 2, 4, 1 c 4, 5, 2, 3, 1

17 a 5 kg b 3 kg c 3 kg d 4 kg

18 2

19 a 100g b 500 g c 300 g

20 a 400 g b 2 kg c 5 kg d 250 g

21 1000

22 a 6 t
b 5 t
c 24 t
d 38 t
e 135 t
f 849 t

23 a 250 kg
b 8750 kg
c 750 kg
d 8250 kg
e 4500 kg
f 500 kg
g 5250 kg
h 1000 kg

24 a 20
b 5
c 1
d 4
e 10
f 2

25

	Grams	Kilograms	Tonnes
a	20 000	20	0.02
b	60 000	60	0.06
c	90 000	90	0.09
d	50 000	50	0.05
e	45 000	45	0.045
f	85 000	85	0.085
g	15 500	15.5	0.0155
h	3 000 000	3000	3
i	7 000 000	7000	7
j	8 200 000	8200	8.2
k	6350	6.350	0.006 35
l	9 110 000	9110	9.11

26 a 375 g
b 45 g
c 725 g
d 35 g
e 490 g
f 440 g
g 5 g
h 10 g

27 a 1 kg for $10.00
b 250 g for $2.45
c 500 g for $9.95
d 250 g for $5.50
e 100 g for $5.50
f 500 g for $8.90
g 1 kg for $3.85
h 300 g for $2.50

8. TIME

O'clock and Half Past

Page 144 – Your Turn

a 

b 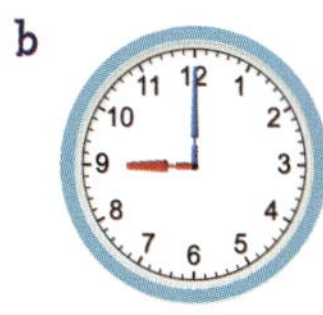

c 

Page 145 – Practice

1 a 

b 

c 

8. TIME CONTINUED

2 a 5:30 b 10:00 c 09:30

3 a half past 4, 4:30, four thirty
b six o'clock, 6:00, six o'clock
c five o'clock, 5:00, five o'clock
d half past 11, 11:30, eleven thirty

4 a 
half past 12
12:30
12 thirty

b 
8 o'clock
8:00
8 o'clock

c 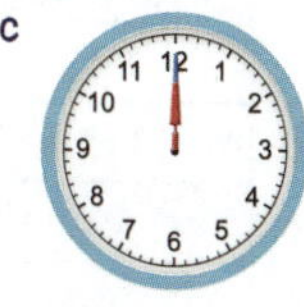
12 o'clock
12:00
12 o'clock

Quarter Past and Quarter To

Page 146 – Your Turn

a 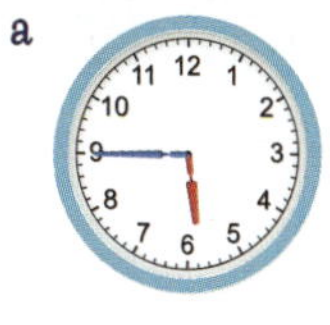

b 

c 

Page 147 – Practice

1 a 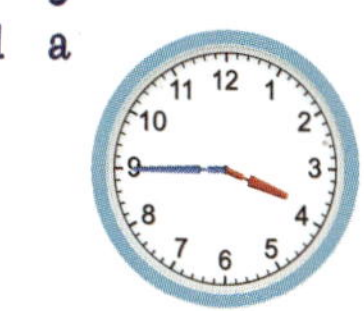

b 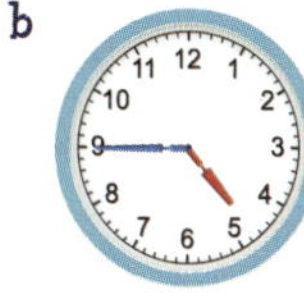

c 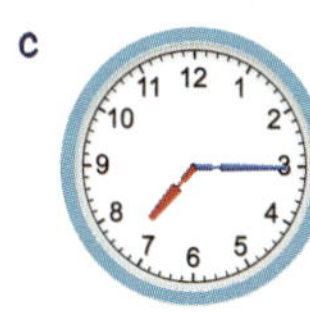

2 a 3:45
three forty-five

b 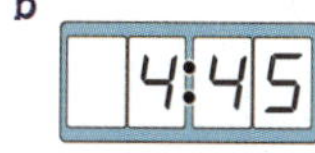

four forty-five

c 

seven fifteen

3 a quarter to 12
11:45
eleven forty-five
b quarter past 10
10:15
ten fifteen
c quarter to 1
12:45
twelve forty-five
d quarter to 5
4:45
four forty-five

4 a 2:30
b 10:30
c 12:00
d 3:00
e 4:00
f 6:30
g 11:30
h 4:35

5 a 2:30
b 11:30
c 4:25
d 6:20
e 1:10

Intervals

Page 148 – Your Turn

1 a 
11:03

b 

c 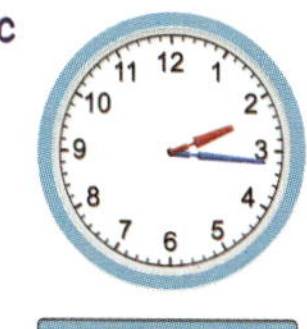

Page 149 – Practice

1 a 24 to 4
3:36
three thirty-six
b 9 past 12
12:09
twelve-oh-nine
c 6 to 7
6:54
six fifty-four

8. TIME CONTINUED

2 a seven-oh-four
b nine-oh-seven
c eleven-oh-one
d ten-oh-three
e six-oh-eight

3

	Digital time	How we read it	What it means
a	6:53	six fifty-three	7 minutes to 7
b	11:14	eleven fourteen	14 minutes past 11
c	4:44	four forty-four	16 minutes to 5

4 a 

b 

c 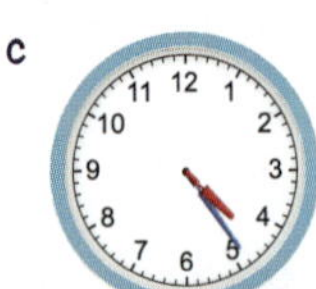

5

	Time	1 minute before	5 minutes before	5 minutes after	13 minutes after
a	11:45	11:44	11:40	11:50	11:58
b	3:21	3:20	3:16	3:26	3:34
c	5:05	5:04	5:00	5:10	5:18

Time Conversions

Page 150 – Your Turn

a ÷, 4 b ×, 300 c ÷, 1

Page 151 – Practice

1

	Seconds	Minutes
a	180	3
b	390	6.5
c	780	13
d	600	10
e	300	5

	Minutes	Seconds
f	7	420
g	11	660
h	9.5	570
i	12	720
j	14	840

2 a 7
b 16
c 11
d 5.5
e 7.5
f 9
g 3.5
h 22
i 18

3 a 150
b 360
c 600
d 750
e 510
f 240
g 1020
h 1140
i 1440

4

	Seconds	Minutes	Hours
a	300	5	0.08
b	600	10	0.17
c	900	15	0.25
d	1800	30	0.5
e	2700	45	0.75
f	3600	60	1
g	10 800	180	3
h	28 800	480	8
i	43 200	720	12
j	86 400	1440	24
k	129 600	2160	36

24-hour Time

Page 152 – Your Turn

Adult to check

Page 153 – Practice

1 a pm
b am
c pm
d pm
e pm
f pm
g am
h am
i am
j pm
k am

2 a 6:16 am
b 6:17 pm
c 10:56 pm
d 7:12 am
e 4:55 am
f 11:48 am
g 3:44 pm
h 8:27 am

3 a 10:54
b 01:27
c 09:10
d 23:12
e 18:43
f 20:48
g 15:07
h 19:26

4 a 20:03, 22:03
b 13:42, 15:42
c 10:20, 12:20
d 14:51, 16:51
e 07:35, 07:59
f 12:46, 13:10
g 05:24, 05:48
h 22:33, 22:57

5 a F
b T
c T
d F
e T
f F
g T
h F
i T

The Calendar

Page 154 – Your Turn

1 a Tuesday b Wednesday c Friday

2 Adult to check

Page 155 – Practice

1 Adult to check

2 Adult to check

3 a 2028
b Saturday, 31st January 2026
c Sunday, 1st March 2026
d 20
e 8

4 a winter b summer

5 a Monday, 23rd February
b Saturday, 28th February
c Wednesday, 18th February

6 a 4
b 4
c 4
d 4
e 4

7 a Sunday
b Thursday
c Tuesday
d Friday
e Saturday
f Monday
g Saturday
h Tuesday
i Friday

8 4

9 a 21st February b 28th February c 24th February

10 a 23rd February b 27th February c 25th February

Timetables

Page 158 – Your Turn

1 a 1
b 2
c 5
d 1
e 4
f 3
g 2

2 a 4
b 1
c 1
d 3
e 4
f 2
g 1
h 1
i 1

Page 160 – Practice

1 a 7 b 11 c 8

2 a 3
b 4
c 5
d 6
e 2

3 a 10:16 b 11:16 c 12:16

4 a 3
b 3
c 6
d 5
e 6

5 a 3 b 4 c 6

6 5

  ISBN: 9781925726190

7 a 55 b 45 c 60 d 60 e 60

8 a Thursday 10:55–11:10 b Wednesday 1:05–3:05 c Tuesday 2:05–3:05

9 a 1:25 b 12:25 c 1:25

10 a 40 mins b 40 mins c 40 mins

11 a 10:05 b 11:10 c 11:05

12 a 24th January 2025 b 8 c 2 pm

13 a 3:32 pm b 5:25 pm c 2:18 pm d 2:00 pm e 5:30 pm

14 a Vie b Yeah! c BTV d ATV e 6mate

15 a 58 mins b 2 hours c 28 mins d 6 mins e 60 mins

16 Start: 9:15, Finish: 3:15

17 a 3 b 4 c 2 d 1 e 1

18 a 105 b 45 c 120

Timelines

Page 164 – Your Turn

1 a Started primary school
b Married Mario
c Competed in National Gymnastics Competition

2 Adult to check

Page 165 – Practice

1 a 1992 b 2004 c 2020

2 a Qantas launched 'Australian Airways'
b Australian Airways stopped operating
c Qantas flight from Singapore to Perth plunged 45 metres after three airspeed sensors malfunctioned

3 Adult to check

4 Adult to check

Time Review Page 166

1 a

b

c

d

2 a

b

c 5:00

d

3 a

quarter to 12

eleven forty-five

b

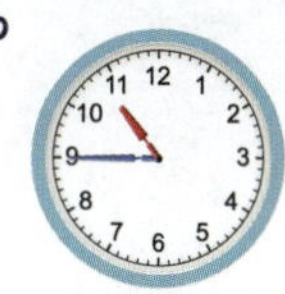

quarter to 11

ten forty-five

c

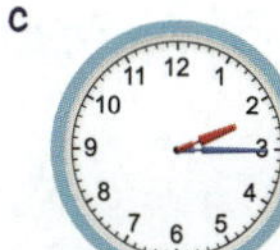

quarter past 2

two fifteen

d

quarter past 6

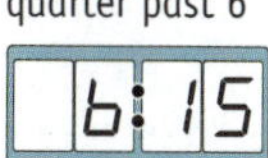

six fifteen

4 a 2:15 b 11:00 c 9:30 d 10:00 e 12:00 f 10:00

5 a

b

c

d

6 a nine thirty-two b eleven thirteen c twelve fifty-eight d seven forty-five

7

	Digital time	How we read it	What it means
a	9:27	nine twenty-seven	27 minutes past 9
b	10:16	ten sixteen	16 minutes past 10
c	4:46	four forty-six	14 minutes to 5
d	7:23	seven twenty-three	23 minutes past 7
e	1:26	one twenty-six	26 minutes past 1
f	12:33	twelve thirty-three	27 minutes to 1

8 a 10:07 b 4:57 c 3:31 d 4:36 e 7:17 f 6:53 g 9:38 h 11:04

9 a 12:06 b 5:20 c 6:43 d 9:32 e 2:57 f 4:51 g 3:59 h 10:15

10 a

half past 11

eleven thirty

b

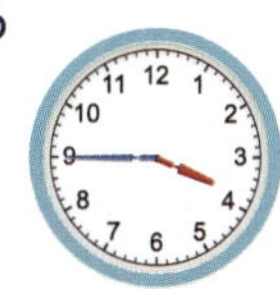

quarter to 4

three forty-five

c

quarter past 8

eight fifteen

d

24 to 5

four thirty-six

ANSWERS

8. TIME CONTINUED

11 **a** 11-oh-3 **b** 12-oh-6 **c** 1-oh-4 **d** 2-oh-2

12

	Seconds	Minutes	Hours
a	300	5	0.083
b	480	8	0.13
c	540	9	0.15
d	2700	45	0.75
e	3600	60	1
f	900	15	0.25
g	1800	30	0.5
h	5400	90	1.5
i	600	10	0.17
j	7200	120	2
k	18 000	300	5
l	39 600	660	11
m	86 400	1440	24
n	172 800	2880	48
o	12 600	210	3.5
p	23 400	390	6.5
q	9000	150	2.5
r	45 000	750	12.5

13 **a** am **b** am **c** am **d** pm **e** pm **f** pm **g** pm **h** pm **i** am **j** pm **k** am **l** pm

14 **a** 05:02 **b** 09:14 **c** 12:52 **d** 14:40 **e** 18:53 **f** 11:31 **g** 16:23 **h** 15:46 **i** 01:01

15 **a** 12:16 pm **b** 5:06 am **c** 10:40 am **d** 10:11 pm **e** 8:28 am **f** 2:27 pm **g** 11:58 am **h** 3:39 am **i** 5:09 pm

16 **a** 06:11, 08:11 **b** 11:03, 13:03 **c** 05:48, 07:48 **d** 19:59, 21:59 **e** 12:32, 14:32 **f** 09:34, 11:34 **g** 10:51, 11:17 **h** 20:40, 21:06 **i** 14:29, 14:55 **j** 09:18, 09:44 **k** 04:39, 05:05 **l** 18:28, 18:54

17 Adult to check

18 Adult to check

19 **a** Sunday **b** Monday **c** Friday **d** Wednesday **e** Monday **f** Friday **g** Tuesday **h** Wednesday

20 **a** Friday, 31st May 2024 **b** Monday, 1st July 2024 **c** summer **d** winter **e** 20

21 **a** Tuesday **b** Saturday **c** Saturday **d** Sunday **e** Sunday **f** Saturday

22 **a** Friday, 21st June **b** Sunday, 16th June **c** Friday, 21st June **d** Monday, 1st July **e** Monday, 24th June **f** Sunday, 9th June

23 **a** 4 **b** 3 **c** 2 **d** 2

24 **a** 4 **b** 2 **c** 1 **d** 3

25 **a** 13:08 **b** 14:08 **c** 15:08 **d** 14:08

26 **a** 3 **b** 3 **c** 3 **d** 1 **e** 2

27 **a** 3 **b** 3 **c** 4 **d** 1

28 **a** 5 **b** 3 **c** 6 **d** 5

29 **a** 6 **b** 4 **c** 3 **d** 3

30 **a** 5:45 am **b** 6:00 am **c** 8:15 am **d** 10:00 am **e** 6:00 am **f** 5:45 am

31 **a** 5:45 pm **b** 9:15 am **c** 5:00 pm **d** 5:45 pm **e** 5:15 pm **f** 5:15 pm

32 **a** 3 **b** 2 **c** 3 **d** 2 **e** 2 **f** 1 **g** 2 **h** 1

33 **a** Spin, Leoni **b** Spin, Bernice

34 **a** 1953 **b** 1969 **c** 1971 **d** 2000 **e** 1986

35 **a** Born **b** Formed a girl group **c** Stars in movie *Grease* **d** Starts clothing label called Koala Blue **e** Stars in a Disney movie

36 Adult to check

9. POSITION

Directions

Page 174 – Your Turn

1 **a** 3 up **b** 4 left **c** 3 up **d** 2 right

2 **a** 1 left **b** 4 down **c** 1 left **d** 3 up **e** 1 left **f** 3 down **g** 1 left **h** 2 up

Page 175 – Practice

1 **a** 3 right, 4 up, 6 right, 2 down
b 3 up, 5 right, 3 down, 1 right, 2 up, 3 right
c 2 right, 1 up, 5 right, 2 up, 1 right, 2 down, 1 right, 1 up

2 **a** Mr Hoff **b** Mrs Toman **c** Mr Callow

3 Adult to check

4 **a** skate park

5 Adult to check

6 Adult to check

7 **a** shopping centre **b** post office

8 Adult to check

The Compass

Page 178 – Your Turn

a north-east **b** north-west **c** south-east

Page 179 – Practice

1 **a**

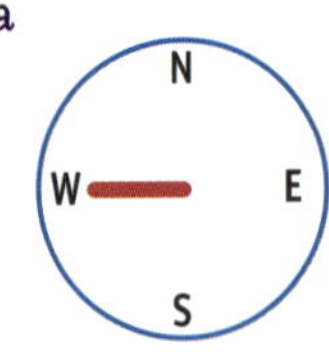

b

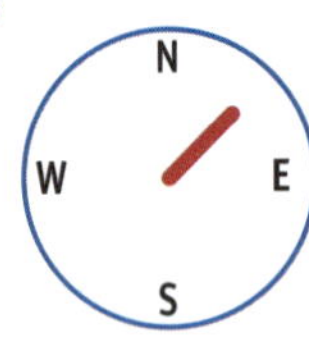

c

N
W
E
S

2 **a** A **b** Z

3

Z	E	B
L	X	A
S	D	K

9. POSITION CONTINUED

4 a east of
b south-east of
c north-east of
d west of
e north-west of
f south of
g south-west of

5

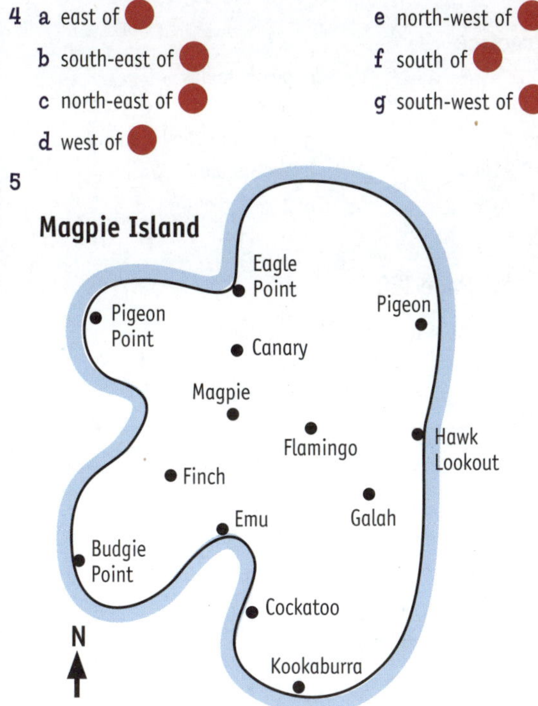

Grid References

Page 180 – Your Turn

1 a (D,2) b (C,3) c (B,1)

2

4	×	×		
3	×		×	
2		×		×
1	×	×	×	×
	A	B	C	D

Page 181 – Practice

1 a (C,5) b (D,3) c (A,7) d (A,4) e (B,2) f (D,6) g (B,6)

2

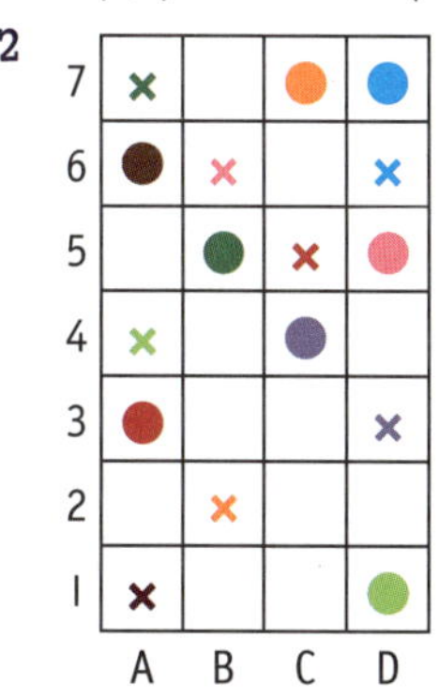

3 a ✔ b ✔ c ✔ d ✔ e ✔ f ✘ g ✔ h ✘ i ✔ j ✘

Coordinates

Page 182 – Your Turn

1

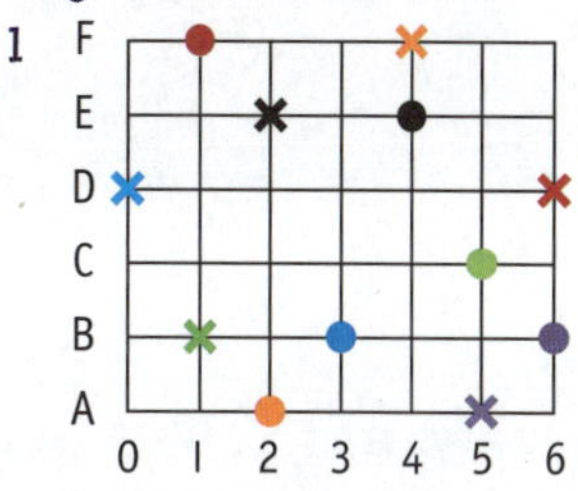

2 a (4,E) b (2,A) c (1,F) d (5,C) e (6,B)

Page 183 – Practice

1 a (B,F), (C,F), (D,E), (C,D), (B,D)
b (H,F), (I,F), (J,F), (J,D), (I,D), (H,D), (H,E), (I,E)
c (B,C), (C,B), (D,C), (B,A), (D,A)
d (E,C), (F,B), (G,C), (F,A)
e (H,C), (J,C), (J,A), (H,A), (I,B)

2

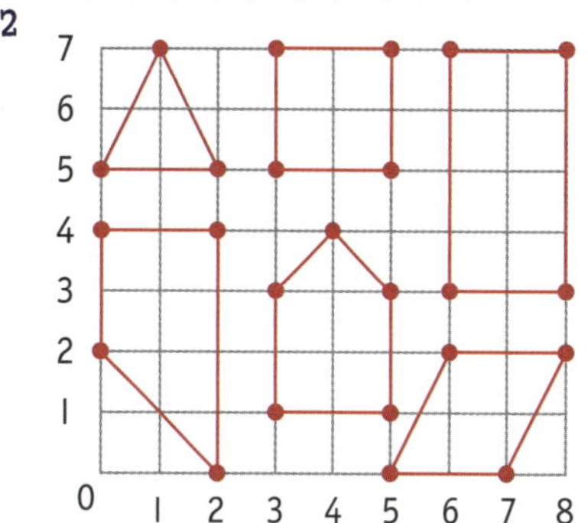

a square
b rectangle
c parallelogram
d trapezium
e irregular pentagon

3 a (5,6) b (2,9) c (4,10)

4 a Mount Natalie b Mount Despair c Mount Pippy

Cartesian Number Plane

Page 184 – Your Turn

A (2,–3) 4 B (1,1) 1 C (–2,4) 2 D (–5,3) 2 E (5,2) 1 F (–2,–4) 3 G (4,–1) 4 H (3,4) 1 I (5,–4) 4 J (–5,–5) 3

Page 185 – Practice

1

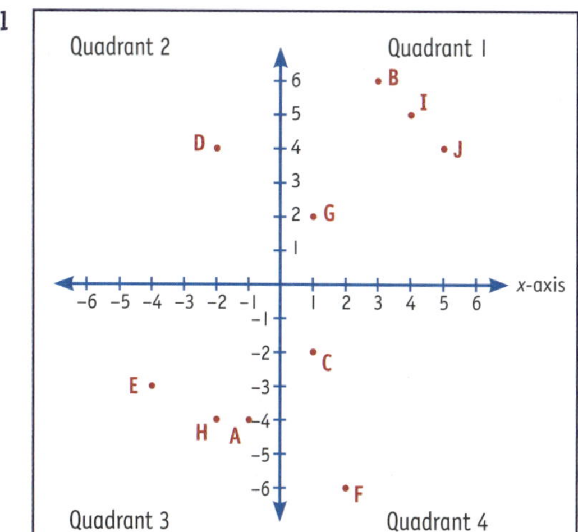

2 A Quadrant 3 B Quadrant 1 C Quadrant 4 D Quadrant 2 E Quadrant 3 F Quadrant 4 G Quadrant 1 H Quadrant 3 I Quadrant 1 J Quadrant 1

9. POSITION CONTINUED

3

	Ordered pair	Quadrant
B	(4,4)	1
Z	(2,–3)	4
P	(–2,–2)	3
D	(5,–1)	4
O	(–4,4)	2
X	(5,2)	1
N	(–7,7)	2
L	(–2,2)	2
Q	(3,–8)	4
R	(3,1)	1
S	(–5,3)	2
A	(–2,–5)	3

4 a triangle
b rectangle
c pentagon
d hexagon
e triangle

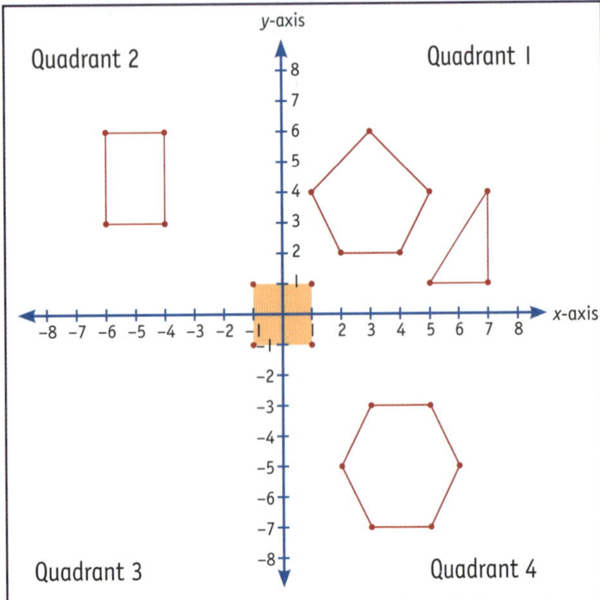

5

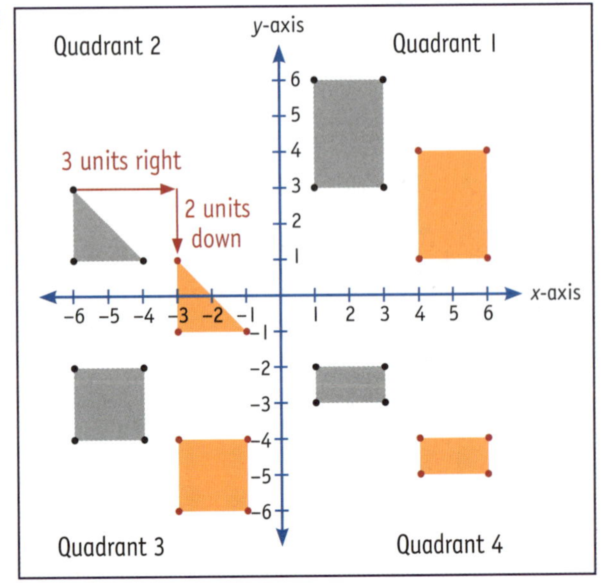

Legends

Page 187 – Your Turn

Adult to check

Page 188 – Practice

1 a mountains
b toilets
c campground
d mountain bike track
e picnic tables
f parkland
g mountain-climbing kiosk
h shower block

2 a 1 b 4 c 2 d 6

3 Adult to check

Position Review Page 189

1 a 3 right, 2 up, 1 right, 1 up, 4 left, 2 up, 2 right
b 1 up, 2 right, 4 up, 4 left, 3 down, 2 right, 1 up
c 2 down, 3 left, 1 down, 4 right, 1 down, 3 left

2

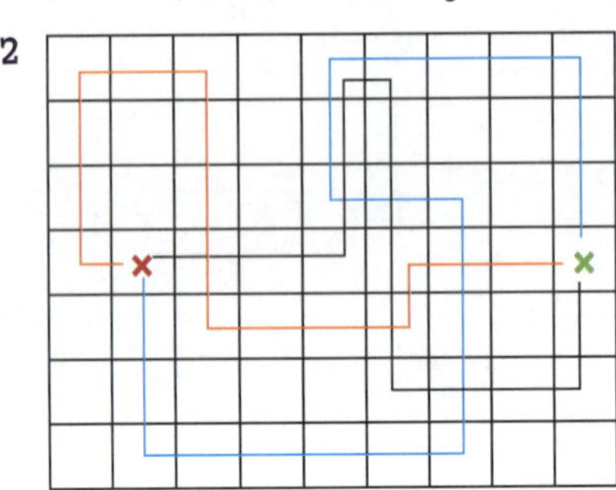

3 Adult to check

4 a Deborah
b Jenai
c Annie
d Tony
e Emily
f Rose
g Kylie
h Bindi
i Shuri

5 Adult to check

6 a the hospital
b the takeaway

7 Adult to check

8 a directions
b cardinal, intercardinal

9

North
North-west
North-east
West
East
South-west
South-east
South

10 a

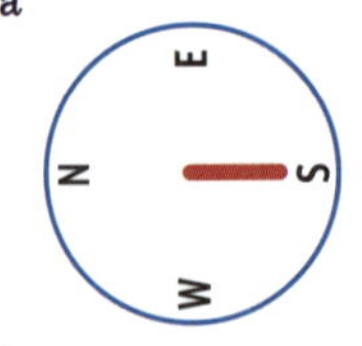

c

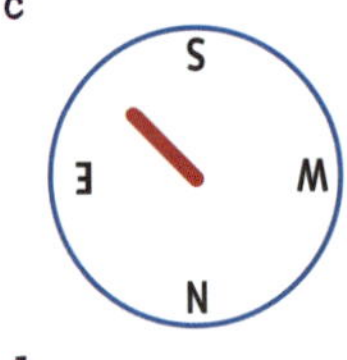

b

d

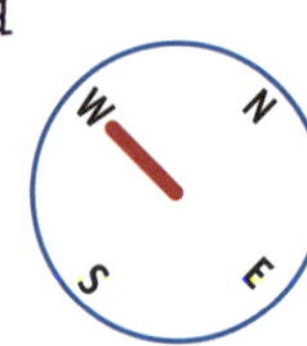

  ISBN: 9781925726190

11

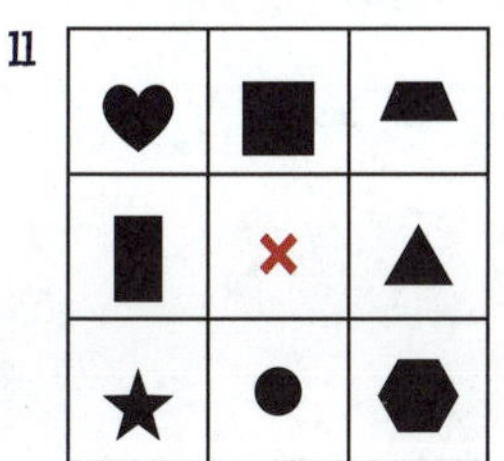

12 a (F,2) c (H,4) e (B,5) g (G,3) i (I,1)
b (C,3) d (K,1) f (A,1) h (J,5) j (D,4)

13

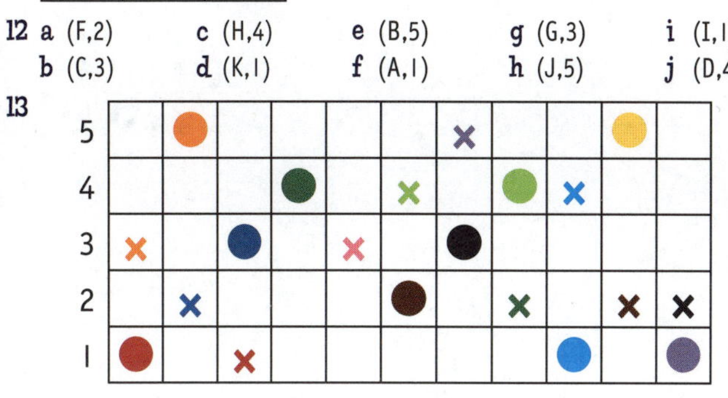

14 a (4,7) c (5,2) e (2,2)
b (2,6) d (3,4) f (6,6)

15 Adult to check

16, 18

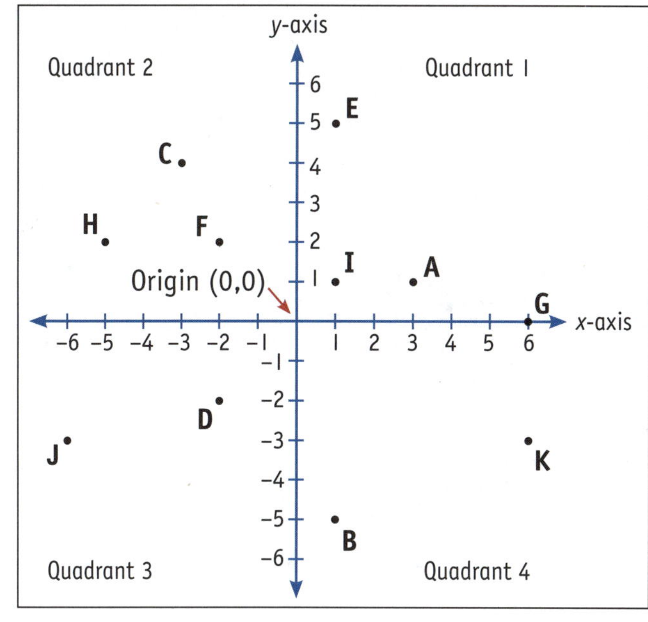

17 A (3,1) C (−3, 4) E (1,5)
B (1,−5) D (−2,−2) F (−2,2)

19

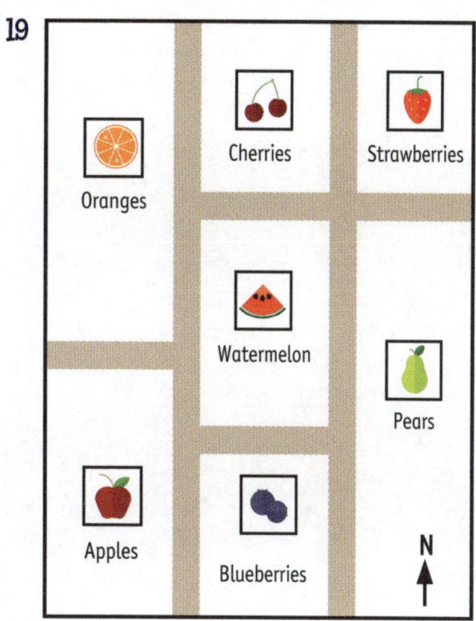

20 a apples
b strawberries
c apples
d watermelon
e strawberries
f pears

Catch-Up Maths Book 6B

ISBN: 9781925726190

Published by Pascal Press
PO Box 250
Glebe NSW 2037
www.pascalpress.com.au
contact@pascalpress.com.au

Design: Janice Bowles
Author: Deborah Frendo-Toman
Publisher: Lynn Dickinson
Typesetter: Ruth Schultz
Editor: Vaishali Batra

Printed in China by 1010 Printing International Ltd.